Space Science and Technologies

Series Editor

Peijian Ye, China Academy of Space Technology, Beijing, China

Space Science and Technologies publishes a host of recent advances and achievements in the field – quickly and informally. It covers a wide range of disciplines and specialties, with a focus on three main aspects: key theories, basic implementation methods, and practical engineering applications. It includes, but is not limited to, theoretical and applied overall system design, subsystem design, major space-vehicle supporting technologies, and the management of related engineering implementations.

Within the scopes of the series are monographs, professional books or graduate textbooks, edited volumes, and reference works purposely devoted to support education in related areas at the graduate and post-graduate levels.

More information about this series at http://www.springer.com/series/16385

Rui You · Wenjun Gao · Chunbang Wu ·
Hongbin Li

Technologies for Spacecraft Antenna Engineering Design

北京理工大学出版社
BEIJING INSTITUTE OF TECHNOLOGY PRESS

Springer

Rui You
Institute of Spacecraft System Engineering
CAST
Beijing, China

Wenjun Gao
Institute of Spacecraft System Engineering
CAST
Beijing, China

Chunbang Wu
Institute of Space Radio Technology
CAST
Xi'an, China

Hongbin Li
Institute of Spacecraft System Engineering
CAST
Beijing, China

ISSN 2730-6410 ISSN 2730-6429 (electronic)
Space Science and Technologies
ISBN 978-981-15-5832-0 ISBN 978-981-15-5833-7 (eBook)
https://doi.org/10.1007/978-981-15-5833-7

Jointly published with Beijing Institute of Technology Press
The print edition is not for sale in China (Mainland). Customers from China (Mainland) please order the print book from: Beijing Institute of Technology Press.

This Springer imprint is published by the registered company Springer Nature Singapore Pte Ltd.
The registered company address is: 152 Beach Road, #21-01/04 Gateway East, Singapore 189721, Singapore

Series Editor's Preface

China's space technology and science research have earned a place in the world but have not been compiled into a series of systematic publications yet. In 2018, the series *Space Science and Technology* edited mainly by me and co-authored by the leading figures in China's space industry was published in China, when China Academy of Space Technology was celebrating the 50th anniversary of its founding. This collection contains 23 volumes in Chinese, only 10 of which have been selected, re-created and translated into English. In addition, each English volume has been re-created at the suggestion of the Springer, by deleting the contents similar to Springer's existing publications and adding the contents that are internationally advanced and even leading, and bear both Chinese characteristics and worldwide universality. This series fully reflects the knowledge and engineering experience recently accumulated by Chinese scientists and engineers in space technology and science research.

As the Editor-in-Chief of this series, I always insist that this collection must be of high quality, either in Chinese version or English version. First, the contents of this series must be condensed and sublimated based on the combination of theory and practice, so as to provide both theoretical value and engineering guidance. Second, the relationships between past knowledge and state-of-the-art and between other people's work and our own new findings should be properly balanced in the book contents to ensure the knowledge systematicness and continuity and to highlight new achievements and insights. Each volume intends to introduce the readers something new. Third, the English version should be customized for international exposure and play a solid supporting role for China to contribute to the world's space field.

This collection consists of ten volumes, including *Spacecraft Thermal Control Technologies, Spacecraft Power System Technologies, Spacecraft Electromagnetic Compatibility Technologies, Technologies for Spacecraft Antenna Engineering Design, Satellite Navigation Systems and Technologies, Satellite Remote Sensing Technologies, Spacecraft Autonomous Navigation Technologies Based on Multi-source Information Fusion, Technologies for Deep Space Exploration, Space Robotics and Manned Spacecraft Technologies.*

Starting from the mission requirements of spacecraft, this volume, namely *Technologies for Spacecraft Antenna Engineering Design*, discusses the characteristics and design methods of various spacecraft antennas and describes the telemetry, track and command (TT&C), data transmission, navigation, inter-satellite link, payload, reliability, testing, the design cases of different antenna systems and the prospect of new technologies sequentially.

The publication of this series adds a new member to the international family of space technology and science publications and intends to play an important role in promoting academic exchanges and space business cooperation. It provides comprehensive, authentic and rich information for international space scientists and engineers, enterprises and institutions as well as government sectors to have a deeper understanding of China's space industry. Of course, I believe that this series will also be of great reference value to the researchers, engineers, graduate students and university students in the related fields.

Peijian Ye
Academician
Chinese Academy of Sciences
Beijing, China

Preface

Since 1957, in which the Soviet Union launched the first man-made satellite, the technologies related to spacecraft have got great success. The radio spectrum has extended from early VHF to L, S and C, X, Ku, Ka, W and even terahertz. China's space technology has also made achievements since 1970, in which China's first satellite DFH-1 was launched successfully into orbit. Up to the end of 2019, China has put more than 300 satellites in different types into orbit, including remote sensing satellites, communication satellites, navigation satellites, planetary exploration satellites, space-based observatory satellites and other satellites. The related antennas mainly include wire antennas, reflector antennas, multi-beam antennas, shaped reflector antennas, large deployed reflector antennas and phased array antennas. There are also advanced measurement facilities including plane scan near field, spherical scan near field, large compact range in CAST (China Academy of Space Technology). We would like to share our technologies and practical experience with worldwide peers to welcome the arrival of space economy.

The main feature of this book is the engineering design of spacecraft antennas. Because of the wide range of topics, the basic principles and technical terms are not addressed in this book. The emphasis of this book is the whole engineering process of spaceborne antennas from their architecture development in accordance with requirements to their practical design and testing.

This book describes the way in which the antenna technology is used in the satellite systems. It is motivated by a belief that the system designers and technology developers would benefit from a systematic view of antenna applications, a description of antenna technology and guidance on evaluation methodologies in practice. On an educational level, this material would be suitable for the academic courses on the application of antenna technology to the systems that have a major importance worldwide.

This monograph is primarily for use as the introduction to newcomers in the field as well as a reference for the advanced practitioners. It can also serve as an intermediate-level course for seniors and graduate students majoring in electric engineering in universities. The technical terms in the book assume that the readers are familiar with basic electromagnetic and microwave knowledge as well as the antennas found in a senior-level engineering course.

This book is divided into 12 chapters. The first chapter introduces the historical development of spacecraft and their composition and classification, the functions of onboard antennas as well as the special needs for space antennas with respect to the environment. At the end of Chap. 1 is the discussion on the system design requirements for both platform antennas and payload antennas. Chapters 2–8 introduce different space antenna designs, including TT&C antennas, data transmission antennas, GNSS antennas, inter-satellite link antennas, communication antennas as well as remote sensing antennas. Chapters 9 and 10 introduce the reliability design and measurement as well as testing of spaceborne antenna. Chapter 11 provides many cases on the spacecraft on orbit in different missions, which were all designed and fabricated by CAST. Chapter 12 provides a historical perspective as well as prospective technologies for space antenna including MEMS and Terahertz, etc. At the end of this book are the basic principles for antenna integrality.

The authors would like to thank Prof. Jiasheng Wang, Prof. Lubin Yan, Prof. Yanping Song and Prof. Shanwei Lv for their great help and careful review of the chapters.

Beijing, China — Rui You
Beijing, China — Wenjun Gao
Xi'an, China — Chunbang Wu
Beijing, China — Hongbin Li
October 2019

Contents

About the Authors

Rui You graduated from Southeast University (Nanjing, China) in 1982. He has been engaged in satellite antenna design for more than 30 years. Many kinds of antennas which he designed have been successfully run in orbit. As a key antenna engineer, his design activities involve spaceborne antenna system for meteorological satellite, remote sensing satellite, communication satellite, space science satellite, etc. He served as the Director of the Antenna Division and Deputy Director of the Technology Commission of Beijing Institute of Spacecraft System Engineering (BISSE), China Academy of Space Technology (CAST). He is also the Project Manager of Hard X-ray Modulation Telescope (HXMT), China's first space telescope for science which was launched on June 15, 2017. His current interests include terahertz payloads and new missions in space.

Wenjun Gao received the B.S., M.S. and Ph.D. degrees in electromagnetic theory and microwave technology from Beihang University, Beijing, China, in 1995, 1998 and 2002, respectively. He is currently a Senior Engineer with Beijing Institute of Spacecraft System Engineering (BISSE), Chinese Academic of Space Technology (CAST). From March 2002 to March 2012, he was a Vice-professor with the Institute of Electronics, Chinese Academic of Sciences. His current research interests include electromagnetic theory, advanced antenna technology and synthetic aperture radar system design. He is the author or co-author of over 30 papers in journals and conferences.

Chunbang Wu graduated from Wuhan University (Wuhan, China) in 1992, is the Director of the Antenna Institute of CAST XI'AN, a member of the Communication and Navigation Professional Group of Science and Technology Committee of CAST, a member of the Science and Technology Committee of CAST XI'AN, academic member of Shanghai R&D Center of CAST XI'AN, an academic member of the Key Laboratory of Antenna and Microwave Technology of National Defense Technology. He participated in many national key projects in the field of Communication, Navigation, Remote sensing and Manned Space, and was awarded several prizes for Science and Technology Progress (STP). He is also a Master's supervisor. His interests include spaceborne large deployable reflector antenna and multi-beam antenna for high-throughput satellite.

Hongbin Li graduated from the University of Electronic Science and Technology (Chengdu, China) in 1996. He has been engaged in satellite antenna design for more than 25 years. Many kinds of antennas which he designed have been successfully run in orbit. As a key antenna engineer, his design activities involve spaceborne antenna system for remote sensing Satellite, communication satellite, etc. He served as the Deputy Director of the Antenna Division of Beijing Institute of Spacecraft System Engineering (BISSE), China Academy of Space Technology. His interests include multifunctional antenna and dual-polarization antenna in space.

Contributors

Rui You, CAST, China
Wenjun Gao, CAST, China
Chunbang Wu, CAST, China
Hongbin Li, CAST, China
Zhi Xu, CAST, China
Xiaoyong Yang, CAST, China
Fan Yang, CAST, China
Xiangni Zhao, CAST, China
Dayuan Sun, CAST, China
Dehong Li, CAST, China
Nan Dong, CAST, China
Gang Chen, CAST, China
Hailiang Li, CAST, China
Yang Si, CAST, China
Weibing Zhang, CAST, China
Yun Liang, CAST, China
Shikun Zheng, CAST, China
Jianlou Zhuang, CAST, China
Changli Ye, CAST, China
Yongfei Mao, CAST, China
Xiaotian Wang, CAST, China
Bing Ma, CAST, China
Zhijia Liu, CAST, China
Chuanqiang Zhang, CAST, China

Acronyms

AIS	Automatic Identification System
AUT	Antenna Under Test
BDS	BeiDou Navigation Satellite System
BFN	BeamForming Network
BISSE	Beijing Institute of Spacecraft System Engineering
BSS	Broadcast Satellite Service
C/I	Ratio of Carrier power to Interference power
CAST	China Academy of Space Technology
CFRP	Carbon Fiber Reinforced Plastic
CMOS	Complementary Metal-Oxide-Semiconductor
CW	Continuous Wave
DBF	Digital BeamForming
DCS	Data Collection System
DOA	Direction Of Arrival
DTA	Data Transmission Antenna
EHF	Extremely High Frequency
EIRP	Equivalent Isotropic Radiated Power
EMC	ElectroMagnetic Compatibility
EMPs	ElectroMagnetic Pulses
FDTD	Finite Difference Time Domain
FOV	Field Of View
FSS	Fixed Satellite Service
G/T	Gain Noise temperature ratio
GEO	Geostationary Earth Orbit
GeoSTAR	Geostationary Synthetic Thinned Aperture Radiometer
GLONASS	GLObal NAvigation Satellite System
GNSS	Global Navigation Satellite Systems
GO	Geometrical Optics
GPS	Global Positioning System
GTD	Geometrical Theory of Diffraction

GTO	Geostationary Transfer Orbit
HEO	High Earth Orbit
HFSS	High-Frequency Simulation Software
HPBW	Half-Power Beam Width
HXMT	Hard X-ray Modulation Telescope
IGSO	Inclined Geosynchronous Satellite Orbit
ISL	Inter-Satellite Link
ISS	Inter-Satellite Service
LDA	Large Deployable Antenna
LEO	Low Earth Orbit
LH	Left Hand
LHC	Left-Hand Circular polarization
LHCP	Left-Hand Circular Polarization
LTCC	Low-Temperature Co-fired Ceramic
MBA	Multi-Beam Antenna
MEMS	Micro-Electro Mechanical System
MEO	Medium Earth Orbit
MFPB	Multi-Feed Per Beam
MMIC	Monolithic Microwave Integrated Circuit
MSS	Mobile Satellite Service
MUSIC	MUltiple SIgnal Classfication
OMT	Ortho-Mode Transducer
PCO	Phase Center Offset
PCS	Personal Communication Service
PCV	Phase Center Variation
PEC	Patch Excited Cup
PIM	Passive Inter-Modulation
PNF	Planar Near Field
PO	Physical Optics
R&D	Research and Development
REV	Rotating element Electric-field Vector method
RF	Radio Frequency
RH	Right Hand
RHC	Right-Hand Circular polarization
RHCP	Right-Hand Circular Polarization
RM	Radiated Mockup
Rx	Receive
SAR	Synthetic Aperture Radar
SNR	Signal-to-Noise Ratio
SSV	Space Service Volume
STAR	Synthetic Thinned Aperture Radiometer
T/R	Transmit/Receive
TDRSS	Tracking and Data Relay Satellite System
TEM	Transverse Electric and Magnetic field
TSV	Terrestrial Service Volume

TT&C	Telemetry, Tracking and Command
TWTA	Travelling Wave Tube Amplifier
Tx	Transmit
UHF	Ultra High Frequency
USB	Unified S-Band
UWB	Ultra Wide Band
VHF	Very High Frequency
VSWR	Voltage Standing Wave Ratio
XP	Cross-polar
XPD	Cross-Polar Discrimination
XPI	Cross-polar Isolation

Chapter 1
Overview

Abstract This chapter introduces the historical development of spacecraft and their composition and classification, the functions of spaceborne antennas as well as the special needs with respect to the environment. General design points of antenna are summarized, including requirements analysis of spacecraft antenna mission, electromagnetic layout, platform and payload antenna design requirements for different spacecraft. Basic space environment is also given.

In 1898, Guglielmo Marconi (1874–1937), the Italian electrical engineer, established the world's first radio equipment factory on Hall Street in Chelmsford, England. In 1902, long-distance maritime communication was realized, and a new era of radio communication opened up. In 1909, the Nobel Prize in Physics was awarded jointly to Guglielmo Marconi and Karl Ferdinand Braun "in recognition of their contributions to the development of wireless telegraphy".

The radio has developed rapidly since it was discovered and put into use. Nowadays, radio technology and its products have become an indispensable means and an important tool all around the world, and its application covers all aspects from civil, military to scientific research, from daily use of individuals to overseas long-distance communications, from astronomy to cosmology. It is no exaggeration to say that the whole Earth is immersed in the ocean of radio electromagnetic fields.

A basic radio system, whose main components are antenna, receiver and transmitter, is shown in Fig. 1.1. In the development process, different improvements and combinations are made to the radio system. For example, by adding a transceiver duplexer and peripheral equipments of display, control, signal processing, etc., a complete radio system is realized. In addition, a radar system can be formed by adding a ranging signal processor.

As shown in Fig. 1.1, whether simple or complex, the basic principles of radio communication system are the same. However, the antenna is always an indispensable part. Antenna theory is based on Maxwell equations established by British scientist Maxwell in the nineteenth century, which originates from four experimental laws: Coulomb's Law, Ampere's Loop Law, Biot-Savart's Law and Faraday's Law of Electromagnetic Induction. In 1887, German physicist Heinrich Rudolf Hertz confirmed the existence of electromagnetic waves by experiments and verified Maxwell's

R. You et al., *Technologies for Spacecraft Antenna Engineering Design*, Space Science and Technologies, https://doi.org/10.1007/978-981-15-5833-7_1

Fig. 1.1 Basic components of a radio system

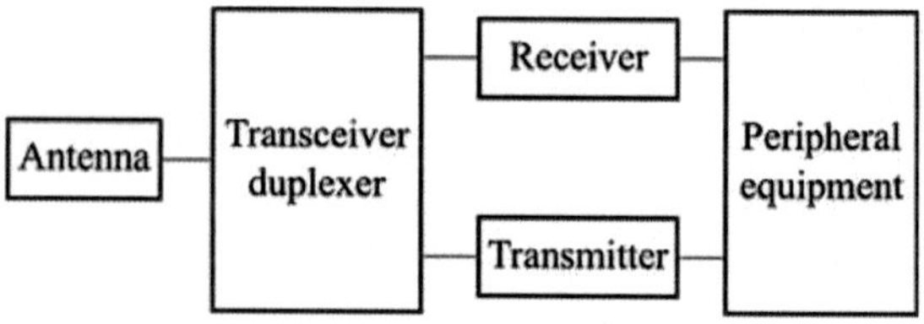

theory through experiments in 1886–1888. He proved that radio radiation has all the characteristics of waves and found that the electromagnetic field equation can be expressed by partial differential equation (commonly called wave equation). This is the theoretical basis equation for antenna analysis.

A spacecraft is a high-speed craft that orbits the planet. Before the laser communication is used (currently still in the experimental stage), the radio is the only means for humans and spacecraft to exchange information of all weather and around the clock. Since electromagnetic waves below the wavelength of 1 m penetrate the ionosphere with less loss, most of the space-to-Earth communication and remote sensing still use microwaves in the radio (the wavelength is in the range of 1 mm–1 m). The microwave antenna technology and products used on the satellite have been greatly developed. From the linear antenna (the first satellite Sputnik-1 in 1957 and "DFH-1", the first satellite of China in 1970) to the 12 m aperture antenna (Inmarsat-F4 mobile communication satellite in 2006), from the earliest frequencies of tens of megahertz to the near-terahertz detection frequency of the Herschel satellite of the European Space Agency (ESA) in 2009, from ordinary parabolic antennas to active phased array antennas for Advanced Extremely High Frequency (AEHF) military communications satellites in the U.S. in 2012, all of which embody the path of technological progress of satellite antennas.

1.1 Components and Types of Spacecraft

On October 4, 1957, the Soviet Union launched the world's first artificial satellite, Sputnik-1 (Fig. 1.2), with an orbital height of 900 km and an inclination of 65°. The satellite is a 58 cm diameter sphere with a mass of 83.6 kg and carries signals of 20 and 40 MHz. Since then, a new industry for spacecraft development has emerged.

A spacecraft orbiting the Earth is generally called a satellite, and a spacecraft far away from the Earth is generally called a space probe, such as Chang'e, a series of lunar probes and Mars probes in China. All of the above-mentioned spacecraft rely on radio waves (mainly microwave) to exchange information with Earth stations, user terminals or other spacecraft on orbit in order to monitor, measure and control the status of the spacecraft or to meet user needs and so on. Therefore, the spacecraft can be regarded as a radio station on orbit. In fact, prior to launch, the spacecraft must obtain a space radio station license issued by the International Telecommunication Union (ITU).

Fig. 1.2 The ground duplication satellite of the world's first man-made satellite. *Note* The satellite is dismantled to show its interior

Generally, spacecraft can be divided into payload and spacecraft common platform. It consists of several subsystems with different functions, including special payload, attitude and orbit control (including propulsion), structure and mechanism, thermal control, TT&C (telemetry, tracking and command), data management, power supply and other subsystems. For recoverable spacecraft, recovery landing subsystem is also included, and for manned spacecraft, environment control and life support, emergency rescue and recovery landing subsystems are inclusive. With the increasing amount of data acquired by spacecraft, many application satellites in low and medium orbits have established independent remote sensing information radio transmission system, which is also known as a data transmission subsystem.

Payload includes all the instruments and equipments that completes a specific space mission and is an important component of the spacecraft. Payload technologies include spacecraft-to-Earth communication, navigation, imaging remote sensing (visible light, infrared, microwave, etc.), electronic remote sensing (reconnaissance of ground-based radar or communication signals), mapping, spacecraft detection (geophysical, space environment, astronomy and planets), etc. The functions of the spacecraft common platform can be summarized as follows: supporting the payload; keeping the payload in a predetermined orbit; keeping the direction required by the payload; maintaining the operating temperature of the payload; providing telemetry, tele-control commands, data storage and transmission functions for the payload; power supply to the payload, etc.

The basic components of a spacecraft are shown in Fig. 1.3. As can be seen from Fig. 1.3, the antenna-related subsystems include three subsystems: TT&C, data transmission and payload.

Early spacecraft was a symbol of national strength for political competition between the U.S. and the USSR and was also a product of political competition between the two countries. Nowadays, it has gradually developed into the areas of human service and scientific exploration. Two large types of satellites, application satellite and scientific satellite, have been developed. For the first time, the U.S.

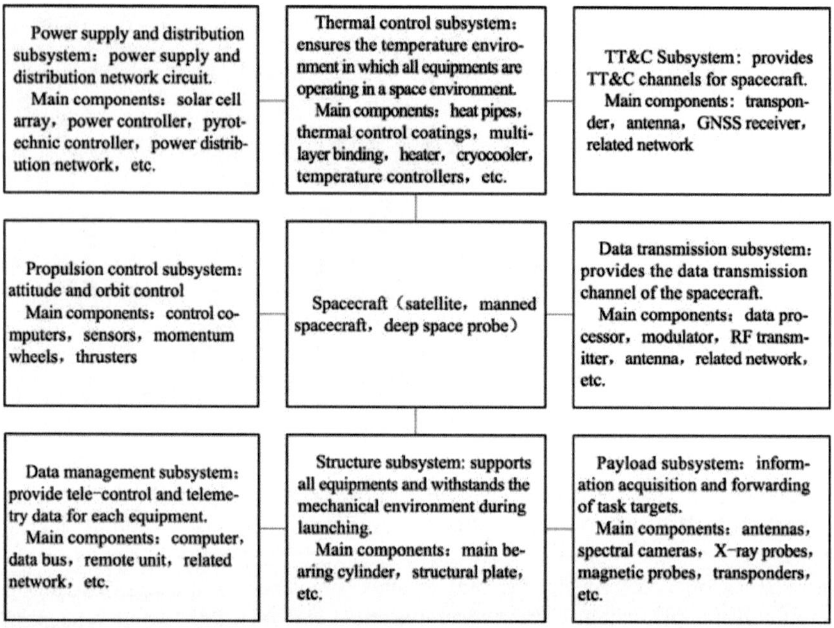

Fig. 1.3 Basic components of a spacecraft

communications satellite "Syncom-3" was used to broadcast the Olympic Games to the U.S., when the 18th Olympic Games was held in Tokyo, Japan in October 1964. In 1972, President Nixon visited China for the first time, bringing two sets of Earth stations (Youdian No. 1 and No. 2 (Youdian in English for Posts and Telecommunications)), which were built in Beijing and Shanghai, respectively. Nixon's visit to China was broadcasted lively via Intelsat-III satellite, which was a precedent for the use of communications satellites in China. According to incomplete statistics, by 2018, more than 2000 satellites of various types were launched worldwide, most of which were launched by the U.S. and Russia (former Soviet Union), and more than 300 by China. Figure 1.4 shows the appearance of several typical satellites. As can be seen from Fig. 1.4, the solar wings panel that provides electrical energy and antennas constitute the main external equipment of the satellite. Obviously, the antenna plays an important role in spacecraft.

Spacecraft can be classified into two categories: application spacecraft and scientific exploration spacecraft. It can be further subdivided according to the payload function. The classification of spacecraft is shown in Table 1.1.

The earliest institution that developed satellite in China was the 651 Design Department of the Chinese Academy of Sciences (now the Beijing Institute of Spacecraft System Engineering, China Academy of Space Technology (CAST), referred to as BISSE), which is famous for the development of DFH-1 (Fig. 1.5). DFH-1 is equipped with a 20 MHz short wave TT&C antenna and a 5 and 10 cm microwave

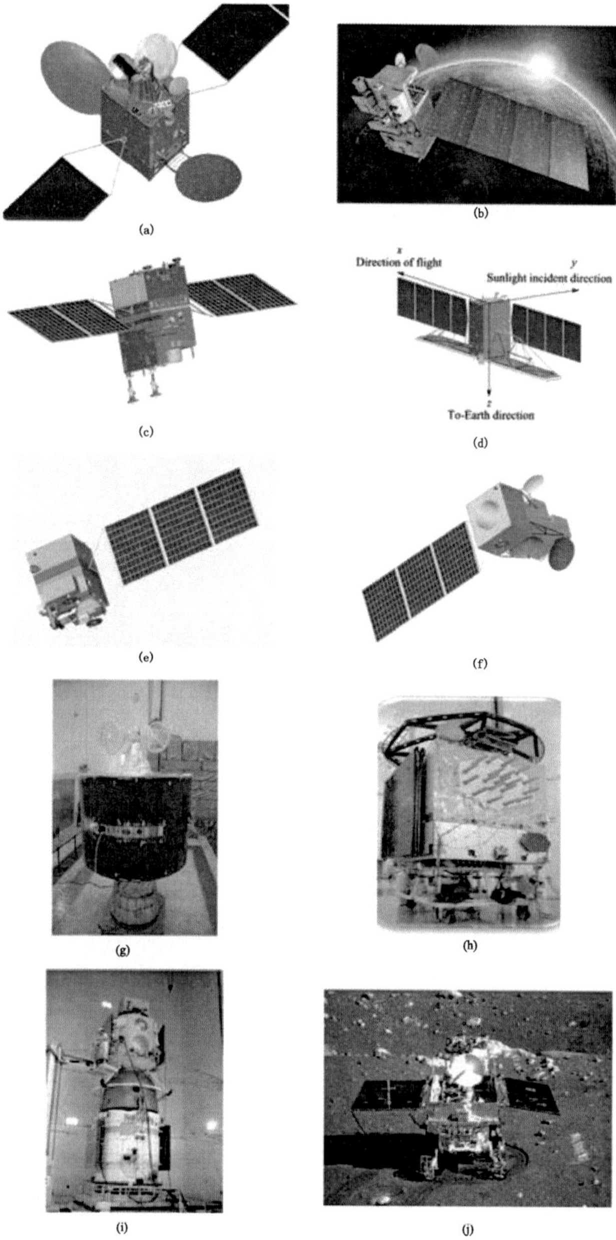

Fig. 1.4 Typical spacecrafts. **a** China communication satellite DFH-4; **b** China meteorological satellite Fengyun-3; **c** China Earth remote sensing satellite GF-2; **d** China radar imaging satellite GF-3; **e** China remote sensing satellite Earth Resources 1; **f** China microwave remote sensing satellite HY-2; **g** China meteorological satellite Fengyun-2; **h** China navigation satellite Beidou; **i** China spacecraft Shenzhou; **j** China lunar probe Chang'e 3

Table 1.1 Main classification of spacecraft

Type of spacecraft	Satellite type	Function	Typical satellites
Application spacecraft	Communication satellite	To-Earth wireless communication	U.S. Intelsat series satellites China satellites DFH-3 and DFH-4
	Remote sensing satellite	Space-to-Earth imaging or atmospheric detection	U.S. ocean observing satellite SeaSat-1, China Earth resources (resources, GF) series satellite
	Meteorological satellite	Detecting and measuring water vapor and temperature in the atmosphere	U.S. GOES, NOAA series, China Fengyun series satellites
	Navigation positioning satellite	Positioning and navigation	Global positioning system (GPS), Galileo satellite system of European Union (EU), Beidou series of China (BDS)
	Recoverable satellite	Early film imaging, scientific experiments, and experimental sample return	Recoverable series of the U.S. and China
	Manned spacecraft	Safeguarding astronauts' activities in outer space	American Apollo lunar landing spacecraft, international space station, Soviet Union spacecraft, China Shenzhou series spacecraft, space station Tiangong, etc.
Scientific exploration spacecraft	Space environment monitoring satellite	Detecting spatial particle ray	U.S. Wilkinson space detector (WMAP) Fermi detector (Fermi), China double satellite 1/2
	Celestial exploration satellite	Exploring extraterrestrial objects	American voyager 1/2, space telescope Hubble, telescope Chandra, telescope James-Weber, China hard X-ray modulation telescope satellite (also known as an in-sight satellite)

(continued)

Table 1.1 (continued)

Type of spacecraft	Satellite type	Function	Typical satellites
	Physics science satellite	Verify some physical theories	American GP-A—verification of gravitational Redshift, American a-magnetometer (AMS)—an exploration of antimatter and dark matter, China satellite Wukong

Fig. 1.5 China's first man-made satellite, DFH-1 (launched in April 1970)

radar transponder antennas. At present, as the main unit of spacecraft development, ISSE is also the main research institute of the satellite platform antenna. Starting from the development of the DFH-2 satellite antenna, CAST Xi'an has developed into the largest microwave payload development base in China and is also one of the few organizations in the world that have R&D capability in large spaceborne deployable reflector antennas.

After more than 50 years of development, the types and quantities of domestic spacecraft (satellites) have been among the best in the world and their performance index and quality (on-orbit service life) have approached the level of developed countries in Europe and America. All the equipments on satellite are developed independently. The development capability of the antenna has also evolved from the original wire antenna and single reflector antenna to the complex phased array antenna, large deployable reflector antenna (more than 10 m in diameter), multi-beam and other antennas. The frequency range is from 20 MHz to 200 GHz, which fully meets the needs of domestic civil, military and scientific spacecraft.

In the field of cosmic background microwave detection in space science, the technology and product capabilities in China have a certain gap with Europe and the U.S., especially for the case of the frequency band above 100 GHz.

1.2 Characteristics and Classification of Spacecraft Antennas

In order to communicate, control and transmit all kinds of information between space and Earth, space and space, inter-constellation and even in deep space, all kinds of radio systems on spacecraft need to load antennas that can establish wireless channels for information transmission on the spacecraft. In aerospace and space communications, spacecraft antenna not only participates in the establishment of wireless channel but also takes part in transmitting and signal processing, especially in the modern space information system. The performance of spacecraft antenna is directly related to the overall system performance and the successful completion of the spacecraft mission.

1.2.1 Characteristics of Spacecraft Antenna

Spacecraft antennas should meet the following specific requirements of their mission: specific electrical specifications, reasonable structural layout, appropriate structural strength and stiffness and proper operation in the space environment.

1. Antenna types vary. Spacecraft antenna is loaded on the spacecraft to perform the scheduled missions. Antenna types that cover almost the entire electromagnetic spectrum vary in performance for different types of missions. The spacecraft antenna does not have a fixed form.
2. Spacecraft antennas must withstand harsh mechanical and space environments. Spacecraft antenna is equipped on a spacecraft and will undergo the entire process of launching, on-orbit flight and even reentry. The performance and functionality of spacecraft antenna are closely related to the carrier environment and the working environment. For spacecraft antenna, besides electrical performance requirements, some special requirements for mechanical structure, material properties and thermal characteristics brought by environmental factors should also be considered. Therefore, spacecraft antenna is a mechanical-thermal-electrical unity.
3. The electromagnetic environment of spacecraft antenna is complex and has special requirements for electromagnetic compatibility. Antennas for different missions are mounted on the spacecraft, which have the characteristics of high density, wide spectrum, strong transmitting and sensitive receiving. In addition, there are scatters and reflectors around the antenna on the spacecraft, and their electromagnetic problems are complex multi-scatter body problem. Therefore, the electromagnetic compatibility design of spacecraft antenna has to be qualified and verified.
4. Due to the on-orbit non-maintainability and long-life requirements, as well as various constraints of spacecraft and the difficulty in achieving the redundant

design of the antenna backup, the requirements for the reliability of spacecraft antennas are greatly restricted. High reliability can be obtained by solving many problems related to materials, components, design and verification, etc.

5. In order to improve the autonomy and survivability of spacecraft, some spacecraft antennas should have the ability of autonomous control and adaption. Therefore, there is a high requirement for the intellectualized spacecraft antenna. In addition, since the performance of a spacecraft antenna is closely related to the system performance, the spacecraft antenna design should focus on the entire system, not just on a single component.

1.2.2 Classification of Spacecraft Antennas

The spacecraft is a craft operating at a certain orbital altitude. It is necessary to exchange information and control the spacecraft via ground equipment, so a TT&C channel is required. Whether it is a remote sensing satellite, a communications satellite or a deep space probe, the goal is to obtain information of the object being detected and to transmit the information detected by the spacecraft to the ground. Therefore, a payload data transmission channel on the spacecraft is an essential component. In addition, as the spacecraft itself uses electromagnetic waves to detect information on space or ground targets, a microwave load channel is also indispensable. Therefore, the general spacecraft must have the most basic configuration of three types of antennas: TT&C, data transmission and payload. Table 1.2 shows the spacecraft antenna categorized by function, including navigation and positioning receiving antenna, inter-satellite link antenna, etc.

Table 1.3 shows the main types of spacecraft antennas and their applications according to the antenna structure.

1.3 Key Points of Spacecraft Antenna Design

1.3.1 Task Analysis and Electromagnetic Layout

The main task of spacecraft antenna engineering design is to ensure that the radiation and impedance characteristics of the antenna meet the overall specification after installing the antenna on the satellite and performing surface thermal control (such as spraying thermal control paint and coating thermal control multilayer). Firstly, the factors such as lightweight, passive intermodulation and multipaction effect are to be taken into account in material selection and structure design. Secondly, spacecraft antennas should operate stably in harsh launch mechanics and space environments (high and low temperatures, electron radiation, particle radiation, cosmic ray radiation) for long periods of time (several years up to 20 years). The antenna design is closely related to materials and manufacturing processes and is a comprehensive

Table 1.2 Main antenna on spacecraft

No	Antenna type	Corresponding functional requirements	Antenna characteristics	Note
1	TT&C antenna	Measure the distance and velocity of the satellite, monitor the internal state of the satellite and upload commands	Omnidirectional coverage is generally required	
2	Data transmission antenna	Transmit the information obtained by the satellite to the ground	Earth-matching beam or spot beam antenna	
3	Payload antenna	Information exchange or acquisition of the detected target, used by the payload	This type of antenna has the richest variety	For example, communication, radar, microwave passive remote sensing, etc.
4	Navigation positioning receiving antenna	Receiving navigation and positioning signals to locate satellites	Generally hemispherical beam	
5	Inter-satellite link antenna	Inter-satellite information exchange	Generally spot beam antenna	

and multi-parameter constraint technical project. By the way, the boundary conditions of the satellite body are the preconditions for the measurement of the electrical properties of the spacecraft antenna.

1.3.1.1 Requirements Analysis of Spacecraft Antenna Mission

1. Electrical performance requirements
   According to the mission analysis, the antenna electrical performance requirement can be converted into characteristic quantities such as radiation pattern, gain, polarization, impedance matching and power tolerance.
2. Structural (and mechanical) performance requirements
   It should be able to withstand various mechanical environmental loads and be compatible with the structure of the satellite body and its related components. That is, there are requirements for structural dimensions, mounting constraints, structural strength and stiffness and mechanical connections. If the antenna moves relative to the carrier (mechanical scanning and tracking), its structure should also satisfy the normal operation requirements in the high-vacuum space environment.

Table 1.3 Application of antennas with different structure types in spacecraft

No	Antenna type		Corresponding functional requirements	Antenna characteristics	Note
1	Wire antenna	Loaded dipole antenna	Large volume at low frequency 243 MHz rescue antenna 406 MHz LEO satellite search and rescue (global satellite search and rescue system)	Early telemetry and tele-control antennas, rescue antennas	Very high frequency (VHF) telemetry, tele-control has been canceled
2		Helical antenna	Wide variety, suitable for ultra high frequency (UHF) to 9 GHz; suitable for circularly polarization; can be designed as medium gain with directional radiation or low gain with omnidirectional radiation	TT&C antenna (S-, X-band); communication antenna (UHF); data transmission antenna (S-, X-band)	
3	Aperture antenna	Horn antenna	Wide variety, suitable for both linear/circular polarization, medium gain, can be designed as narrow band/wideband	Medium-gain (area coverage) communication antennas (C, Ku, Ka); data transmission antennas (X, Ku, Ka); feed source of reflector antenna	

(continued)

Table 1.3 (continued)

No	Antenna type		Corresponding functional requirements	Antenna characteristics	Note
4	Aperture antenna	Front-fed parabolic antenna	Simple structure, suitable for both linear and circular polarization	Narrowband high-gain communication antenna (C, X, Ku); radar antennas (C, X)	The structure of the parabolic reflector is a wide variety. Main types are fixed reflector (diameter under 3 m) and deployable reflector (diameter upon 3 m and up to 20 m and more)
5		Cassegrain antenna	High gain, suitable for both linear and circular polarization	Communication antenna (C, Ku, Ka, millimeter wave, terahertz), active/passive remote sensing antenna	
6		Gregorian antenna	High gain, suitable for both linear and circular polarization	Communication antenna (C, Ku, Ka, millimeter wave, terahertz), active/passive remote sensing antenna	
7		Offset paraboloid antenna	Linear polarization and multi-beam	Regional communication, regional broadcasting antenna, remote sensing	
8		Resonant backfire antenna	UHF, narrow band, medium gain, low side lobe	Data collection antenna	
9	Slot antenna	Rectangular waveguide slot antenna	Above C-band, linear polarization is dominant and can form large arrays	Radar antenna and passive remote sensing antenna	
10		Circular waveguide slot antenna	Capable to form a circumferential symmetrical radiation pattern, above C-band	TT&C antenna, especially for spin-stabilized satellites	

(continued)

Table 1.3 (continued)

No	Antenna type		Corresponding functional requirements	Antenna characteristics	Note
11	Microstrip antenna	Single-layer patch antenna	Narrowbands, large arrays can be formed	Above L-band, radar antenna, passive remote sensing antenna	
12		Multi-layer patch antenna	More than 10% broadband, large arrays can be formed	Above L-band, radar antenna, passive remote sensing antenna	
13	Array antenna	Linear array	Directional pattern widths of two orthogonal planes are different	Above L-band, radar antenna, passive remote sensing antenna	
14		Planar array	High gain, beam variable	Above L-band, mobile communication antenna, radar antenna, passive remote sensing antenna, navigation antenna, etc.	Generally, antennas are required to be deployed
15		Conformal array	Conformal with carrier without affecting aerodynamic characteristics	TT&C, radar response, navigation and positioning, etc., of missile, recoverable satellite	

3. Thermal design requirements
   Determine the operating temperature range of the antenna, and perform various thermal protection designs to ensure antenna performance and function under various operating environment temperatures.
4. Space environmental protection requirements
   In the orbital environment, spacecraft antennas may face various space conditions such as high-vacuum space electromagnetic radiation, various charged particle radiations, atomic oxygen bombardment, multipaction breakdown effect, electrostatic charge accumulation and dense plasma envelopment in the orbital environment. Therefore, the spacecraft antenna design should be suitable for the orbital environment.
5. EMC requirements
   Spacecraft is an integration of highly dense instruments and equipment, while the antennas on the spacecraft are an aggregation. For the antennas on spacecraft, not only the interaction between antennas, antennas and carriers (including spacecraft body and solar wing, etc.) but also the electromagnetic interference between antenna RF equipment and RF transmission lines, spacecraft internal equipment and cables have to be considered in order to ensure the performance, function and normal operation of the respective antennas.
6. High-reliability requirements
   Considering the overall life of the spacecraft and the operating mode of the antenna, the requirements of the antenna to withstand the mechanical environment of the launch and the reliability requirement in the orbital high-low temperature and space radiation environment are put forward.

1.3.1.2 Electromagnetic Layout Analysis of Spacecraft Antenna

For the layout design of the spacecraft antenna on the carrier, first, various constraints of structural installation and layout should be satisfied to achieve its structural compatibility, and then its optimal electromagnetic layout should also be considered. The optimization means that the performance should still be optimized under the consideration of various effects of spacecraft carrier environment (such as inter-antenna, antenna and satellite, antenna and solar wing). To reach this goal, an antenna electromagnetic layout technology is developed by means of computer numerical simulation based on the theory of electromagnetic field.

1.3.1.3 Verification of Electromagnetic Layout of Spacecraft Antenna

Under the premise that the performance of single antenna meets the requirements, the configuration size of the spacecraft and the installation position of each antenna on the spacecraft are decided, a full-scale Radiating Mockup (RM) test was performed. Full-scale RM is one of the technological states of the whole satellite. The whole satellite technological state also includes Structural Mockup (SM), Electrical Mockup

(EM) and Thermal-control Mockup (TM). Full-scale RM is an engineering model whose high-frequency electromagnetic performance is equivalent to the real satellite. After all the satellite antennas are installed on the Mockup as required, the electrical performance of each antenna is tested to verify whether their predetermined requirements are still met; the isolation between antennas and the interference fields of some sensitive points are tested to verify whether the requirements of electromagnetic compatibility (EMC) of the whole satellite are met. The test can also be used to observe the effect of the solar wing on the antenna radiation pattern if possible. During the test, the solar wing simulates the orbital operation state and tests whether the antenna radiation performance meets its predetermined requirements during the entire operation. Generally, due to the large size of the solar wing and limited by the test conditions, it is impossible to determine the influence of the test. Therefore, the numerical analysis and high-frequency approximation simulation can be used instead of the test. The purpose of the RM test is to verify whether the spacecraft antennas meet the design requirements (including the EMC design) standing at the system level and then determine the installation modes and optimal layout of the spacecraft antennas.

1.3.2 Essential Factors for Adaptive Design of Space Environment

Spacecraft operate in the space areas of several hundreds to tens of thousands kilometers above the Earth, while deep space probes operate even higher. Therefore, in order to operate properly, spacecraft should adapt to the space environment of the orbit. The antenna is generally installed outside the spacecraft, which is directly affected by the space environment. Therefore, in the initial stage of antenna design, various factors of the space environment must be considered. For detailed space environment and adaptive design, the reader can refer to relevant professional books and literature.

In the space environment adaptive design for spacecraft antenna, the launching environment (sufficient mechanical strength), space vacuum environment, thermal radiation environment, space charged particles and plasma environment are mainly observed. The main effects are shown in Table 1.4.

1.3.3 Platform Antenna Design

Platform antenna, whose function is to ensure the basic information exchange and command of the satellite, generally includes TT&C antenna, data transmission antenna, navigation and positioning antenna, etc. TT&C antenna together with its transponder always works from launching to satellite-rocket separation until the end

Table 1.4 Space environment and requirements of spacecraft antenna

No	Space environment factor	Characteristic	Requirements for antennas
1	Launch environment	Mechanical vibration and acceleration overload	Sufficient strength and stiffness are required. Specific conditions are given in the mechanical conditions of the whole satellite
		Air flow washout and pneumatic heating	These considerations must be taken into account for satellites without fairing
2	Space vacuum environment	Low-pressure gas and high-vacuum environment	Low-pressure discharge and multipaction effects should be considered for antennas with transmitting power over 1 W. The low-pressure and vacuum adaptability design of antennas should be verified by ground simulation test
3	Thermal radiation environment	The surface temperature of the antenna can reach up to 100 °C when there is sunlight, and it can reach below −100 °C without sunlight, or even lower	Thermal analysis and thermal design are to be performed. In addition to considering its orbital illumination, the transmitting antenna must consider its own electrothermal loss. The antenna is required to be qualified by the ground thermal and vacuum simulation test

(continued)

Table 1.4 (continued)

No	Space environment factor	Characteristic	Requirements for antennas
4	Space charged particles and plasma environment	Earth radiation belt 1. The main compositions of the orbital space below 10,000 km are energetic protons and electrons 2. The main compositions of the orbital space between 10,000 and 40,000 km are electrons 3. The South Atlantic Anomaly over the South Atlantic Ocean can reach the lowest altitude of 200 km from the surface, mainly composed of energetic protons and electrons 4. The main compositions of geomagnetic bipolar space are energetic protons, electrons and plasma	The antenna materials used (especially dielectric materials) must be able to withstand some radiation dose of proton and electron radiation in space. For active antennas, the Complementary Metal–Oxide–Semiconductor (CMOS) transistor devices used should be higher than a certain threshold of single event Linear Energy Transfer (LET), generally more than 15 meV cm^2/mg and anti-single-event design should be carried out in the circuit
		Galactic cosmic rays, mainly composed of energetic protons and heavy ions	Anti-single-event design of components
		The solar cosmic rays can be neglected when the sun is quiet. When the sun breaks out, the intensity of the solar cosmic rays will increase greatly. The main compositions are energetic protons and heavy ions	Total dose resistant shielding and anti-single-event design
		Plasma: 1. More than 20,000 km above the equator, during geomagnetic storm/geomagnetic substorm, thermal plasma above keV energy level may occur in orbital space such as Geosynchronous Orbit (GEO) and Medium Earth Orbit (MEO). Charge and discharge effects may occur on the surface of the satellite 2. At a distance of less than 20,000 km from the Earth surface, especially below a few hundred kilometers, the space plasma is mainly cold-thick ionospheric particles or auroral precipitation particle, which acts on the low-altitude spacecraft to produce surface charge and discharge effects	Surface equipotential design, charge leakage channel design

(continued)

Table 1.4 (continued)

No	Space environment factor	Characteristic	Requirements for antennas
5	Internal charging effect environment	Space energetic electrons (relativistic electrons) deposit into materials or devices, which results in high potential difference and may cause internal discharges	The antenna surface and the bracket should have a conductive path with the satellite casing. The active circuit of the antenna should be shielded and grounded to avoid creating a large area of the isolated conductor

of satellite on-orbit life, so it has the highest reliability requirements. Navigation and positioning antenna together with its receivers are usually used from the separation of satellite and rocket to the end of satellite life. TT&C antenna and navigation and positioning antenna are characterized by broadband design and narrow band operation, medium or low power (depends on orbit height) and continuous operation. Data transmission antenna, which is connected with the data transmission system, is used to transmit payload data (such as image data) and some platform state telemetry data. It is characterized by broadband operation and high power, but intermittent operation. Generally, low-orbit satellite transmits data no more than 15 min per orbit and high-orbit satellite does no more than 20 min per hour. Among these three types of antennas, the failure of either TT&C antenna or data transmission antenna will lead to the failure of the whole spacecraft function, so the reliability of these both types of antennas should be designed to be close to 1.

It can be seen from the above features that the design of the platform antenna has a certain specialization. The features of the three types of antennas are shown in Tables 1.5, 1.6 and 1.7, which is the primary consideration in the design of the platform antenna.

Table 1.5 Design characteristics of spacecraft TT&C antenna

No	Orbit	RF characteristics of the data transmission antenna			Remarks
		Polarization	Radiation pattern	Power	
1	Low Earth orbit	Circular	Near 4π spatial coverage	≤1 W	
2	GEO orbit	Circular	Nearly 4π spatial coverage in the launch phase, and switching to directional antenna after the fixed point on orbit	≤1 W before the transfer orbit (GTO), ≥5 W for the GTO orbit, ≤1 W after the fixed point on orbit	Multipaction effect should be considered in the case of 2 W or more
3	Interstellar orbit	Circular	Cover by nearly 4π space from launch to tens of millions of kilometers, switch to the directional high-gain antenna in long distance or insert a medium-gain antenna (about 10 dBi) between them	Omnidirectional <1 W; medium gain >5 W; high gain >5 W	Depending on the combination of specific designs

Table 1.6 Design characteristics of spacecraft data transmission antenna

No	Orbit	RF characteristics of the data transmission antenna			Remarks
		Polarization	Radiation pattern	Power	
1	Low Earth orbit	Circular	For medium to low code rates, the pattern is an Earth-matched beam; for high code rates (above 300 Mb/s), a double-polarized, gimbal high-gain (≥20 dBi) antenna is typically used	Earth-matched shaped beam antenna 10–100 W; mobile high-gain antenna ≤20 W	Power capacity and multipaction effect factors are to be considered
2	GEO orbit	Circular	18° medium-gain antenna or gimbal high-gain (≥25 dBi) antenna	10–100 W, depending on the code rate	
3	Interstellar orbit	Circular	Gimbal high-gain antennas (≥25 dBi)	≥20 W, depending on the orbital distance	

Table 1.7 Design characteristics of spacecraft navigation and positioning antenna

No	Orbit	RF characteristics of the data transmission antenna			Remarks
		Polarization	Radiation pattern	Phase	
1	Low Earth orbit	Circular polarization	Generally half-space coverage, or special design for special space-oriented satellites	Continuous phase within ±80°	For satellites with high-precision orbit determination requirements, the phase center stability must be less than a certain value (e.g., 10 mm) within ±80°
2	GEO orbit	Circular polarization	The field angle is generally ±30°, preferably the middle concave beam of the Earth	Continuous phase within ±30°	Receiving the limb signals, the side or back lobe signals of the navigation satellite

As described in Table 1.5, the main features of the TT&C antenna are near-omnidirectional coverage and circular polarization, which are mainly to adapt to reliable TT&C of spacecraft in arbitrary attitude (including attitude in case of spacecraft failure), so it is generally composed of two half-space covered antennas. The specific antenna form can be divided into two-arm spiral, quadrifilar helix, microstrip patch, horn, waveguide open-ended port, etc. Off-the-shelf products have been available in relevant institutes in China. The design is mainly related to the antenna system such as element selection, antenna layout design. The specific design of the TT&C antenna is shown in Chap. 2.

As described in Table 1.6, the most important feature of the data transmission antenna is the Earth-matched shaped beams in LEO. Different orbits have different field angle. For example, on 550 km orbit, the satellite's field angle to the Earth is $\pm 67°$ when the Earth station operates at an elevation of 5°. The maximum gain and polarization characteristics of this angle should be calculated when designing antenna elements. For medium- and high-gain antennas, horn and paraboloid antenna are preferred; for antennas used at high code rate, double circular polarization isolation (XPD) is an important criterion. For the design of data transmission antenna, see Chap. 3.

From Table 1.7, the main features of navigation and positioning antenna are half-space coverage and circular polarization. In a spacecraft with high-precision positioning, the antenna phase center stability index is required, which is closely related to the spacecraft boundary condition (such as the reflection of the satellite surface and the solar wing). Therefore, in the antenna design, the side lobes and the back lobes should be reduced as much as possible, while in the installation layout, it is necessary to avoid the influence of large-area scatterers such as the solar wing, in order to ensure the continuity of the amplitude pattern and the phase pattern at low elevation. For the specific design of the navigation positioning antenna, see Chap. 4.

1.3.4 Payload Antenna Design

The payload antenna, which is equipped with dedicated payload to complete user's task, can be classified into two categories: one is information acquisition type payload antenna, such as microwave radiometer antenna, the microwave scatterometer antenna and the synthetic aperture radar (SAR) antenna of microwave remote sensing satellite; the other is information exchange transmission payload antenna, such as communication antenna for communication satellite, navigation payload antenna for navigation satellite, etc. The common characteristics of these antennas are large bandwidth, high power of transmitting antennas (ranging from tens to several kilowatts), long-term continuous operation (such as communication satellite antenna, navigation payload antenna) and short-term intermittent operation (such as SAR antenna). For microwave radiometer antenna, the equivalent noise temperature and beam efficiency are also specified. Due to the variety of satellite missions, the payload antenna is the fastest-growing antenna with the most types, most complex technology in the spacecraft antenna.

In terms of electromagnetic layout, the electromagnetic coupling between the payload antenna and the satellite platform is relatively poor, and it can be independently designed, but the structure and thermal index are closely related to satellite platform. The electromagnetic boundary formed by the payload antenna has a great influence on the weak directional platform antenna. Therefore, for the system design of the platform antenna, the boundary of the payload antenna must be considered. For frequency reuse antennas through orthogonal polarization (generally parabolic antenna and its variants), the influence of the antenna rod and the edge scattering effect of the outer cladding on the polarization isolation of the dual-polarization feed must be considered.

The information acquisition type payload antenna includes microwave passive remote sensing antenna and microwave active remote sensing antenna. The design characteristics are shown in Table 1.8. For microwave passive remote sensing antenna, large aperture paraboloid is preferred, with a focus on multi-band, dual-line polarized feeds. Active remote sensing, such as synthetic aperture radar (SAR), is characterized by the fact that the beam should meet the requirements of various operating modes. Therefore, planar phased array or multi-feed (for beam switching) offset parabolic antennas are two typical forms. Due to the long distance, the aperture of the antenna is large. Whether it is a planar phased array or a multi-feed offset parabolic antenna, they all adopt deployable structure design. For the specific design, see Chap. 7.

Payload antennas for information exchange and transmission (Table 1.9) include communication broadcast antennas and navigation payload antennas. Communication broadcast antenna is categorized into fixed communication and mobile communication satellite antenna. For fixed communication and broadcasting services, the antenna aperture of Earth station is larger, so the antenna on satellite can be relatively compact (e.g., about 2 m in diameter), and is generally a solid surface offset parabolic antenna. In order to achieve dual-polarization frequency reuse with one paraboloid, a dual grid reflector and multi-feed design are generally used to form the coverage of the service area, while the regional television broadcasting antenna uses the combination of parabolic reflector and single-feed source to achieve the coverage of the area. For mobile communication satellite antennas, high-power flux is required to support the communication of mobile terminals. Therefore, large-diameter (10 m or more in diameter) mesh deployable paraboloid plus multi-feed is used to achieve high Equivalent Isotropic Radiated Power (EIRP). For specific design, see Chaps. 5 and 6.

The spacecraft antenna engineering design is not only the electrical design of the isolated antenna but also related to the spacecraft and other carriers, so the orbit, attitude, mechanics and space environment and related transceiver characteristics should be considered. From the perspective of the system level, the development of spacecraft antenna begins with the mechanical, electrical and thermal integrated design. The spacecraft antenna engineering design is a very complicated process, so advanced design methods must be adopted. At present, computer numerical analysis

Table 1.8 Design characteristics of information acquisition type payload antenna

No	Load type	RF characteristics of the payload antenna			Remarks
		Polarization	Radiation pattern	Power	
1	Microwave passive remote sensing	Single/double linear polarization	The frequency coverage is wide, the lowest is 1 GHz, the highest is 1 THz and the bandwidth is wide, which can reach several octaves Sharp beam, beam efficiency (>0.9); generally, the area scanning function is required, and mainly mechanical rotation scanning	No requirement	Large aperture paraboloid is the main type of antenna, for cosmic microwave background detection, low-temperature operation is also required
2	Microwave active remote sensing	Single/double linear polarization	Generally, the highest frequency is not more than 30 GHz, and the bandwidth is several hundred MHz, the antenna aperture is large, the gain is high (more than 30 dBi), at the same time, the beam width is variable and the beam can be moved	High RF power, the peak value of the pulse can reach more than 10 kW and the average power is several hundred watts to kilowatts	The antenna form mainly includes planar phased array and multi-feed large-aperture parabolic antenna

and simulation technology are important means in the design of spacecraft antenna. In addition, to ensure the correctness of the design scheme, it is necessary to test and verify some important and key performances in the design process. In summary, various validation tests are necessary and effective methods for spacecraft antenna research, also the most reliable and intuitive means.

Table 1.9 Information exchange transmission type payload antenna

No	Load type	RF characteristics of the payload antenna			Remarks
		Polarization	Radiation pattern	Power	
1	Communication broadcast	Fixed communication and broadcasting services adopt single/double linear polarization; mobile communications adopt circular polarization	Wide frequency coverage, minimum bandwidth can be determined on demand. Generally, shaped beams for area coverage, and high-gain point beams for emergency or military communications	High power, from tens of watts to nearly kilowatts	Large aperture paraboloid is the main type, including front-fed single reflector, offset reflector and its shaped, dual-grided reflector, multi-feed and so on
2	Navigation payload	Circular polarization	L-band, narrow bandwidth, medium gain (within 16 dBi), a matching beam of Earth's field of view	Medium RF power, continuous operation mode	The antenna form mainly includes planar phased array and multi-feed large-aperture reflector antenna

Chapter 2
Spacecraft TT&C Antenna

Abstract The TT&C antenna of spacecraft is used to receive command signals and to transmit telemetry signals between spacecraft and earth station. Based on the introduction of the characteristics of spacecraft TT&C antennas, this chapter also provides several commonly used TT&C antennas, including S-band dual-arm helical antenna, S-band quadrifilar helical antenna, S-band cavity-backed dipole antenna, C-band coaxial-slot double-cone-disk antenna, and other TT&C antennas, which are used in both low and high-orbit satellites.

Spacecraft TT&C antenna is the core component of spacecraft TT&C subsystem. Together with ground TT&C antenna, it can establish a stable and reliable radio transmission channel to meet the predetermined requirements and to ensure the transmission, reception and transmission of TT&C information. In short, the TT&C antenna of spacecraft is used to receive command signals from Earth station to spacecraft or to transmit telemetry signals from spacecraft to Earth station.

Space TT&C system is an indispensable part of space engineering. In the early stage of the development of space TT&C system, a decentralized system was adopted. In this system, multiple devices are required to complete the overall TT&C of a spacecraft. Each device performs different functions and assigns different independent carriers. In this way, spacecraft not only has a lot of equipment with large weight and high power consumption but also has poor reliability, and ground equipment is very complex. With the development of technology and the urgent need of flexible and high precision for TT&C system in space missions, it is urgent to upgrade the technology of TT&C system. In the mid-1960s, NASA built a Unified S-Band (USB) space TT&C system, which was successfully used in the Apollo lunar landing program. Then, the multifunctional unified microwave TT&C system has replaced the single-function decentralized space TT&C system. On this basis, the U.S. has built more than 20 S-band integrated TT&C stations with better performance, forming a global space TT&C network. Countries such as Europe, Russia, Japan and India are sparing no effort to develop their own microwave unified TT&C systems, which are responsible for the engineering TT&C and on-orbit TT&C of satellites and other spacecraft, and basically belong to the unified C-band or S-band TT&C system.

R. You et al., *Technologies for Spacecraft Antenna Engineering Design*, Space Science and Technologies, https://doi.org/10.1007/978-981-15-5833-7_2

Microwave unified TT&C system is an integrated space radio TT&C system, which can modulate multiple signals such as ranging, velocity measurement, TT&C, data onto a carrier and accomplish multiple functions at the same time. This system is used for TT&C and communication of various aircraft, such as missiles, satellites, spacecraft and other space probes. The microwave unified TT&C system equipment is mainly composed of two parts, which are located in TT&C equipment of the ground TT&C station and spacecraft TT&C equipment.

With the development of space technology and satellite application, space-based TT&C resources based on Satellite Navigation System BDS (Beidou Navigation Satellite System), GPS (Global Positioning System) and other global navigation and positioning satellite systems have been applied. At present, most satellites have used global positioning and navigation system as a supplement to satellite TT&C, and satellite-borne navigation and positioning antenna has also been used on most satellites as platform antenna. The spacecraft navigation and positioning receive antenna will be introduced in Chap. 4.

In order to meet the requirements of various application satellite constellation and satellite autonomous control, space-based TT&C resources based on TDRSS are also developing and improving. The successful application of inter-satellite link TT&C technology of Tianlian series to the docking test between Tiangong-1 and Shenzhou-8 indicates that a complete "space-based" TT&C system has been established in China. The corresponding inter-satellite TT&C technology is constantly improving. See Chap. 5 for details.

Based on the introduction of the characteristics of spacecraft TT&C antennas, this chapter also introduces several commonly used TT&C antennas, including dual-arm helical antenna, quadrifilar helical antenna, cavity-backed dipole antenna and coaxial slot antenna.

2.1 Main Characteristics of Spacecraft TT&C Antenna

Spacecraft TT&C antenna is used to receive tele-control signals from Earth station to spacecraft and to transmit remote measurement signals from spacecraft to Earth station. It can be divided into telemetry antenna, command antenna, orbit measurement antenna and unified system TT&C antenna (compatible with telemetry, command and orbit measurement function).

The common working frequency bands of TT&C antenna include S-band, C-band, X-band and Ku/Ka-band. The uplink and downlink coherent carriers are used, and the forwarding ratio is specified. For example, the S-band TT&C antenna shall operate in 2000–2300 MHz with a forwarding ratio of 221:240; the C-band TT&C antenna is required to be operated in 5925–6425 MHz for uplink and 3700–4200 MHz for downlink. Spaceborne TT&C system mostly uses the common antenna for receiving and transmitting, and a few adopt the separate antenna, which depends on the requirements of the flight mission.

Wide-beam antenna is generally used in TT&C antenna, and narrow-beam antenna can be used in special cases. For the wide-beam antenna, helical antenna and dipole antenna can be used, and for the narrow-beam antenna, reflector antenna and horn antenna can be used.

For TT&C antenna, a near-omnidirectional coverage beam shall be formed in order to meet the requirements of continuous tracking and TT&C during the launch of spacecraft and normal TT&C in any attitude of the satellite when the satellite fails. For three-axis stabilized satellites, especially communication satellites in geostationary orbit, directional antennas can be switched to reduce transmission power and suppress malicious interference or omnidirectional and directional antennas can be used simultaneously. For deep space exploration satellite, the TT&C antenna also requires a high gain to meet the transmission link requirements.

2.1.1 TT&C Antenna Array Mode

TT&C antennas have different array modes for different satellites. For S-band and C-band TT&C antennas, the two-element antenna array is used to form near-omnidirectional covering beams, which meets the requirements of satellite TT&C links in the whole life cycle. The block diagram of TT&C antenna array system is shown in Fig. 2.1. TT&C antenna 1 is generally installed on the surface of satellite that is Earth-facing, while TT&C antenna 2 is installed on the surface of satellite that is away from Earth. The pattern gain formed by the combination of two TT&C antennas and feed network is called composite pattern gain. Feed network generally includes high-frequency cable assembly, power synthesizer/divider, etc. The length of high-frequency cable assembly is determined according to the satellite layout. The TT&C antenna usually works in circular polarization, and the two pairs of TT&C antennas can be arranged in a co-polarization or cross-polarization array mode.

The advantage of co-polarization array is that normal TT&C can be achieved under one polarization state of Earth station antenna. However, due to the interference phenomenon of the two antennas in the beam overlap area, that is, there is a waist (overlap area) interference area behind the antenna array, the location and range of the waist interference area need to be properly controlled to minimize the influence on the TT&C link. Generally, it is required that the angle range of the pattern in the

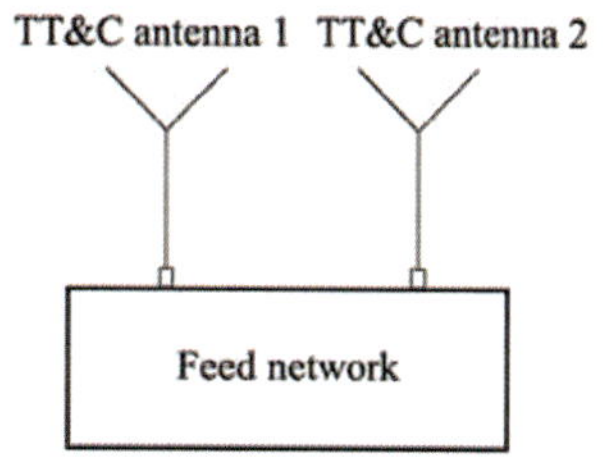

Fig. 2.1 Block diagram of TT&C antenna array system

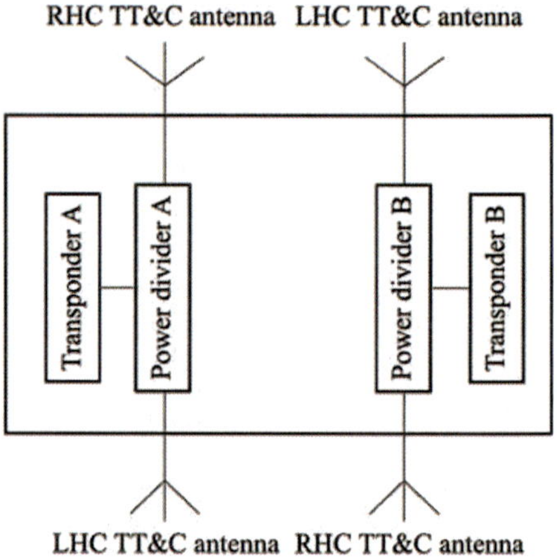

Fig. 2.2 Block diagram of TT&C antenna reverse-rotating array system

waist interference area which is continuously lower than the threshold level is not more than 1°. Possible approaches include selecting fore antenna with a relatively wide beam and aft antenna with a relatively narrow beam to reduce interference. In addition, the coverage of the toward-Earth TT&C antenna can be further increased by adjusting the power-dividing ratio of the feed network.

The block diagram of the TT&C antenna cross-polarization array system is shown in Fig. 2.2. Only one transponder works when transmitting signals and two transponders work simultaneously when receiving signals (the working mode of the two transponders can be determined by the software of the satellite-borne computer). When the telemetry signal is transmitted, the interference based on the physical characteristics of co-frequency and co-source of radio frequency signals, between two antennas connected to the same transponder, can be avoided in the overlapping area of the beams by the orthogonal polarizations. The single circular polarization pattern of the antenna array is basically the same as that of the unit antenna, which makes full use of the advantages of wide-beam antenna coverage area. When receiving command signals, according to the inherent characteristics of the same frequency and different sources, the same polarization antennas connected to the two transponders are used to form the beam coverage in the whole space so as to ensure that the command signals arrived from any direction to the satellite can be successfully received. For the downlink telemetry signals transmitted by satellites, two orthogonal polarizations are transmitted simultaneously in space, which form 4π full-space coverage. The ground station can receive only one or two polarization signals in diversity to ensure that the satellites can reliably transmit telemetry signals in any attitude as long as they enter the field of view of ground TT&C station.

The TT&C antenna of the early tri-axis stabilized satellite in the MEO and LEO adopts a co-polarization or cross-polarization array to form near-full-space coverage, and the flare angle of the ground coverage area is shown in Fig. 2.3a. The design method of the TT&C antenna array is as described above.

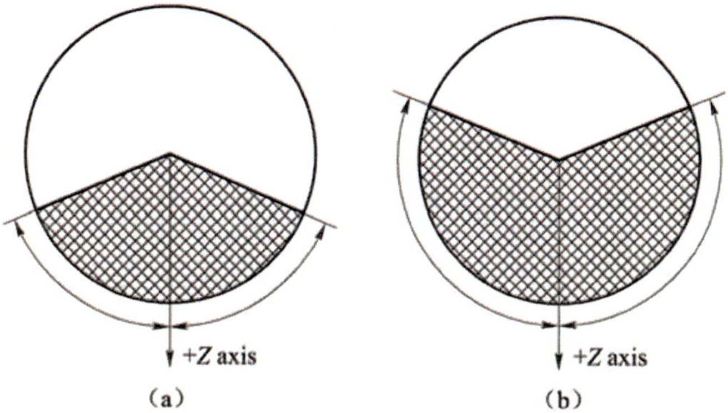

Fig. 2.3 Satellite-to-Earth coverage requirements. **a** Flare angle of TT&C antenna between satellite and Earth; **b** flare angle of TT&C antenna between side pendulum mode satellite and Earth

With the development of space technology, some satellites have added side pendulum maneuver mode, which requires the ground coverage to be extended in the Y direction (perpendicular to the satellite flight direction). The flare angle of the ground coverage area is shown in Fig. 2.3b. Because the coverage of satellite to the ground is required to be expanded gradually, the coverage of Earth TT&C antenna also needs to be widened accordingly.

For high-orbit satellites, to reach the intended orbit, multiple apogee maneuver and attitude adjustments are required. During this period, a stable and reliable RF channel is to be established by the TT&C system with the ground. After the satellite establishes the apogee orbital maneuver, the satellite $+Z$-axis points to the thrust direction and $-X$-axis points to the center of Earth. In order to make the TT&C antenna installed on the satellite $+z$-plane cover the Earth, its beam angle width shall be no less than 102°. In order to solve the TT&C problem in the state of apogee orbital maneuver, an identically polarized antenna array scheme with two TT&C antennas shall be used, in which the $+z$-direction TT&C antenna is inclined to install. To this end, the opposite polarized array method can also be used. Moreover, a combination of three TT&C antennas is also a feasible solution, and different TT&C antenna can be switched by a microwave switch to achieve full-space TT&C beam coverage.

2.1.2 Main Technical Requirements of TT&C Antenna

In this section, the main technical requirements of TT&C antenna are illustrated by taking S-band near-omnidirectional covering beam TT&C antenna as an example.

Operating frequency and bandwidth: The operating frequency band of the S-band TT&C antenna should be 2000–2300 MHz. According to ITU regulations, the ratio of uplink and downlink of TT&C frequency is 221:240. A transceiver antenna is often used, however, as an alternative, two separate antennas, a receiving antenna and a transmitting antenna can also be used.

Operating mode: on-orbit continuous operating mode.

Pattern and gain: using the antenna spherical coordinate system shown in Fig. 2.4, gain of the typical TT&C antenna unit and coverage requirements are

The antenna +Z-axis (usually toward-Earth) is the rotation axis. When θ is within the range of 0°–±68°, the gain is ≥−4 dBi.

Using the satellite spherical coordinate system shown in Fig. 2.5, the gain and coverage requirements of typical TT&C antenna array are as follows:

With the satellite +Z-axis as the rotating axis and θ in the coverage range of 0°–±68°, the gain is ≥−10 dBi.

With the satellite −Z-axis as the rotating axis and θ in the coverage range of 0°–±68°, the gain is ≥−10 dBi.

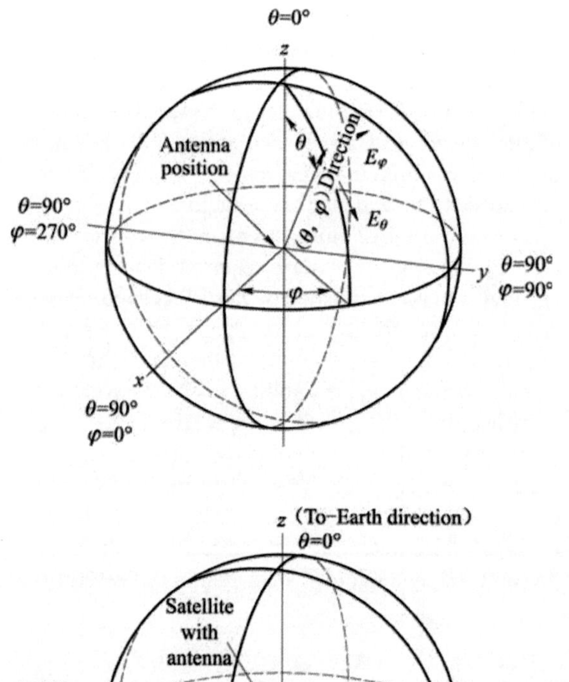

Fig. 2.4 Unit antenna spherical coordinate system

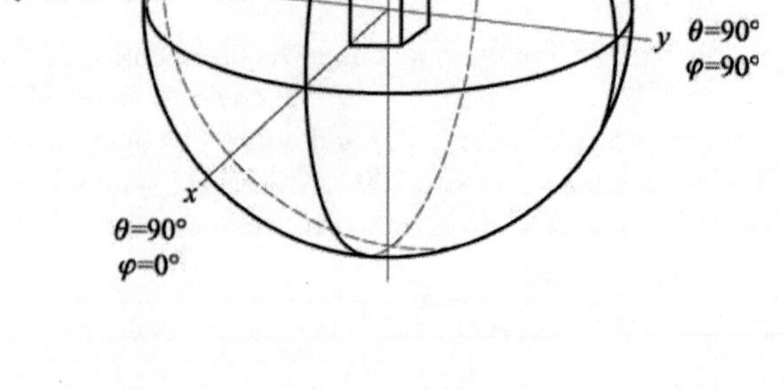

Fig. 2.5 Satellite spherical coordinate system

Polarization and axial ratio: Use left-handed circular or right-handed circular polarization, usually with an axis ratio of less than 3 dB.

Voltage standing wave ratio: Generally not more than 1.5.

Power capacity: Depending on the specific mission, the power capacity of low-orbit spacecraft generally does not exceed 1 W, while that of high-orbit spacecraft or deep space probes is generally 10–25 W. Because the TT&C antenna operates continuously in orbit, it is necessary to prevent multipaction and low-pressure discharge.

Space environment adaptability: TT&C antenna is located outside the cabin, and its space environment is relatively harsh. EMC, mechanics, heat, radiation, etc., shall be taken into account in the design. In addition, the product should still be able to operate normally after the corresponding test and assessment.

Reliability and product life: The TT&C antenna is usually on-orbit without backup, and cannot be maintained on-orbit, so its reliability is required to be higher, and the reliability at the end of on-orbit life is not less than 0.999.

2.2 S-Band Bifilar Conical Helical TT&C Antenna

The S-band bifilar conical helical TT&C antenna is derived from the bifilar cylindrical helical antenna. It features excellent broadband impedance, miniaturization, lightweight and high reliability. This antenna adopts the concept of conical equiangular spiral antenna, which can expand the operating bandwidth of the antenna. The structure of conical equiangular spiral antenna is completely determined by the angle. A continuous scaling antenna similar to the original structure can be obtained when the angle changes. Its electrical performance is independent of frequency and belongs to the frequency-independent antenna. Therefore, the pattern and impedance characteristics of this antenna have a broadband characteristic.

The geometric parameters of the S-band bifilar conical helical TT&C antenna are shown in Fig. 2.6. A schematic diagram of monofilar helical line TT&C antenna is given in Fig. 2.6, where $\delta = 0°$.

The helix is wound on a conical surface with a half-cone angle θ_0. The helix equation is as follows:

$$r = \begin{cases} r_0 e^{b'\varphi} & \theta = \theta_0 \\ 0 & \theta \neq \theta_0 \end{cases} \tag{2.1}$$

where r_0 is the distance from the vertex to the starting point of the helix along the tapered surface; r is the distance from the vertex to the helix along the tapered surface.

The angle between the helix and the conical generatrix is α,

$$\tan\alpha = \sin\theta_0 / b' \tag{2.2}$$

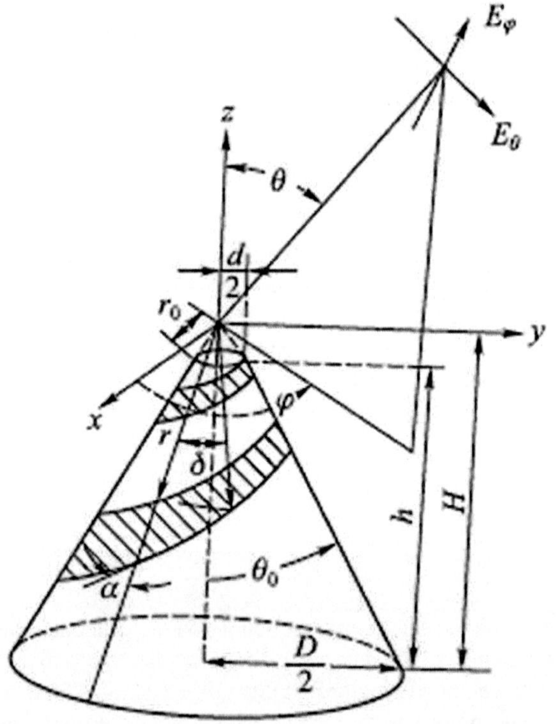

Fig. 2.6 Schematic diagram of S-band bifilar conical helical TT&C antenna winding

Similar to a planar helix, the helix angle remains unchanged when it travels from the starting point to the end, so it is also called an equiangular helix. The value of parameter b' reflects the degree of tightness of the helix. When the cone angle is fixed, the larger the b' is, the smaller α is, which indicates that the helix is looser. The spiral wire moves around the cone at a greater distance along generatrix.

The S-band bifilar conical helical TT&C antenna adopts a balanced feed structure. The helix of this structure consists of two symmetrical helical arms. The equations of these two helixes are as follows

$$r_1 = r_0 e^{b'\varphi}$$
$$r_2 = r_0 e^{b'(\varphi - \pi)} \tag{2.3}$$

The effective radiation area of the tapered helical TT&C antenna is substantially in an area whose circumference is about one wavelength. When the frequency changes, the effective radiation area moves forward and backward along the axis. When the frequency is high, it moves toward the top of the cone. When the frequency is low, it moves toward the cone bottom, so that the antenna has broadband characteristics. The antenna full-cone can be roughly divided into three parts: the area from the feeding end to $a/\lambda \approx 0.12$ area of the tapered helical TT&C antenna is known as transmission area. The frequency area of $a/\lambda \approx 0.12$–0.18, called the effective radiation area, is substantially in an area whose circumference is about one wavelength. The frequency area of $a/\lambda > 0.18$ to feeding end, where the electric field intensity is rather weak, is therefore called the terminal non-excitation area. If the amplitude of the electric field falls to 15 dB below the peak level in this area and the antenna is truncated, there is no significant influence on the electrical performance of the antenna. Similarly, if the antenna is truncated where the electric field amplitude in the antenna transmission area drops from peak level to 3 dB, the electrical performance of the antenna will not be significantly affected. The electric field distribution along the cone provides a numerical basis for truncating cone.

The application example of the product is given below. The S-band bifilar conical helical TT&C antenna includes two polarization modes: LH circular polarization and RH circular polarization. It can provide half-space (LH or RH) circular polarization beam coverage. The antenna is a common receiving/transmitting antenna and is widely used in high-, medium- and low-orbit satellite TT&C subsystems.

The antenna consists of a radome, helix, an outer conductor, an inner conductor and a flange. Thermal control white paint is sprayed on the outer surface of the radome. The antenna is shown in Fig. 2.7. According to the layout of the whole satellite, the antenna can be supported at different altitudes to improve the radiation performance of the antenna.

Typical technical specifications of the S-band bifilar conical helical TT&C antenna are shown in Table 2.1.

High Frequency Simulation Software (HFSS) is used to simulate the electrical performance of S-band bifilar conical helical TT&C antenna. The gain pattern of an antenna unit is shown in Figs. 2.8 and 2.9. The measured pattern is shown in Figs. 2.10

Fig. 2.7 Diagram of an S-band bifilar conical helical TT&C antenna

Table 2.1 Typical technical specifications of the S-band bifilar conical helical TT&C antenna

No	Item	Technical specification requirement
1	Operating band (MHz)	2000–2300
2	Polarization	RHC or LHC
3	Voltage standing wave ratio in bandwidth	≤1.5
4	Gain pattern	Half-cone angle 0°–68°, gain ≥ −4 dBi
5	Electrical interface	TNC socket
6	Operating temperature (°C)	−172 to +112
7	Reliability	0.999
8	Life (year)	(In orbit) 8, (ground test, storage) 5

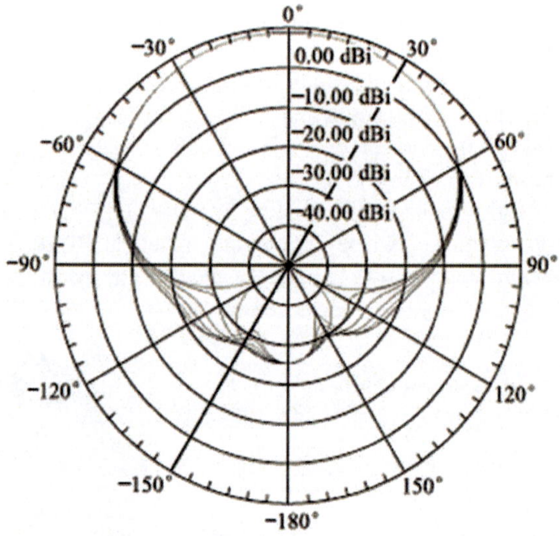

Fig. 2.8 Gain pattern simulation results (uplink)

and 2.11, and the specification line is given as a reference. Also, the pattern curves of different sections are shown in Figs. 2.8, 2.9, 2.10 and 2.11.

2.3 S-Band Quadrifilar Helical Antenna

Kilgus C. C. first invented the resonant quadrifilar helical antenna in 1968. The structure of the antenna is shown in Fig. 2.12. The antenna consists of four helical arms and metal support rods. Each arm has an integral multiple of a quarter of

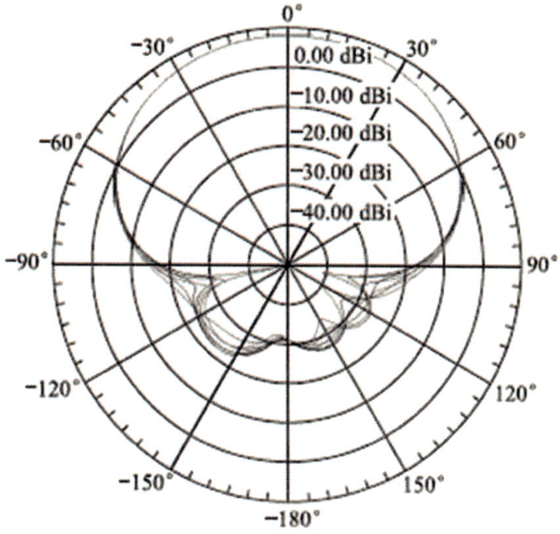

Fig. 2.9 Gain pattern simulation results (downlink)

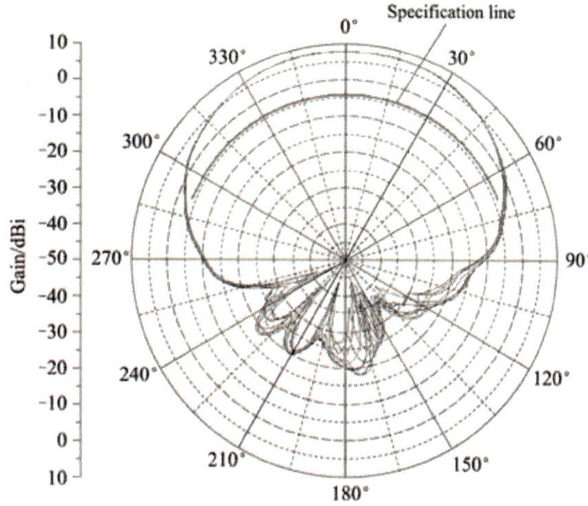

Fig. 2.10 Gain pattern measurement results (uplink)

the operating wavelength. When the multiplier relationship is odd, the helical arm terminal is an open end; otherwise, the helical arm terminal is short-circuited. The four helical arms have the same feeding terminal current and each phase differs by 90° and is 0°, 90°, 180° and 270°, respectively. The circular polarization can be realized by this feeding method.

The quadrifilar helical antenna parameters are shown in Fig. 2.13. It can be regarded as two sets of bifilar helix fed by spatially orthogonal 90°. The helical-shaped design can form a cardioids-shaped circular polarization pattern, and the maximum direction of the radiation pattern is the axial direction.

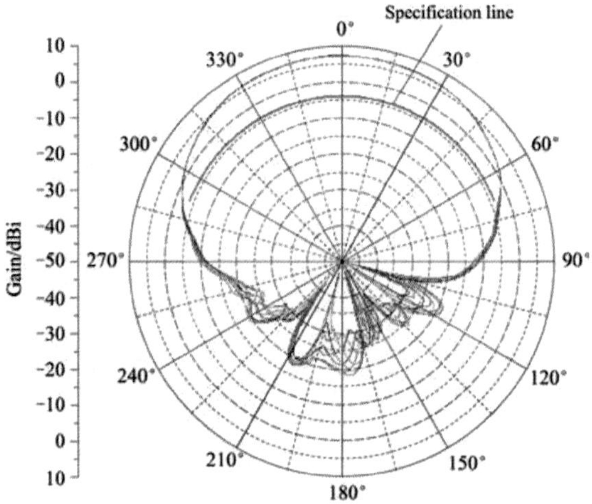

Fig. 2.11 Gain pattern measurement results (downlink)

Fig. 2.12 Schematic diagram of resonant quadrifilar helical antenna structure

According to the annular dipole model, the working principle of the antenna can be analyzed theoretically. The radiation field of the bifilar helix formed by units 1 and 2 can be expressed as

$$E_{\theta 1,2} = A_1 \mathrm{e}^{-\mathrm{j}kr}(-\cos\varphi + \mathrm{j}\sin\varphi\cos\theta) \tag{2.4}$$

$$E_{\varphi 1,2} = A_1 \mathrm{e}^{-\mathrm{j}kr}(\sin\varphi\cos\theta + \mathrm{j}\cos\varphi) \tag{2.5}$$

Fig. 2.13 Quadrifilar helical antenna parameter

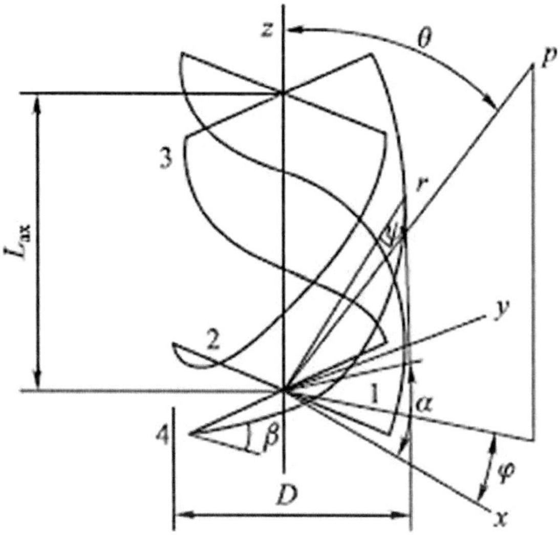

Similarly, the radiation field of the bifilar helix formed by unit 3, 4 can be expressed as

$$E_{\theta 3,4} = A_2 e^{-j(kr-\pi/2)}(\sin\varphi + j\cos\varphi\cos\theta) \tag{2.6}$$

$$E_{\varphi 3,4} = A_2 e^{-j(kr-\pi/2)}(\cos\varphi\cos\theta - j\sin\varphi) \tag{2.7}$$

When $A_1 = A_2$, the total radiation field of the quadrifilar helix can be expressed by the following equation

$$E_\theta = E_{\theta 1,2} + E_{\theta 3,4} = Ae^{-jkr}(1+\cos\theta)e^{-j\varphi} \tag{2.8}$$

$$E_\varphi = E_{\varphi 1,2} + E_{\varphi 3,4} = Ae^{-jkr}(1+\cos\theta)e^{-j(\varphi-\pi/2)} \tag{2.9}$$

Circular polarization can be achieved when E_φ and E_θ are equal in amplitude and the phase difference is 90°.

The key factor of a quadrifilar helical antenna is to feed the four arms with equal amplitude and 90° difference in sequence. It can be realized by using phase-shifting network or self-phase-shifting structure.

At present, the backfire resonant quadrifilar helical antenna is widely used and has good performance. It consists of four helical arms and two straight arms at the top. The rotation angle of each helical arm is generally 90°, and the length of which is approximately λ/4. Its radiation performance is less affected by the satellite, so it is especially suitable for spacecraft installations with high antenna density. As long as it is properly designed, the free-space pattern of the quadrifilar helical antenna is

Fig. 2.14 S-band four-arm helix antenna

not much different from that mounted on the carrier. In addition, the antenna is small in electrical size and has excellent circular polarization characteristics; therefore, it is also the preferred antenna solution on satellite.

An application is given below: S-band quadrifilar helical antenna, which provides half-space circular polarization (LHC or RHC) beam coverage, and transmit-receive sharing, is suitable for low-orbit satellite TT&C systems.

Generally, the S-band quadrifilar helical antenna consists of a radome, helical strips, a reflecting dish and a phase shifter. The radome and reflecting disk are coated with thermal control white paint. The antenna is shown in Fig. 2.14. According to the layout of the satellite, the antenna can be supported at different altitudes to improve the radiation performance of the antenna.

Typical technical specifications of the S-band quadrifilar helical antenna are shown in Table 2.2.

Table 2.2 Typical technical specifications of S-band quadrifilar helical antenna

No	Item	Technical specification requirement
1	Operating band (MHz)	2000–2300
2	Polarization	RHC or LHC
3	VSWR in bandwidth	$\leq$1.5
4	Gain pattern	Half-cone angle 0°–68°, gain ≥ -4 dBi
5	Electrical interface	TNC socket
6	Operating temperature	−100 to +100°C
7	Reliability	0.999
8	Life (year)	(On orbit) 8, (ground test, storage) 5

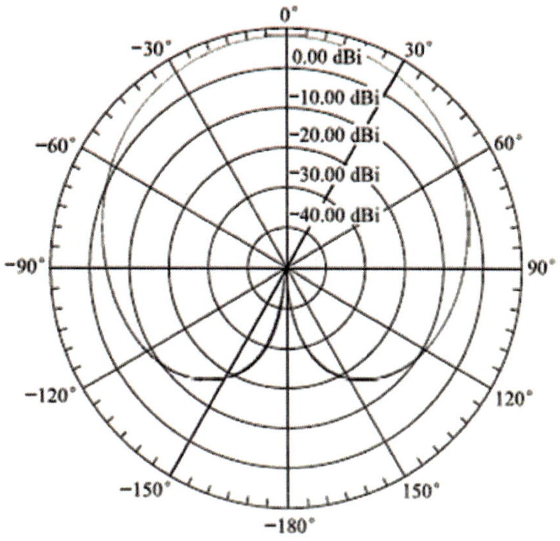

Fig. 2.15 Gain pattern simulation results (uplink)

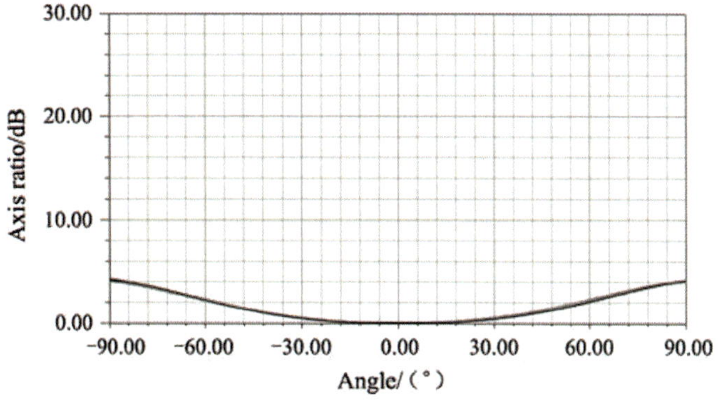

Fig. 2.16 Axis ratio simulation results (uplink)

High Frequency Simulation Software (HFSS) is used to simulate the electrical performance of the S-band quadrifilar helical antenna. The gain patterns and axis ratio curves of antenna units are shown in Figs. 2.15, 2.16, 2.17, 2.18, 2.19, 2.20, 2.21 and 2.22.

2.4 S-Band Cavity-Backed Dipole TT&C Antenna

For the TT&C subsystem of a high-orbit satellite, in order to ensure the requirement of TT&C in the active segment orbit maneuver of the satellite, the pattern shall satisfy

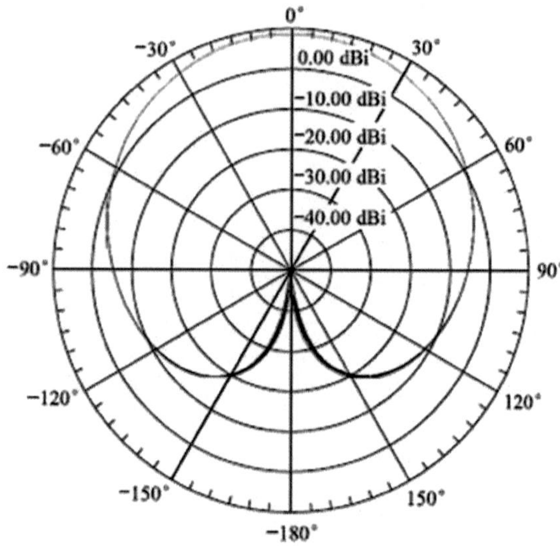

Fig. 2.17 Gain pattern simulation results (downlink)

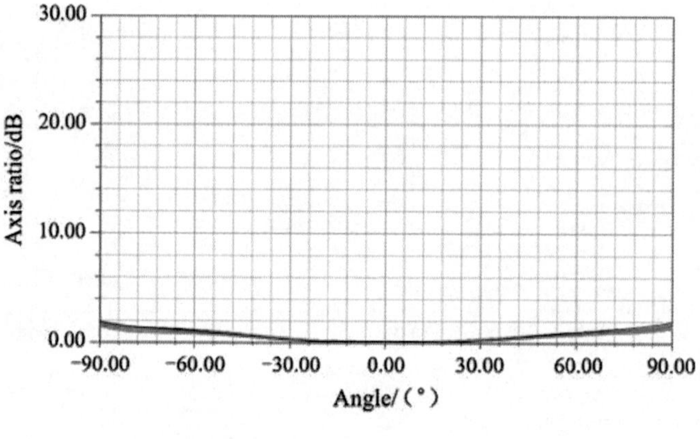

Fig. 2.18 Axis ratio simulation results (downlink)

the coverage requirement near half-space under complex boundary conditions. The cavity-backed dipole TT&C antenna is a wise choice. The antenna has the advantages of hemispheric radiation with low back lobe, small size, lightweight and can transmit large power.

The S-band cavity-backed dipole TT&C antenna is composed of radiation dipoles, back-cavity, choke ring and feed line, as shown in Fig. 2.23. The design parameters of the antenna include the diameter of the back cavity, the height of the back cavity, the diameter of the choke ring, the height of the choke ring, the length of the two groups of radiation dipoles and the distance between the radiation dipole and the bottom of the back cavity. Because of the function of the back cavity reflector, the antenna

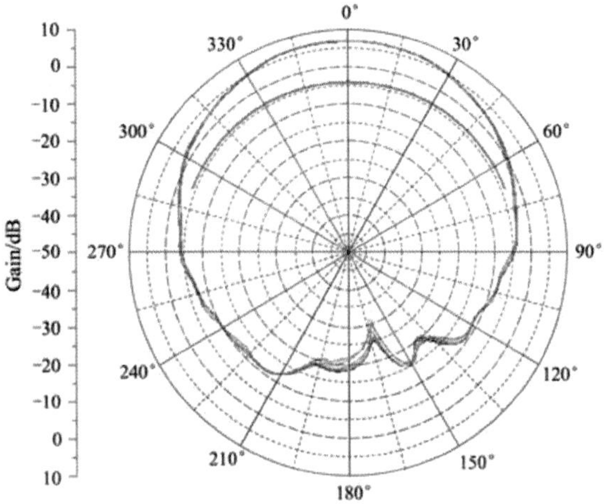

Fig. 2.19 Gain pattern measured results (uplink)

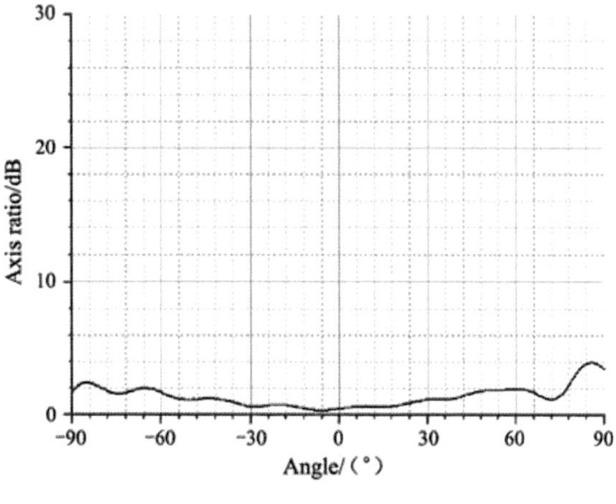

Fig. 2.20 Measured results of axis ratio (downlink)

realizes unidirectional radiation. In the design, when the height of the back cavity is about $\lambda/4$, the diameter of the back cavity is λ. The choke ring is used to reduce the back lobe of the antenna. Two groups of orthogonal unequal-length radiation dipoles realize circular polarization radiation, feeding Balun realizes 90° difference between two groups of dipoles, and the relative position between feeding Balun and dipole depends on the antenna rotation direction. Furthermore, the antenna is fed by coaxial mode.

By optimizing the size of the dipoles and the back cavity, the wide beam coverage of the antenna can be obtained. The antenna is machined in one piece without solder joints and has high reliability. The antenna technical specifications are shown in Table 2.3.

HFSS is used for electrical performance simulation of the S-band cavity-backed dipole TT&C antenna, and the simulation gain patterns of the antenna unit are shown

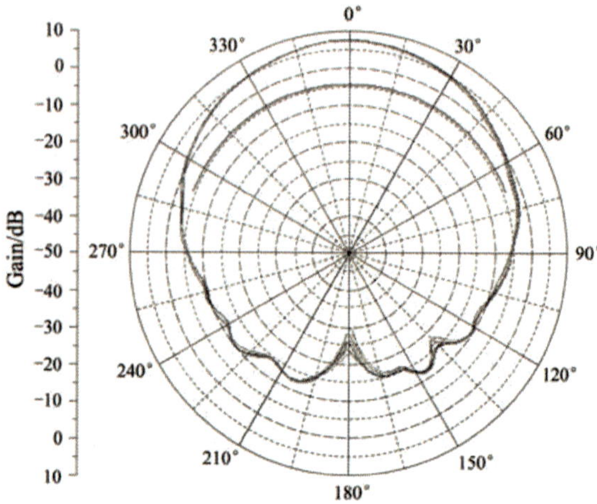

Fig. 2.21 Measured results of gain pattern (downlink)

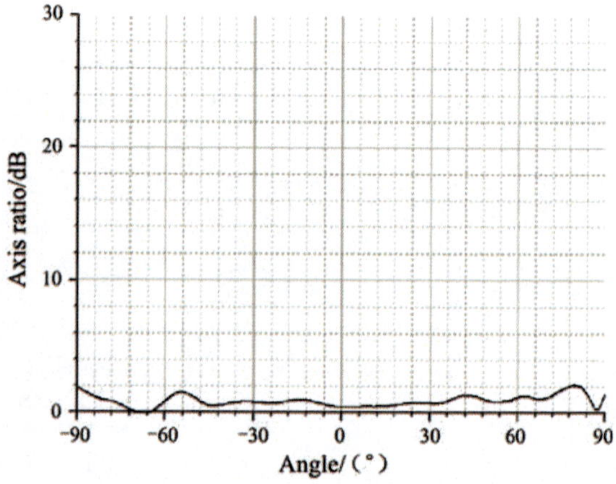

Fig. 2.22 Measured results of axis ratio (downlink)

in Figs. 2.24 and 2.25. The measured patterns of the antenna unit are shown in Figs. 2.26 and 2.27. The comparison of the simulated and measured patterns of an orbital satellite is shown in Figs. 2.28 and 2.29. It can be concluded that antenna performance meets the requirements.

2.5 C-Band Coaxial-Slot Double-Cone-Disk TT&C Antenna

Antennas introduced in the above three sections are suitable for three-axis stabilized satellite. For the spin-stabilized satellite TT&C subsystem, in order to avoid channel

Fig. 2.23 An S-band back cavity dipole TT&C antenna

Table 2.3 S-band back cavity dipole TT&C antenna

No	Item	Technical specification requirement
1	Operating mode	Transmit-receive sharing
2	Operating band (MHz)	2000–2300
3	Polarization	RHC or LHC
4	VSWR in bandwidth	$\leq$1.5
5	Gain pattern	Half-cone angle 0°–90°, gain ≥ -7 dBi
6	Electrical interface	TNC socket
7	Operating temperature	-180 to $+100$°C
8	Total mass (kg)	0.2
9	Power capacity (W)	20
10	On-orbit life (year)	$\geq$12

level fluctuation caused by satellite spin, the radiation pattern of TT&C antenna is required to be axisymmetric. Antenna forms such as double-cone or double-disc are usually adopted.

Figure 2.30 shows a C-band coaxial slot double-cone-disc TT&C antenna, which consists of a telemetry transmitting antenna and a command receiving antenna. The upper part is a tele-control antenna (receiving), and the lower part is a telemetry antenna (transmitting). The two antennas have the same structure and are installed coaxially. Each antenna is fed by a coaxial tube. The outer wall of the coaxial tube is connected with a pair of cone discs. There are eight slots with an inclination of 45° on the outer wall of the coaxial tube between the cone discs. The inclination direction depends on the circular polarization rotation direction. The outer diameter of the cone disc is about 2λ, and the size of the opening is about 0.8λ, forming a ring

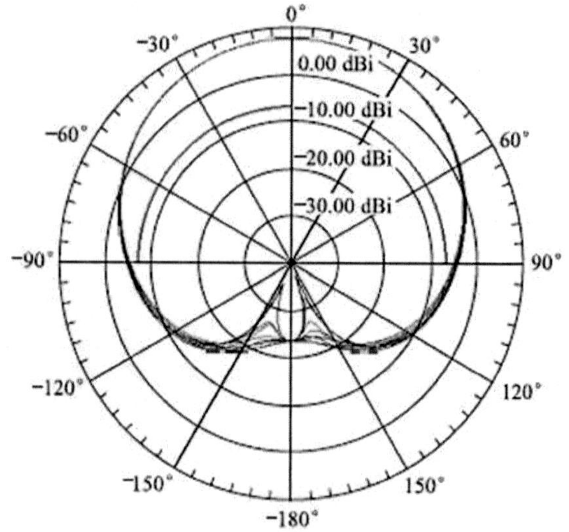

Fig. 2.24 Gain pattern simulation results (uplink)

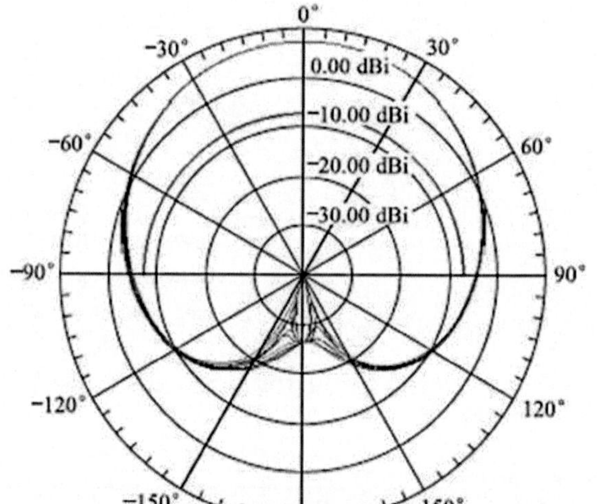

Fig. 2.25 Simulation results of gain pattern (downlink)

beam with a half-power beamwidth of about ±30°. In order to match the impedance of the antenna well, the inner diameter of the coaxial tube with slots has a diameter of 0.7λ, and a step transition is used between the coaxial tube and the input port.

This antenna is fed by a coaxial transverse electric and magnetic field (TEM) waves. The coaxial inner and outer conductors are extended to coaxial waveguides which can excite and transmit H_{11} modes. The mode excites the circularly distributed inclined slot array of coaxial cylinders. The radiated electric field of the inclined slot is decomposed into two components of parallel and vertical discs. By adjusting

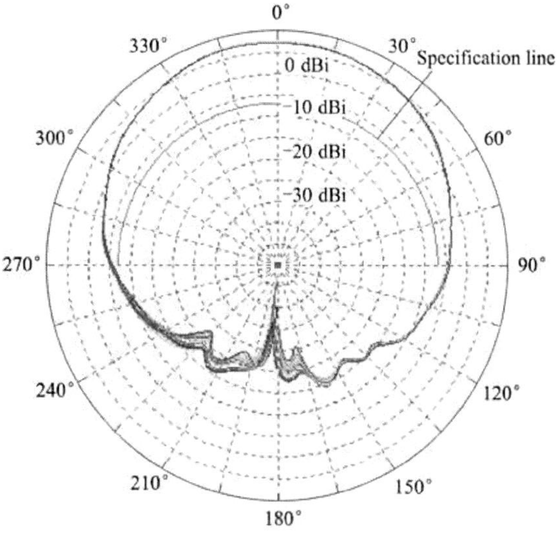

Fig. 2.26 Measured results of gain pattern (uplink)

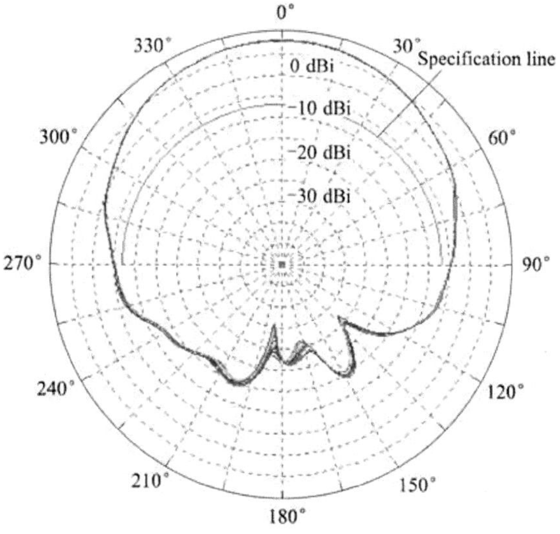

Fig. 2.27 Measured results of gain pattern (downlink)

the spacing between the two discs, the two-component electric fields have different phase shifts, so that the antenna aperture satisfies the circular polarization radiation condition. In order to ensure the electromagnetic isolation between the receiving and transmitting channels, the inclined slot directions of the upper and lower disk antennas are orthogonal, so that they radiate counter-rotational circular polarization. The satellite telemetry is LH circular polarization and the remote control is RH circular polarization. This dual-disk antenna has been widely used in synchronous orbit spin-stabilized satellites.

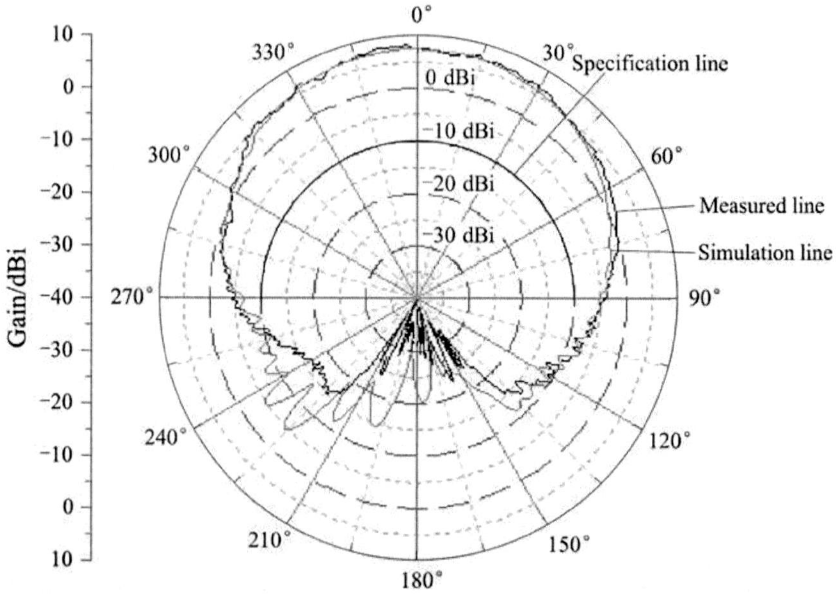

Fig. 2.28 Simulation and measured results of satellite pattern (uplink)

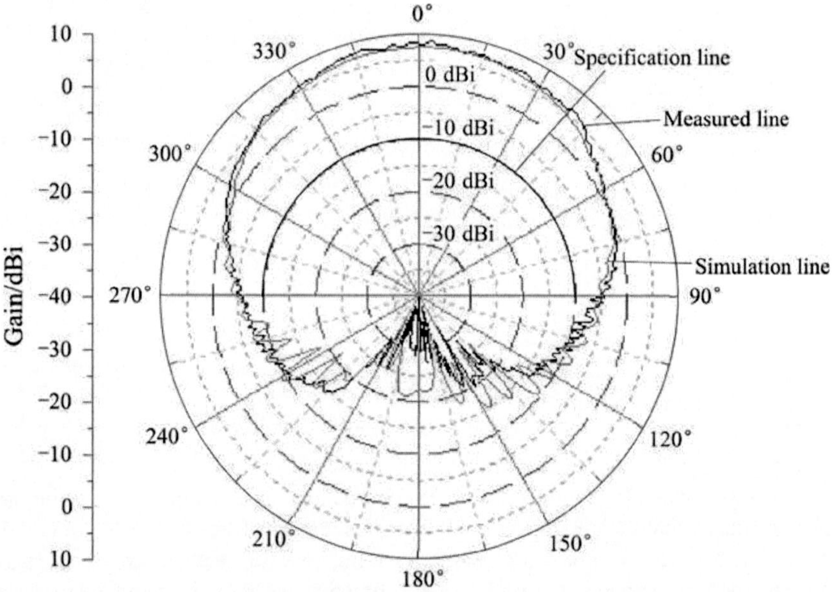

Fig. 2.29 Simulation and measured results of satellite pattern (downlink)

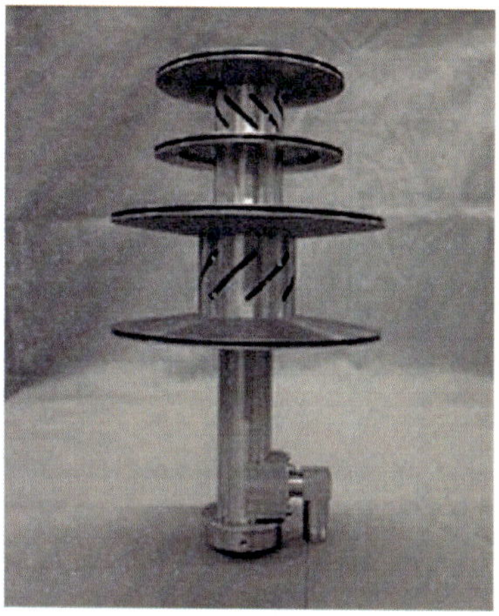

Fig. 2.30 C-band coaxial slot double-cone-disk antenna

The technical specifications of the antenna are shown in Table 2.4.

Figures 2.31, 2.32, 2.33 and 2.34 show the test gain of the antenna in the satellite environment. The *xy*-plane is the equatorial plane and the *xz*-plane is the meridian plane. The specification lines are given in Figs. 2.31, 2.32, 2.33 and 2.34 for reference.

Table 2.4 Technical specification of C-band coaxial slot double-cone-disk antenna

No	Item	Technical specification requirement
1	Operating band (MHz)	Receiving: 5926–5965 Transmitting: 4170–4192
2	Polarization	Receiving: RHC Transmitting: LHC
3	VSWR in bandwidth	$\leq$1.2
4	Gain pattern	Circle pattern in equatorial plane, 8-shaped in meridian plane Receiving: at the maximum point of the meridian surface 2.5 dBi, with this as the center within $\pm 30° > -0.5$ dBi Transmitting: at the maximum point of the meridian surface 3 dBi, with this as the center within $\pm 30° > 0$ dBi
5	Electrical interface	TNC socket
6	Operating temperature (°C)	−180 to +100
7	Total mass (kg)	0.85
8	Dimension ($\phi \times L$) (mm)	ϕ160 × 281
9	Power capacity (W)	10

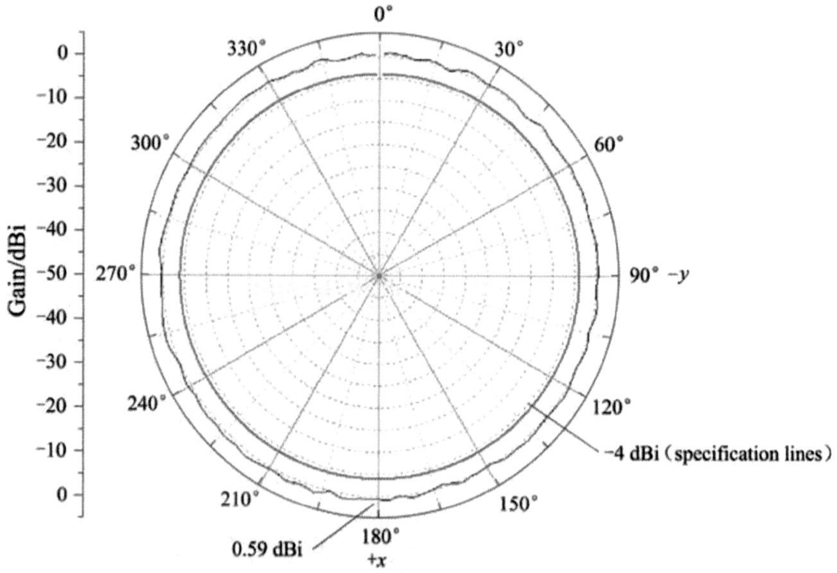

Fig. 2.31 *xy*-plane pattern and gain (downlink, measured)

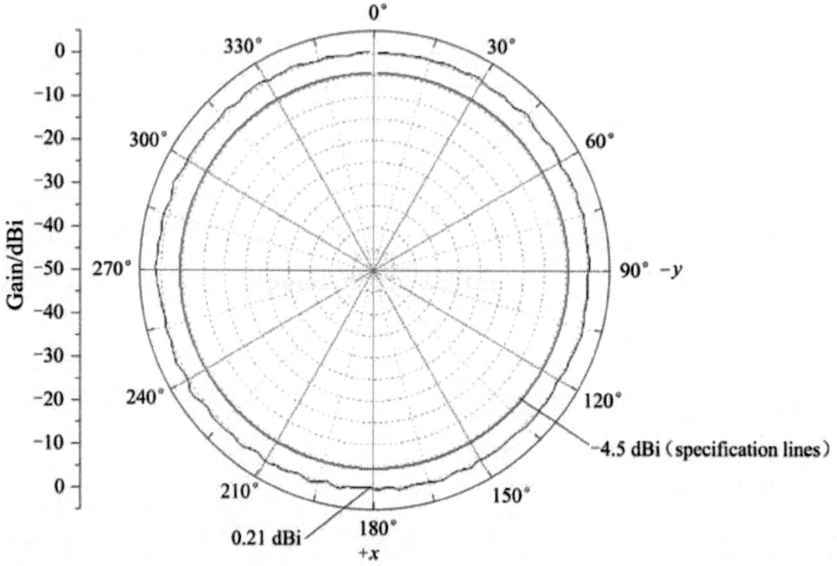

Fig. 2.32 *xy*-plane pattern and gain (uplink, measured)

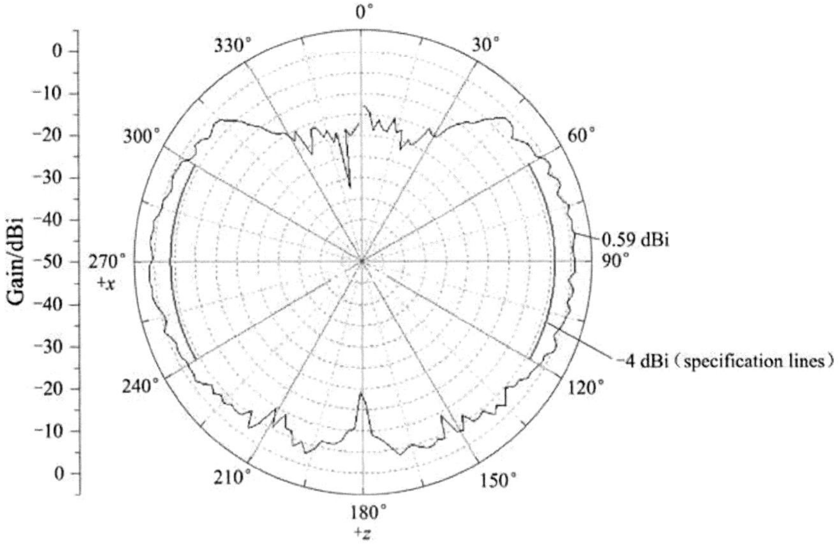

Fig. 2.33 *xz*-plane pattern and gain (downlink, measured)

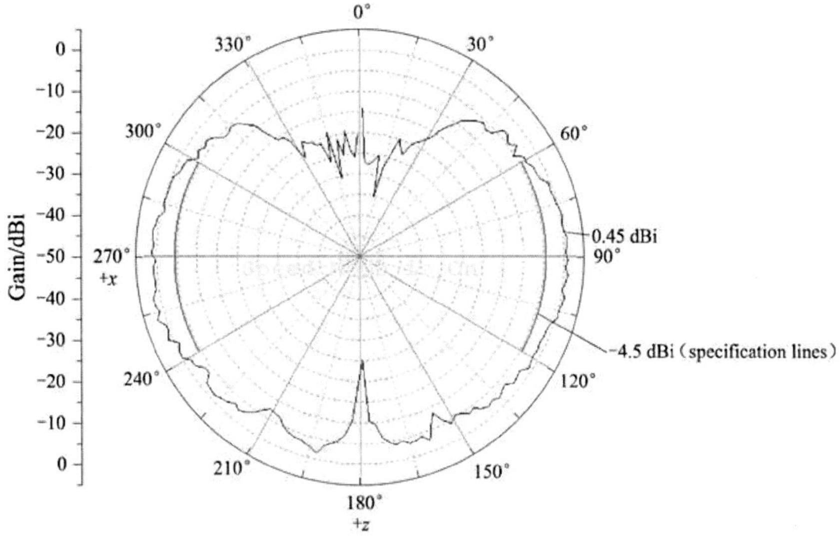

Fig. 2.34 *xz*-plane pattern and gain (uplink, measured)

2.6 Other TT&C Antennas

According to the regulations of ITU, the TT&C frequency of communication satellite can use the frequency points in the communication frequency band of the communication satellite. Communication satellites used C-band in the early stage globally, and then gradually extended to Ku- and Ka-bands. Although the frequency bands are different, the technical requirements are basically the same as those of the same altitude orbit satellites. The two TT&C antennas shown in Figs. 2.35 and 2.36 are both TT&C antennas used by China's communications satellites in geostationary orbit. They operate in C-band and can be divided into linear polarization and circular polarization. They receive (command) and transmit (telemetry) separately. The circular waveguide aperture antenna is adopted. The outer choke is to suppress the back lobes and reduce the influence of the satellites. The side lobes and the back lobes outside the range of $\pm 100°$ are below -20 dB and the main specifications are shown in Table 2.5. The X-band TT&C antenna is also used in deep space exploration spacecraft. Please refer to Chap. 11 for more details.

In a word, there are various types of TT&C antennas. According to different satellite sizes, structures, layouts and the feature of the mission, different antenna

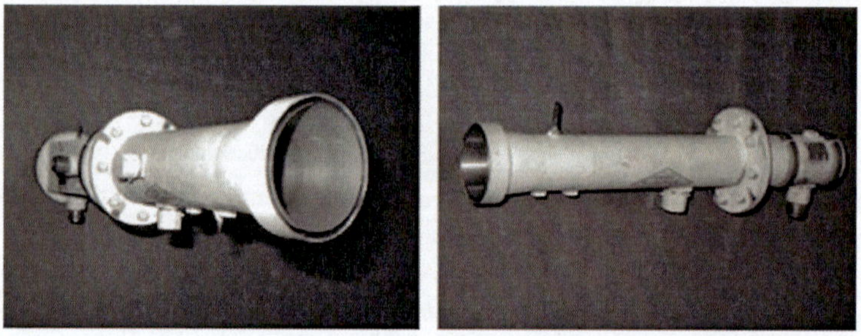

Fig. 2.35 C-band linear polarization telemetry and tele-control antenna for three-axis stabilized satellite platform in geostationary orbit

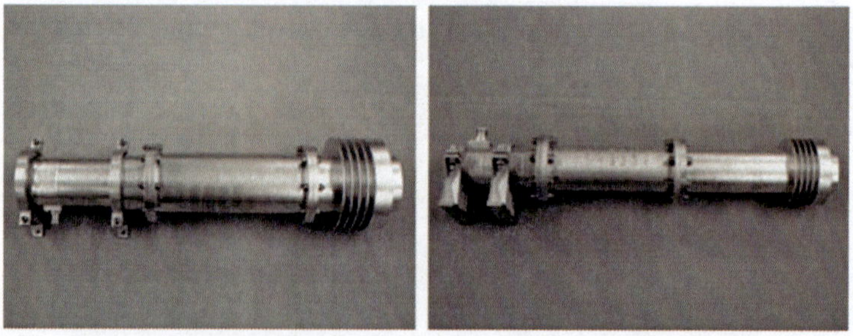

Fig. 2.36 C-band circular polarization telemetry and tele-control antenna for three-axis stabilized satellite platform in geostationary orbit

Table 2.5 Technical specifications of C-band TT&C antenna for communication satellites in geostationary orbit

No	Item	Technical specification requirement
1	Operating mode	Separate mode of receiving and transmitting
2	Operating band (MHz)	Receiving: select two frequency points from 3700 to 4200
		Transmitting: select two frequency points from 3700 to 4200
3	Polarization	Right circular polarization (RCP) or left circular polarization (LCP) or linear polarization (LP)
4	VSWR in bandwidth	≤ 1.5
5	Gain pattern	Half cone angle 0°–90°, gain ≥ -7 dBi
6	Electrical interface	TNC socket
7	Operating temperature	-180 to $+100$°C
8	Total mass (kg)	≤ 0.8
9	Power capacity (W)	20
10	On-orbit life (year)	≥ 12

unit design schemes or their combination forms can be selected. In addition, low-cost microstrip patch antennas are widely used in microsatellites, which will be briefly introduced in Chap. 8.

Chapter 3
Spacecraft DTA

Abstract Spacecraft data transmission antenna (DTA) refers to the antenna for transmitting remote sensed data to earth station. This chapter provides the design of two types of earth-matched beam antennas and spot beam dual circularly polarized DTA.

Spacecraft data transmission antenna (DTA) refers to the antenna for remote sensing data transmission of spacecraft, which can be divided into three categories according to the types of spacecraft orbits: MEO and LEO remote sensing satellite DTA, HEO remote sensing satellite DTA and deep space exploration satellite DTA. According to the transmission target, it can also be divided into satellite-to-ground antenna and inter-satellite antenna. Satellite-to-ground DTA is used to transmit data from satellite-to-Earth station, which is generally known as Data Transmission Antenna.

Early remote sensing satellites were all recoverable satellites, and the remote sensing data could not be processed and applied until the satellite returned to the ground. With the development of space technology, radio transmission systems of remote sensing information to the Earth were equipped in many application satellites on LEO and MEO, which are commonly called transmission-type remote sensing satellites. Because of their low orbits, most of these satellites have a large field of view (about 120°) in the altitude range of more than 1000 km above the ground. In the range of satellite-to-ground field of view, the transmission loss of satellite-to-ground link varies greatly. Generally, the level change caused by the spatial attenuation of transmission paths at the sub-satellite point and the edge of the coverage section is about 10 dB. During the whole operating period, in order to ensure that the ground receiving station has a good reception effect on satellite remote sensing information, especially the digital image information, it is required that the satellite downlink data transmission signal has essentially the same level in the coverage area. Therefore, the gain distribution of satellite-borne DTA in the coverage area should compensate for the level difference caused by transmission path attenuation. This type of beam is often referred to as Earth-matched beam.

Data transmission antennas of typical transmission-type remote sensing satellites, such as Spot in France, Landsat-D of NASA and ERS-1 of ESA, use shaped reflector antennas to realize the Earth-matched beam design.

R. You et al., *Technologies for Spacecraft Antenna Engineering Design*, Space Science and Technologies, https://doi.org/10.1007/978-981-15-5833-7_3

Earth-matched beam-shaped antenna for MEO and LEO satellite has been studied and applied for a long time in China. In October 1999, China-Brazil Resources 1 satellite (CBERS-1) had been successfully launched, and in September 2000, transmission-type Earth observation remote sensing satellite YG-3 had been successfully launched. Because of the limitation of installation space and some technical factors, shaped reflector antennas were not used as the X-band DTA. Although the X-band DTA adopted backfire dual-line helical antenna or waveguide cross-slot-array combination antenna, the performance is very satisfactory. This kind of antenna has many outstanding advantages, such as compact structure, small size, lightweight, high reliability and low development cost. In the future, such antennas will be more widely used, especially on the small satellites which are flourishing recently.

Besides, a DTA with shaped reflector, which has the advantages of high gain and large power capacity and meets the increasing transmission rate requirements, has been also developed in China. With the further improvement of data transmission rate of remote sensing satellite, the bandwidth requirement of DTA has been increased to more than 400 MHz. However, fixed-beam Earth-shaped beam antenna can no longer meet the mission requirements. X-band spot beam dual circularly polarized DTA is developed at the right moment. This antenna uses a two-axis driving mechanism to realize the pointing of the antenna to the Earth station. The Equivalent Isotropic Radiated Power (EIRP) of the data transmission system on the satellite is greatly improved, while the transmission power is reduced.

With the development of remote sensing technology, the amount of data is further increased (data rate up to 1–2 Gbps). Some LEO remote sensing satellites have adopted Ka-band data transmission system, thus achieving larger transmission bandwidth and higher EIRP. The Ka-band DTA also uses a two-axis drive mechanism to realize that the DTA points to the Earth station. Besides, Ka-band data transmission system is widely used in high-orbit remote sensing satellites.

Deep space exploration satellites all adopt movable spot beam DTA to realize data transmission between satellite and Earth. The operating frequency band of antenna has changed from early S-band in Chang'e-1 to X-band at present.

With the continuous development of data relay satellite application in China, some LEO remote sensing satellites are equipped with relay data transmission system. The satellite remote sensing data is transmitted to the data relay satellite through relay DTA, and then to Earth station by the data relay satellite. This greatly improves the data transmission capability of the spacecraft system.

3.1 Main Technical Characteristics of Spacecraft DTA

At present, most remote sensing satellites (code rate below 300 Mbps) use DTAs with fixed matched beams, which cover the field of view of the satellite. For LEO remote sensing satellites, the beam coverage can reach over 130°. Therefore, the gain of Earth-matched beams cannot be too high. With the increase of remote sensors, the improvement of the remote sensors resolution and the operation of new remote sensors on the satellite, the amount of data required to be transmitted is increasing, which puts forward higher requirements for data transmission rate. However, this also requires an increase in the EIRP at the transmitting end. EIRP is an important indicator of the data transmission subsystem, which can be expressed as the product of the transmitter output power and antenna gain. Increasing the antenna gain means narrowing the beam and forming spot beam coverage. Therefore, beam scanning technology should meet the requirement of wider beam coverage. In the data transmission of HEO remote sensing, deep space exploration and inter-satellite links, the spot beam DTA with beam scanning capability is also used.

3.1.1 Earth Shaped Beam

The Earth shaped beams of LEO satellites are usually rotationally symmetrical, so the pattern function $P(\theta, \varphi)$ can be simplified to the form of $P(\theta)$ that only related to θ. The data transmission links depicted in the plane composed of satellite, Earth station and Earth center are shown in Fig. 3.1.

As shown in Fig. 3.1, point S stands for the current position of the satellite, point G represents the position of the Earth station, and point O represents the position of the Earth's center. Point D is the corresponding sub-satellite point, θ is the coverage angle, ε is the elevation angle of the Earth station, h is the altitude of the satellite orbit, and R_e is the average radius of the Earth.

According to the plane geometry, there is a vertical intersection point V in the plane, then

$$\begin{aligned} R(\theta) &= \overline{\mathrm{SV}} - \overline{\mathrm{GV}} \\ &= \overline{\mathrm{SO}}\cos(\theta) - \sqrt{R_e^2 - \left(\overline{\mathrm{SO}}\sin(\theta)\right)^2} \end{aligned}$$

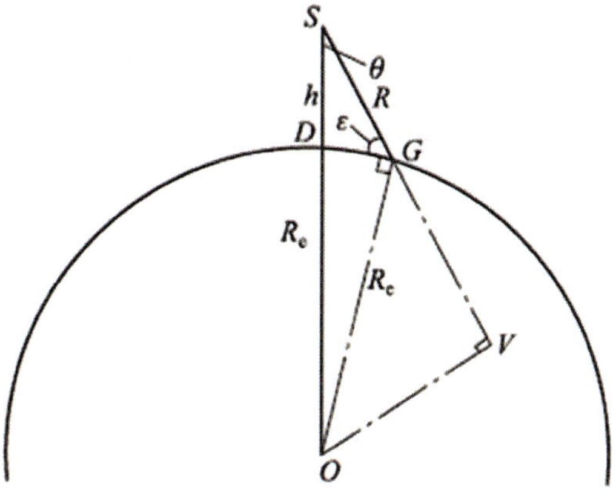

Fig. 3.1 Plane schematic of LEO satellite data transmission links

$$= (h + R_e)\cos(\theta) - \sqrt{R_e^2 - [(h + R_e)\sin(\theta)]^2} \tag{3.1}$$

As shown in Fig. 3.1, the relationship between the elevation angle of Earth station ε and coverage angle of antenna beam θ is as follows:

$$\frac{h + R_e}{\cos(\varepsilon)} = \frac{R_e}{\sin(\theta)} \tag{3.2}$$

When the elevation angle of Earth station ε reaches the minimum (e.g., 5°), the coverage angle θ reaches the maximum value of θ_{max}, and the corresponding $R(\theta)$ reaches the maximum value R_{max}, then the normalized power pattern function $P(\theta)$ when $\theta < \theta_{max}$ is as follows:

$$P(\theta) = \frac{R^2(\theta)}{R_{max}^2} = \frac{R^2(\theta)}{R^2(\theta_{max})} \tag{3.3}$$

The mathematical model of the Earth matching beam in case of $\theta < \theta_{max}$ can be obtained by transforming the above equation in decibel form.

$$P(\theta) = 20\lg\left(\frac{R(\theta)}{R_{max}}\right) \tag{3.4}$$

Taking a satellite with an orbital altitude $h = 1000$ km as an example, the minimum receiving elevation angle of the Earth station is 5°, and the maximum pointing gain is 5 dBi. According to the above mathematical model, the Earth shaped beam specification curve is calculated, as shown in Fig. 3.2.

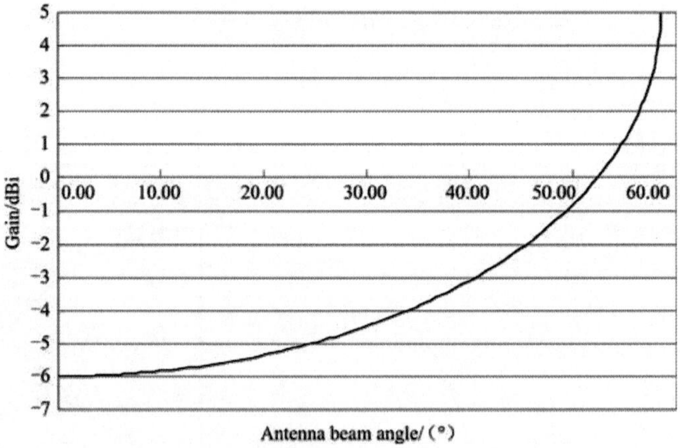

Fig. 3.2 An example of calculation of the data transmission antenna gain specification curve

3.1.2 *Main Characteristics of DTA*

1. Broadband characteristics

With the increasing of remote sensing data, the bandwidth requirements of spacecraft DTA are increasing. At present, remote sensing satellites generally use X-band for satellite-to-ground data transmission. According to the regulations of ITU, X-band can be used for transmitting remote sensing satellites from 8025 to 8400 MHz with a bandwidth of 375 MHz. In order to meet the higher transmission rate requirement, the transmission frequency is further extended to Ka-band with an available frequency range of 25–27.5 GHz and a bandwidth of 2.5 GHz.

2. High-power characteristics

The high EIRP requirement of spaceborne data transmission channel dictates that the spacecraft DTA must generally work at high power. The Earth shaped beam antenna for low- and medium-orbit spacecraft, which works at 50 W input power, is most widely used in China. Special requirements are put forward for its power tolerance in on-orbit vacuum environment and anti-multipactor capability in the high-power state. Emphasis should be placed on antenna product design, and the power tolerance and anti-multipactor capability of each product should be tested and verified.

The interference to the near microwave receiving system on the spacecraft surface is increased by the high-power operation of the DTA. Therefore, in the layout design of the antenna system, it is necessary to consider whether the isolation characteristics between the DTA and other microwave receiving antennas meet the system compatibility requirements.

3. High-reliability requirement

DTA is a key component of data transmission subsystem. The failure of DTA is fatal to satellite mission. For this reason, the backup design is generally adopted for DTA. Typical block diagram of data transmission system is shown in Fig. 3.3. DTA a and DTA b are backup configurations, and cross-backup of the X-band solid-state amplifier and the DTA are realized by switching matrix.

3.2 X-Band Bifilar Backfire Helical Antenna

Since there are many kinds of remote sensing payloads installed on the spacecraft surface toward-Earth and the room is very limited, a backfire bifilar helical antenna with small size and lightweight is used, as shown in Fig. 3.4. By utilizing the leaking wave radiation characteristics of the helical antenna, the main lobe of radiation is changed from back direction to broadside direction, thus forming a miniaturized Earth-matched beam-shaped antenna. The total height of the antenna is less than 300 mm and the total weight is less than 300 g. Because of its high-power operation, the design of its power capacity is a hard job.

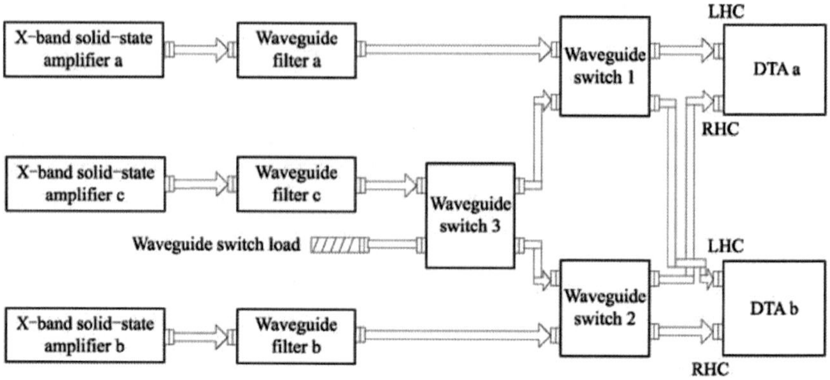

Fig. 3.3 Block diagram of data transmission channel backup design

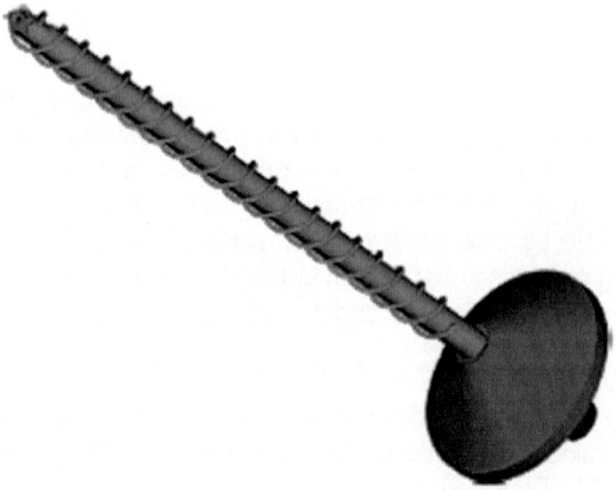

Fig. 3.4 Stereoscopic effect of backfire bifilar helical antenna

3.2.1 Radiation Mode of Helical Antenna

A helical antenna is a widely used circularly polarized broadband antenna. As a traditional radiation mode, the axial and normal modes of helical antennas are the main radiation modes that have been widely used. In the 1960s, Mittra and Klock studied the wave mode characteristics of the helical antenna and firstly pointed out the existence of complex propagation constant wave on the helix. This characteristic has attracted extensive research. Until the 1990s, Peterson and Mittra further studied and summarized the wave mode characteristics of the helical antenna, pointing out that helical antenna has not only axial attenuation-free propagation wave but also the surface wave and leaking wave radiation. The leaking wave radiation characteristics related to complex propagation constants are different from the normal mode and axial mode radiation of the classical helical antenna. When the operating frequency changes, the radiation main lobe will also change from backfire direction to broadside direction. The complex propagation constant β for helix can be expressed as

$$\beta = \beta_r - \mathrm{j}\beta_i \tag{3.5}$$

Equation (3.5) is put into the equation of propagation factor $\mathrm{e}^{-\mathrm{j}\beta z} = \mathrm{e}^{-\beta_i z}\mathrm{e}^{-\mathrm{j}\beta_r z}$. When the wave is strongly excited, its maximum radiation direction deviates from backfire direction, and its maximum directional angle approximates the following formula:

$$\theta_m = \cos^{-1}(\beta_r / k) \tag{3.6}$$

where $k = 2\pi/\lambda$ is the wave number in free space; β_r is the real part of the helix mode propagation constant; θ_m stands for the conical angle between the broadside direction and the antenna axis; β_i represents the exponential attenuation of the wave along the axis, that is, the characterized guided wave energy is converted into radiation rate. If the real part of the complex propagation constant is properly controlled, the maximum beam pointing desired can be approached, while the imaginary part of the complex propagation constant can be controlled to meet the desired lobe shape and gain requirements.

3.2.2 *Radiation Pattern of Complex Transmission Mode*

The helix parameters are selected according to different operating frequency bands. The model of X-band backfire bifilar helical antenna is shown in Fig. 3.4. The Earth shaped beam pattern is obtained by simulation and optimization, as shown in Fig. 3.5. The U-shaped curve in Fig. 3.5 is the Earth-shaped beam specification curve. As shown in Fig. 3.5, the gain margin of the antenna in the maximum pointing direction is smaller, while the margin in the central region is larger.

The maximum pointing direction of the pattern shifts outward with the increase of frequency, contrariwise, the maximum pointing direction of the pattern shifts inward. This determines the antenna pattern bandwidth. The maximum application bandwidth of the antenna is about 400 MHz in X-band.

3.2.3 *Multipactor Margin Design for DTA*

Multipactor margin design of DTA mainly refers to *the summary report of measurement and research technology of multipactor effect* completed by the microwave equipment division of European Space Research and Technology Center (ESTEC). The 6 dB test tolerance is used in the design of multipactor.

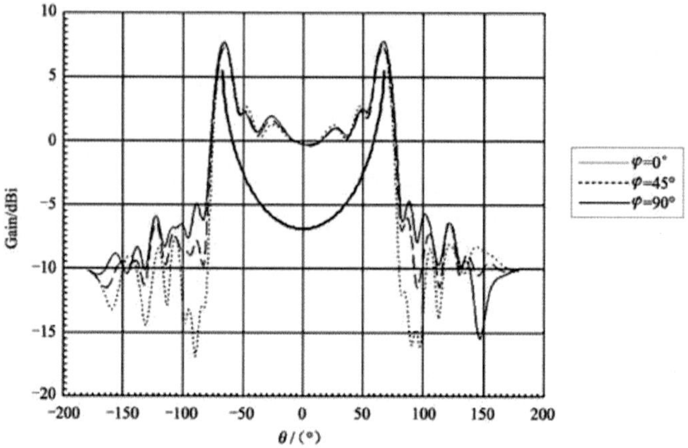

Fig. 3.5 Earth shaped beam patterns

The data transmission antenna is a high-power component. Therefore, in addition to satisfying the power capacity, the multipactor margin requirement is required. According to the practical development experience, 6 dB is generally taken to implement the requirements for in-orbit use. Table 3.1 shows the 6 dB distribution estimation.

The multipactor of DTA is simulated and analyzed using the software HFSS. The antenna design parameters are optimized according to the requirements of electrical performance, while the current and voltage distribution of the helical antenna are adjusted to improve the multipactor threshold. The specific process is as follows.

Because the helical antenna has periodic structural characteristics, and the voltage and the current are continuously tapered, so the voltage and current shall be calculated for different parts of the antenna during the analysis. The sampling part includes the first, the second and the third turn, in which the current and voltage are relatively concentrated, and three points are sampled in the first turn. In the design process, an appropriate structural adjustment should be made to the first two turns of the helical antenna to ensure that the multipactor thresholds of the first and the second turn are greatly improved under the condition that the antenna port inputs a continuous wave of 50 W, and are greater than that of the third turn. The multipactor threshold of the third turn is 16.5 dB and exceeds the standard design threshold by more than 10 dB.

Table 3.1 Design tolerance distribution for multipactor

Declining factor	Tolerance decibels (dB)
The worst case of VSWR 2:1	2.5
Oxidation	0.5
Pollution	2.0
Pollution migration	1.0
Total	6

Then, the multipactor analysis is performed on the narrowest slot of the model (i.e., the quarter-wave slot), and there the voltage is concentrated and gradually increased along the radius direction. After the analysis of the maximum voltage point, the design threshold of multipactor at this point is also higher than 13.8 dB, which is larger than the standard design threshold of 7.8 dB. The analysis curve for the multipactor threshold is shown in Fig. 3.6, and the analysis curve for the multipactor margin in consideration of the 6 dB multipactor evaluation standard is shown in Fig. 3.7.

Here, the analysis is performed at an operating frequency of 8 GHz. As the operating frequency increases, the multipactor threshold should also increase. The analysis results show that the DTA has enough multipactor design margin.

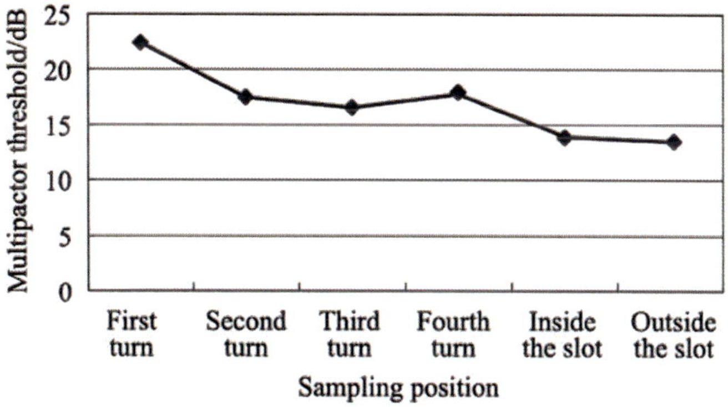

Fig. 3.6 Analysis curve of multipactor threshold

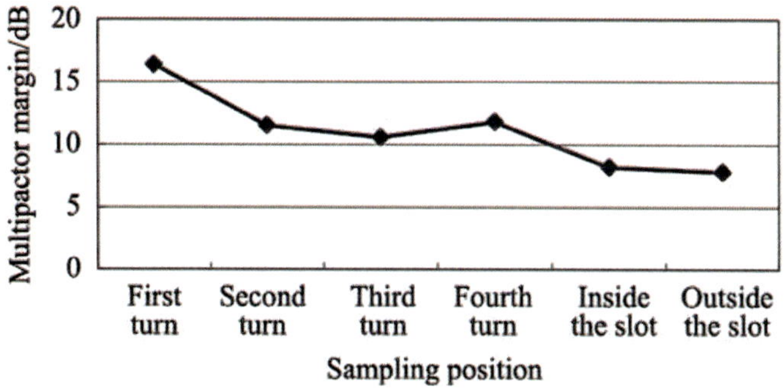

Fig. 3.7 Analysis curve of multipactor margin

3.2.4 Engineering Application Design of DTA

3.2.4.1 High-Power Structure Design

In order to meet the power capacity requirement of X-band backfire bifilar helical DTA on satellite, considering the on-orbit working environment, the following structure design is further used on the antenna body:

1. Helix uses a hollow medium to realize unsupported structure, thus greatly improving power capacity.
2. The internal feeding structure of the antenna is designed with an anti-deformation connection to release the high-temperature thermal stress caused by high-power transmission on orbit.
3. The interior dielectric of the socket part adopts the outgassing design.

3.2.4.2 Anti-satellite Body Environmental-Impact Design

Backfired bifilar DTA has a good Earth-shaped beam characteristic. However, due to its wide coverage, weak directivity, there is a large radiation level outside the coverage area. In the satellite environment, the radiation level outside the coverage area, which is projected onto the surface of satellite or other large structures and reached the main radiation area after reflection, will bring obvious interference to the radiation level in the coverage area. So, it is essential to attach importance to this factor in the practical application of the antenna.

The pattern of backfire bifilar DTA unit is shown in Fig. 3.8, and the pattern of backfire bifilar DTA after installation under the interference of satellite environment is shown in Fig. 3.9. The pattern of satellite-mounted DTA fluctuates sharply in the coverage area, which has an obvious effect on the amplitude-frequency characteristics in the data transmission channel and results in error code of transmission data. Therefore, the antenna should not only meet the requirement of the Earth-shaped beam gain but also minimize the similar sharp fluctuation of the pattern to ensure the quality of the data transmission channel.

There are two ways to improve the amplitude-frequency characteristics of DTA in the satellite environment:

1. The DTA is arranged on the edge of the satellite to minimize the energy projected onto the surface of the satellite.
2. A cup-shaped structure (Fig. 3.10) is added to the antenna body to block the path of the back lobe of the DTA projecting onto the satellite. Its effect is shown in Fig. 3.11. That is, the coverage area pattern fluctuates obviously gently, thus meet the requirements of the data transmission system.

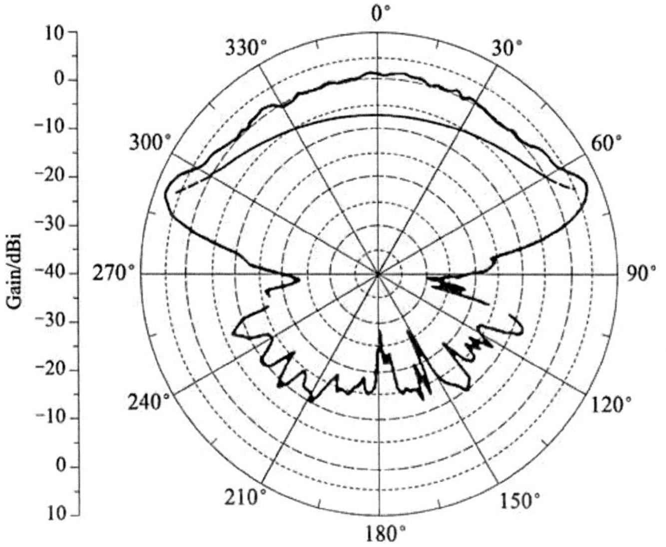

Fig. 3.8 Pattern of DTA unit

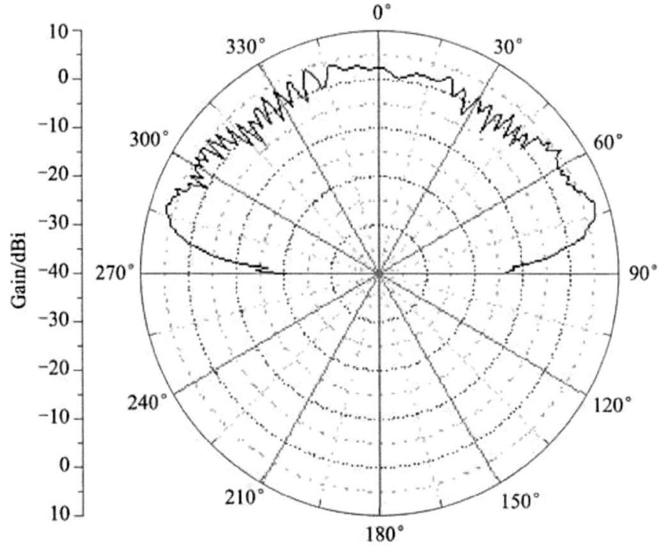

Fig. 3.9 Pattern of DTA interfered by satellite environment

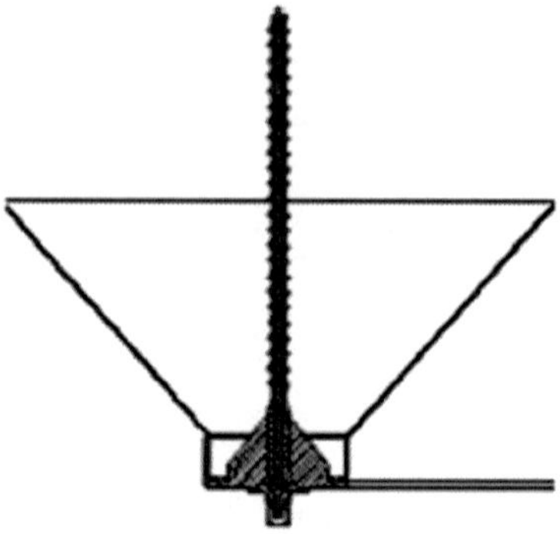

Fig. 3.10 Schematic diagram of anti-interference cuplike structure of DTA

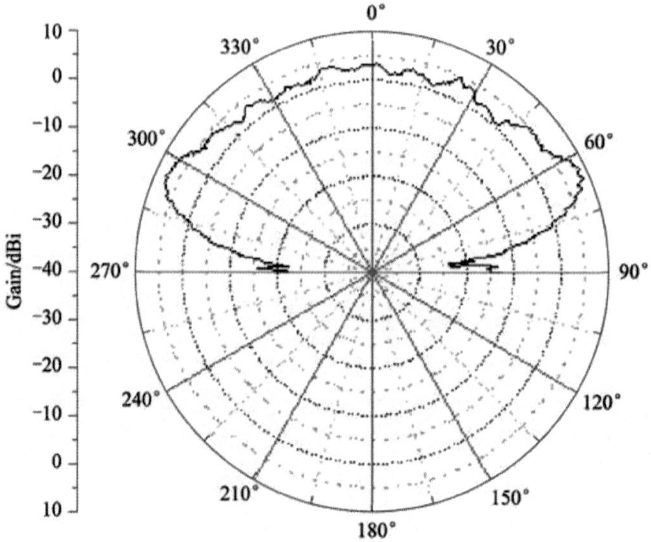

Fig. 3.11 Satellite pattern of DTA with additional anti-interference cuplike structure

3.3 X-Band-Shaped Reflector Antenna

Earth-shaped reflector antenna consists of shaped reflector and feed. In order to achieve the desired gain and coverage angle, the diameter of the shaped surface is generally about 600 mm. The geometric size of the reflector is much larger than the wavelength, so it is suitable for geometric optics approximation design. Because the required beam coverage pattern is circularly symmetrical and the shaped reflector should adopt a circularly symmetrical structure, the shape design of the reflector can be reduced to the design of the transversal line. Once the transversal line is

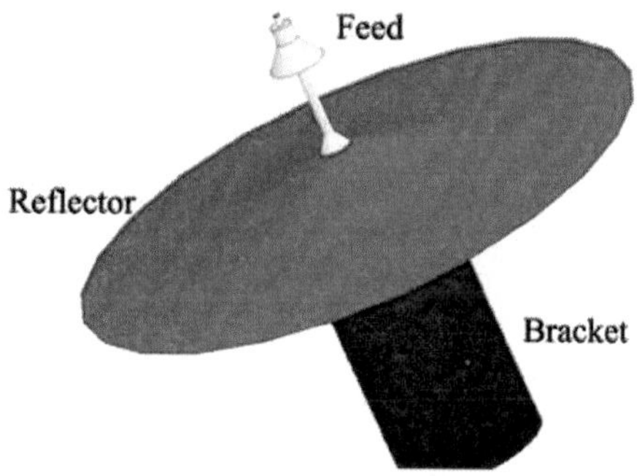

Fig. 3.12 Schematic diagram of X-band-shaped reflector data transmission antenna composition

determined, the shaped reflector can be obtained by rotating the line around the axis. The specific design methods can be found in the relevant literature [1].

Moreover, a coaxial multi-mode matching feeder or a cross-dipole-plus-reflecting-cavity feeder can be used, and the former features high radiation efficiency, circular symmetry, low cross-polarization and low side lobe level. However, the feeder aperture blockage is large, and the pattern of antenna coverage area has an obvious ripple, resulting in large level fluctuations. Although the latter is not as ideal as the former, it has fewer blockages to the reflector and better overall performance, so it is generally preferred.

X-band-shaped reflector antenna can provide Earth-shaped beams to meet the requirements of fixed beams for ground data transmission of LEO and MEO satellites. It features high power handing and high edge gain. The X-band-shaped reflector data transmission antenna consists of a cross-dipole feed, a shaped reflector and a bracket, as shown in Fig. 3.12.

3.3.1 Design of Cross-Dipole Feed

3.3.1.1 Cross-Dipole Radiation Principle

The structure of the cross-dipole antenna consists of the dipole 1 and the dipole 2, which are placed perpendicular to each other and the geometric centers coincide, is shown in Fig. 3.13. The two dipoles are fed in the same phase, and the circular polarization is achieved by changing the relative length and thickness between the two dipoles.

In the Cartesian coordinate system, the resultant electric vector of two electric field vectors, which are perpendicular to each other in space with equal amplitude, same frequency and 90° phase difference, rotates at the same angular rate with time. And the mode of the resultant vector is constant, i.e., the magnitude of the resultant vector is equal in all directions, and the rotation trajectory of the vector end is a circle. Thus, a circularly polarized wave is formed.

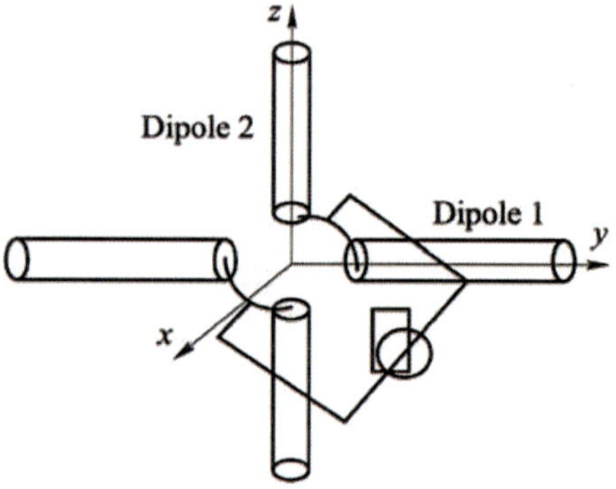

Fig. 3.13 Cross-dipole antenna structure

According to the self-impedance analysis of the dipole, the impedance of the dipole is inductive, when the electric length of the two dipole arms is greater than $\lambda/2$. Conversely, the impedance of the dipole is capacitive, when the electric length of two dipole arms is less than $\lambda/2$. Correspondingly, when the electric length of the two dipole arms is greater than $\lambda/2$, the phase of the current is leading. Conversely, the phase of the current lags and the polarization direction is rotated from leading to lagging.

3.3.1.2 Structure and Performance of Cross-Dipole with Reflector as Feed

The feed consisting of a cross-dipole and a reflector is shown in Fig. 3.14. The cross-dipole with simple and reasonable structure is fed by a coaxial waveguide,

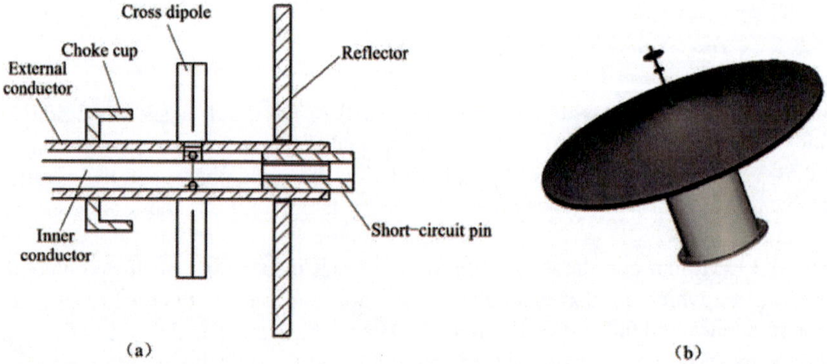

Fig. 3.14 Feed consisting of a cross-dipole and a reflector. **a** Feed structure with cross-dipole and reflector; **b** configuration of shaped reflector antenna

and its performance is adjusted and matched by the choke cup and short-circuit pin. The simulation radiation pattern of the feed and the measured pattern of the shaped reflector antenna are shown in Figs. 3.15, 3.16 and 3.17, respectively. It can be seen that the performance of shaped reflector antenna with feed consisting of cross-dipole and reflector is not satisfied, and the radiation pattern gain fluctuates sharply due to the influence of complex satellite environment. Analytically, the structure has two disadvantages:

1. The equalization performance (symmetry of each section) of the feed is poor because the feed adopts a balanced feeding structure with coaxial slotting. Besides, the equalization performance of the feed is deteriorated by the slot radiation.

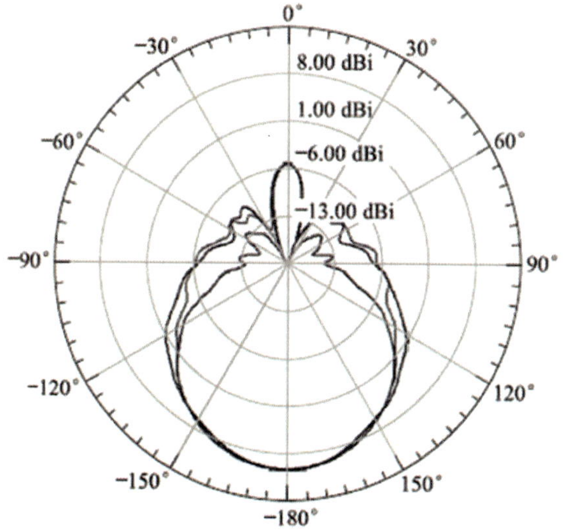

Fig. 3.15 Simulation results of the radiation pattern of cross-dipole and reflector feed

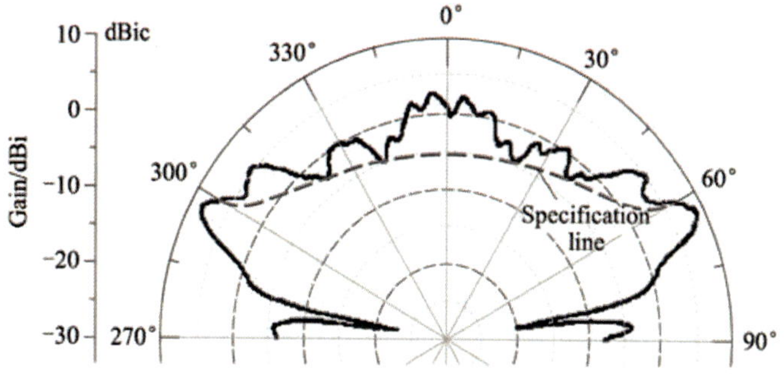

Fig. 3.16 Measurements of the shaped reflector antenna with cross-dipole and reflector feed

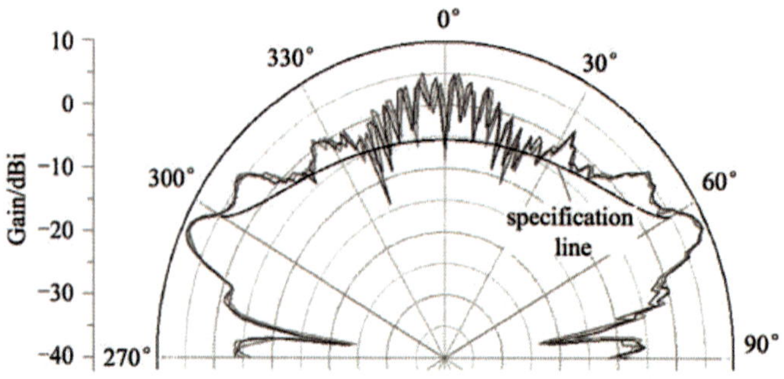

Fig. 3.17 Test results of the shaped reflector antenna with cross-dipole and reflector feed on satellite

2. The cross-polarization level of the feed is high, and the radiation of the cross-dipole is bidirectional, while the ability of the reflector to suppress the backward radiation is insufficient.

It is these disadvantages of the feed that cause the poor performance of the antenna.

3.3.1.3 Feed Structure and Performance of Pyramid Horn Excited with Cross-Dipole

In order to improve feed performance, a pyramid horn fed by the cross-dipole is adopted, as shown in Fig. 3.18. The simulation results of the feed pattern are shown in Fig. 3.19. It can be seen that the pyramid horn feed suppresses the radiation of the balanced feed slot, improve the backward radiation performance of the cross-dipole, and reduce the cross-polarization radiation component of the dipole.

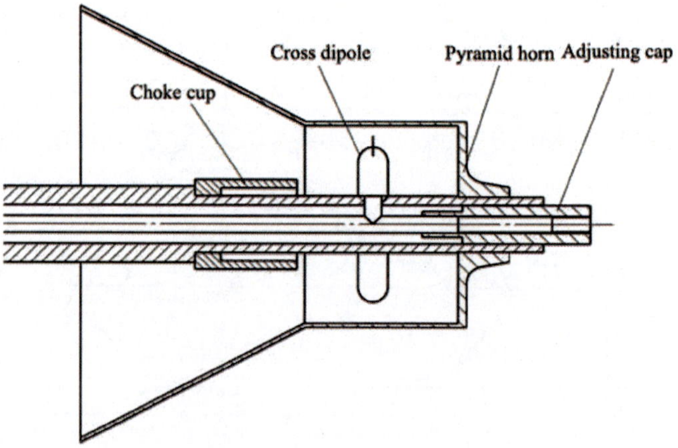

Fig. 3.18 Feed structure of the pyramidal horn with cross-dipole feed

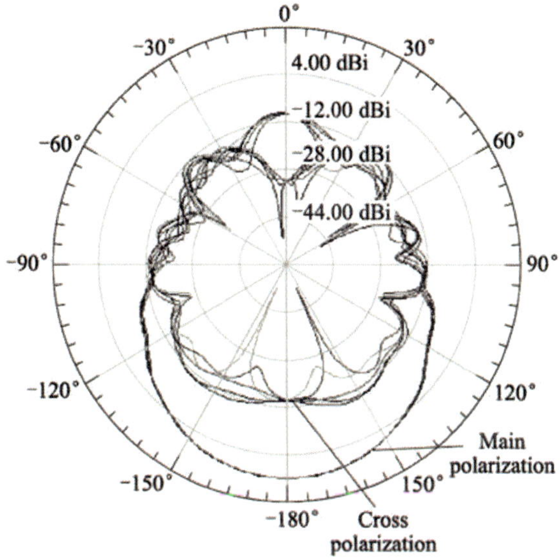

Fig. 3.19 Feed pattern of the pyramidal horn with cross-dipole feed

3.3.2 Performance of Shaped Reflector Antenna

3.3.2.1 Power Tolerance and Multipactor Test

The cross-dipole pyramidal horn feed has passed 100 W continuous-wave power tolerance test under normal temperature and pressure, 100 W average power electron-radiated multipactor test at normal temperature and vacuum and 100 W average power with 2.2 dB pulse power (166 W), electron-radiated multipactor test under normal temperature and vacuum conditions. All these indicate that the feed has large power operation capability.

3.3.2.2 Performance After the Combination of Feeder and Reflector and on Satellite

The pattern test results of the single shaped reflector antenna with pyramid horn fed by crossed dipole and that after it installed on the satellite are shown in Figs. 3.20 and 3.21. It can be seen that the radiation pattern of the single unit has better-shaped beams, and the gain fluctuation is also significantly reduced after installing on satellite.

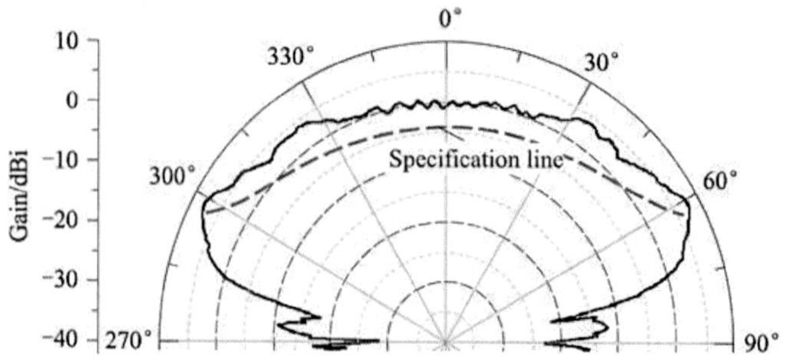

Fig. 3.20 Test result of the single-shaped reflector antenna with pyramid horn feed

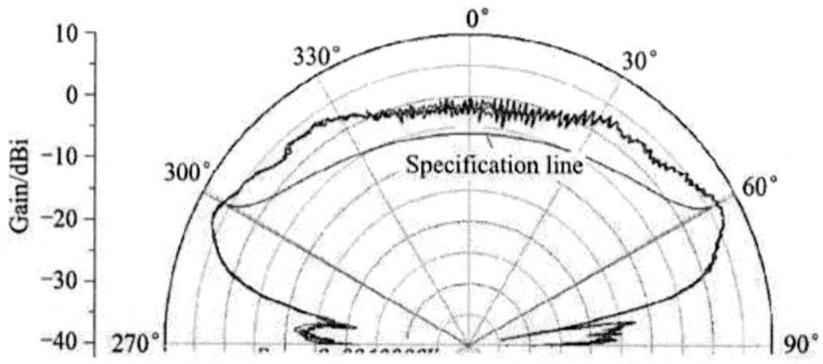

Fig. 3.21 Test result of the shaped reflector antenna with pyramid horn feed on satellite

3.4 Spot Beam Double Circularly Polarized DTA

3.4.1 Demand and Task Analysis

There are two main ways to improve the code rate of transmitting sub-system: one is to improve the EIRP of the antenna coverage area; the other is to use frequency reuse technology. Because the output power of satellite-borne microwave power source is limited, EIRP can only be realized by increasing the antenna gain. High-gain beam can only cover a very small area on the Earth, i.e., spot beam coverage. A dual-axis drive mechanism for beam scanning is equipped in the antenna to cover a larger area. The so-called frequency reuse refers to the simultaneous transmission of multiple signals by using different antenna polarizations at the same carrier frequency. Both of the orthogonal circular polarizations are generally used to transmit two signals. Such antenna is also called the double circularly polarized antenna, which can transmit two data-transmission signals at the same frequency without changing the EIRP of the antenna, thus doubling the data transmission rate.

An important design of spot beam double circularly polarized DTA is to determine the form of scanning mechanism. The medium-low-orbit satellite with a high-speed

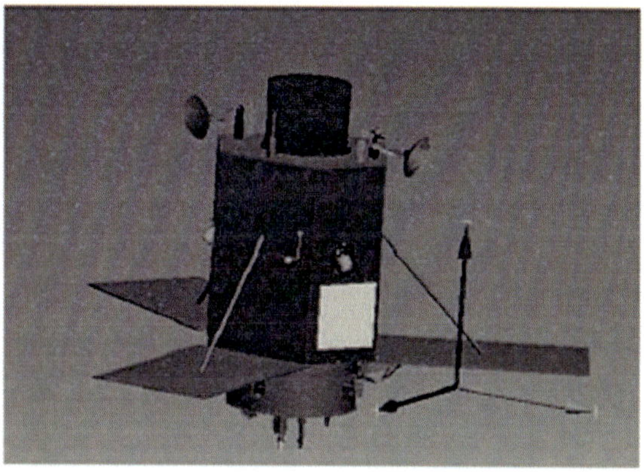

Fig. 3.22 Installation layout of a spot beam double circularly polarized DTA on satellite

data transmission to the Earth usually requires side maneuver, so that the scanning range of DTA must exceed hemispheric space.

For example, a satellite orbital altitude H is 600 km, the operation initial elevation angle of Earth station δ is 5°, and the radius of the Earth R is 6378 km. Viewing from the satellite, the half taper angle of the view field of the Earth is as follows:

$$\theta_m = \arcsin \frac{R \cos \delta}{R + H} \approx 65^\circ \tag{3.7}$$

When the satellite's side swing (pitching and rolling) is ±45°, the actual half taper angle coverage of the DTA should reach 110°, which exceeds the rotational scanning capability of a single two-axis driving mechanism. Therefore, two identical DTAs are generally used on the satellite, as shown in Fig. 3.22. A single antenna driven by a two-axis mechanism has half-space scanning capability. The two antennas tilted on both sides of the satellite to Earth can cover the field of view with a half taper angle of 110°.

In order to maintain the transmission of microwave signal during scanning, a dual-channel microwave rotating joint is installed on each rotating axis of the driving mechanism, and a circularly polarized signal is transmitted through each channel. Since the working mode of the two-axis driving mechanism is reciprocating swing, not 360° circular rotation, two single-channel rotating joints can be combined to form the two-channel microwave rotating joints, as shown in Fig. 3.23. For the single-channel rotating joint technology, see Sect. 5.2 for details.

3.4.2 *Stepped Diaphragm Double Circular Polarizer Design*

One core component of spot beam double circularly polarized DTA is a dual circular polarizer. A stepped diaphragm double circular polarizer is usually used, i.e., a

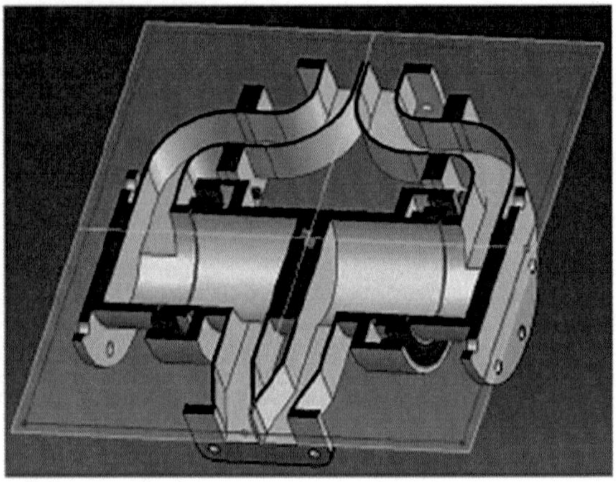

Fig. 3.23 Sectional view of dual-channel rotating joint

stepped diaphragm is inserted into the symmetrical plane of a square waveguide, one end of which is two identical rectangular waveguides, and the other end is a square waveguide. The rectangular waveguide is a feed port, and the square waveguide is connected with the radiator. Left- or right-handed circularly polarized beams can be obtained by feeding different rectangular waveguides, and double circularly polarized beams can be formed by feeding at the same time. In addition to being used for antennas, stepped diaphragm double circular polarizer is also commonly used in polarization measurement equipment, etc.

As shown in Fig. 3.24, the stepped diaphragm double circular polarizer can be regarded as a four-port network. Both rectangular waveguide ports transmit only TE_{10} main mode, each representing one port, while two modes are transmitted simultaneously in a square waveguide: TE_{10} mode and TE_{01} mode, each mode represents one port.

The working principle of stepped diaphragm double circular polarizer can be analyzed by odd-even mode theory. During even mode excitation (Fig. 3.24), the

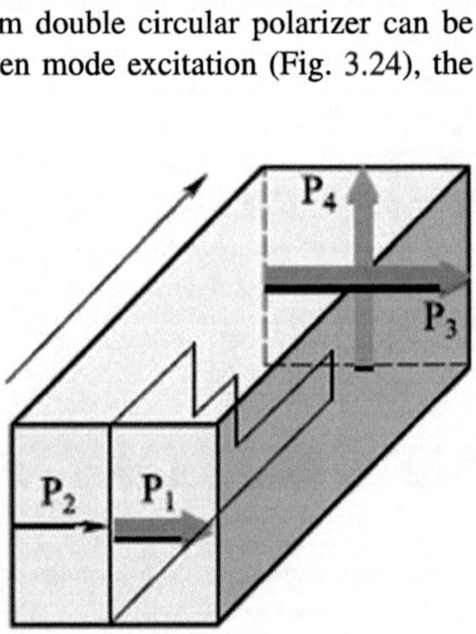

Fig. 3.24 Stepped diaphragm double circular polarizer

electric field orientation of port P_1 and port P_2 is the same (e.g., both are →), currents on both sides of the diaphragm are in the opposite direction and can bypass the diaphragm at the step to maintain continuity, so the structure of the field is not affected. The TE_{10} modes of port P_1 and port P_2 are fused into the TE_{10} mode of port P_3. By odd mode excitation, the electric field of port P_2 is opposite to that of port P_1, and the current direction of both sides of the diaphragm is the same. It cannot be bypassed at the step, and the continuity of the current is cut off. Therefore, the field structure is disturbed and the TE_{01} mode of port P_4 is excited.

According to the basic theory of waveguide transmission mode, TE_{10} mode of port P_3 is propagated in a rectangular waveguide, TE_{01} mode of port P_4 is propagated in a ridge waveguide. Under the condition of the same wide-edge size, the cut-off wavelength of ridge waveguide is longer (i.e., permitting lower frequency electromagnetic waves to pass through), and the phase velocity of TE_{01} mode is lower than that of TE_{10} mode, the appropriate step size can be designed to make the phase of TE_{01} mode lag 90° than that of TE_{10} mode on the square waveguide interface.

When even mode and odd mode are added, it is equivalent to feeding only to port P_1, and port P_2 has no input. Please refer to the direction of electric field shown in Fig. 3.24. The TE_{01} mode electric field (P_4) with direction↑ lags 90° than the TE_{10} mode electric field (P_3) with direction →, resulting in a left-handed circular polarization. When the even mode and the odd mode are subtracted, it is equivalent to feeding only to port P_2, and port P_1 has no input, thus forming a right-handed circular polarization.

When stepped diaphragm double circular polarizer is used in horn antenna fed by circular waveguide, square-circular transformation section is usually used to transition between square waveguide and circular waveguide, as shown in Fig. 3.25. The geometry of inner cavity of the square-circle transformation section is an intersection of a quadrilateral platform and a circular platform. The two bottom planes of the quadrangular platform are the square waveguide port and the circular waveguide

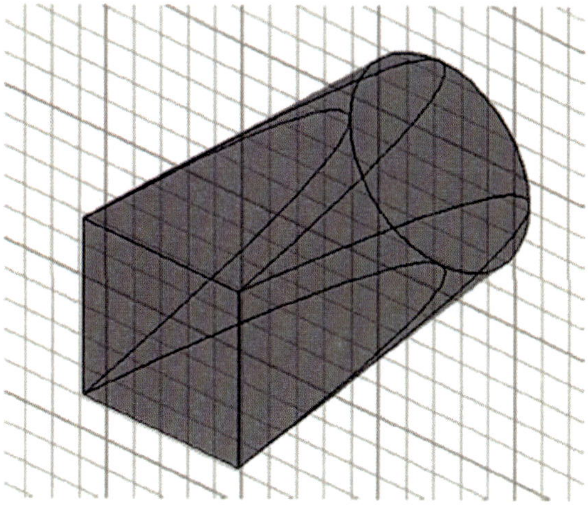

Fig. 3.25 Square-circle transformation section

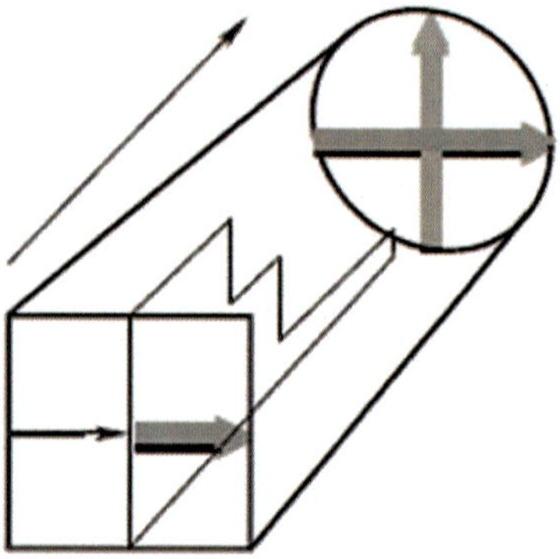

Fig. 3.26 Stepped diaphragm double circular polarizer with square-circle transformation section

circumscribed square, respectively. The two bottom planes of circular platform is the square waveguide port inscribed circle and the circular waveguide orifice, respectively. Such geometry can not only guarantee good standing wave characteristics, but also facilitate mechanical processing, so it is also a common microwave device for antenna feeding.

If an independent square-circle transformation section is used, the length of the feed structure will be increased. In fact, the effect of double circular polarization can also be achieved by inserting stepped diaphragms directly into the square-circle transformation section (Fig. 3.26), the principle of which is like that of the abovementioned stepped diaphragm double circular polarizer.

The length of square waveguide is generally equal to one waveguide wavelength of TE_{10} mode. The generation of high-order modes other than TE_{10} mode (and TE_{01} mode) should be suppressed by the selected size of waveguide. The selection of stepped diaphragm thickness should take into account the electrical size, structural strength and the difficulty of process realization. The optimization design of step size mainly based on the standing wave ratio and isolation of ports P_1 and P_2, as well as the phase difference of ports P_3 and P_4, and it is better that the design is simulated with the antenna part together, and observe whether the cross-polarization isolation level of the antenna meets the application requirements. Generally, for a frequency reuse antenna, the cross-polarization isolation of linear polarization should be no less than 30 dB, and that of circular polarization should be no less than 27 dB within its effective beam width. An effective measure to prevent the deterioration of cross-polarization isolation is to control the size of the feed port of the antenna so that it can only transmit the main mode. For example, when the feed port of the antenna is a circular waveguide, its inner diameter should be taken as 0.59–0.76 λ to ensure that only TE_{11} mode transmission is allowed in the circular waveguide.

Note that the cross-polarization isolation level is not the level of port P_1 and port P_2, but the ratio of two orthogonal polarization components of the wave beam in a certain direction in the field of view of antenna.

3.4.3 Double Circularly Polarized Antenna Design

There are many types of double circularly polarized antennas, such as horn antenna or reflector antenna with a dual circular polarizer, and sinusoidal antenna, a variant of the planar helical (spiral) antenna. For satellite-borne data transmission applications, the frequency is generally X-band or Ka-band, so using horn antenna or reflector antenna has the advantages of simple feeding structure, large power capacity and easy to form spot beam.

In general, if larger beam coverage is required, the horn antenna with lower gain can be adopted; if the required beam coverage is small, the reflector antenna with higher gain must be used. The boundary line between higher gain and lower gain is approximately 25 dBi. The advantages of horn antenna are small size, lightweight and low pointing accuracy requirement. However, this is also the disadvantage of reflector antenna, which means larger size and weight, and a higher requirement for pointing control accuracy. In practical engineering applications, antenna aperture should be reasonably selected according to specific EIRP requirements, and a compromise should be made among power consumption and weight of microwave power source and envelope size and weight of the antenna.

For circularly polarized horn antennas used for data transmission, the main requirements are broadband, good equalization of beam E-plane and H-plane, as well as a good axis ratio, such as multi-mode conical horn, corrugated horn. The main principle is to form an ideal electric field distribution on the radiation aperture. For multi-mode conical horns, more than two modes (TE_{11}, TM_{11}, TE_{12}, etc.) are generally superimposed on the aperture surface. For corrugated horns, a mixed HE_{11} mode is formed inside the horn. Comparatively, the corrugated horn has better broadband characteristics, an easy-to-process structure, and a mature design method. A double circularly polarized corrugated horn for data transmission with a diameter of 200 mm is shown in Fig. 3.27, and the simulation results of its pattern are shown in Fig. 3.28. At 8–9 GHz, the peak gain of the pattern is greater than 22 dBi.

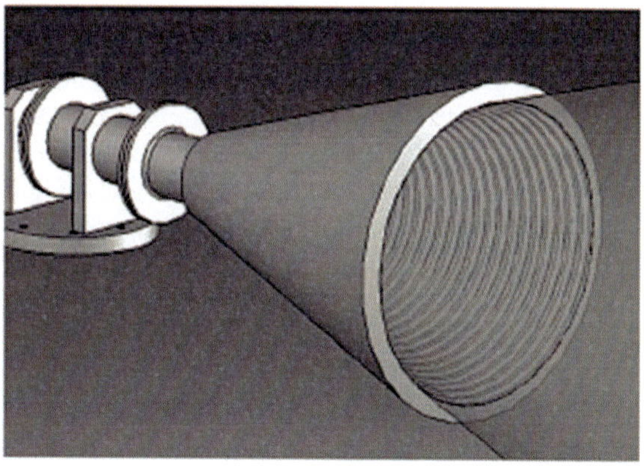

Fig. 3.27 Double circularly polarized corrugated horn antenna

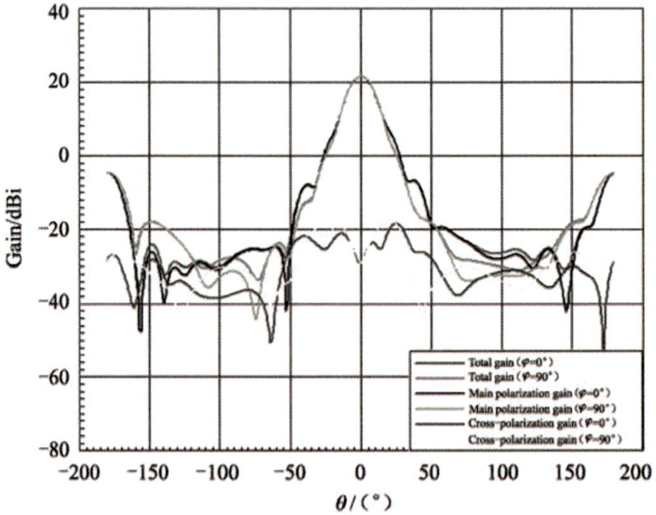

Fig. 3.28 The simulated pattern of double circularly polarized horn antenna (8.0 GHz)

There are many types of double circularly polarized reflector antennas for data transmission, and there is no difficult to select a reflector type. However, as mentioned above, 25 dBi is a demarcation point for selecting horn or reflector antenna, while the general requirement of satellite-borne data transmission in X-band is limited by envelope size, and usually, only Φ 400 mm diameter aperture is allowed. This is just a dilemma that horn antenna cannot achieve and reflector antenna cannot play a high-efficiency advantage. If the double circularly polarized reflector antenna is designed under such condition, the ring-focus reflector antenna is the best choice considering the efficiency of the antenna. This is because the sub-reflector has less blockage to the main reflector in the ring-focus reflector antenna, and it reflects the concentrated energy of the primary beam center to the edge of the main reflector, which makes the aperture field distribution closer to the equal-amplitude and same-phase distribution, thus achieving higher efficiency.

The electrical performance design of the ring-focus reflector antenna can be referred to as the relevant literature [2]. Figure 3.29 shows a ring-focus spot beam double circularly polarized antenna with the main reflector, a sub-reflector, a double circularly polarized horn feed, waveguide rotating joints, and waveguide transmission lines, etc. The antenna pointing to Earth station is realized by using a two-axis driving mechanism. The simulation results of its pattern are shown in Fig. 3.30.

However, in Ka-band, the size constraints are not as harsh as those in X-band. There are more configurations available for reflector antenna of 400 mm in diameter. For example, feed forward single reflector is a preferred scheme. Moreover, it can achieve better double circular polarization characteristics in the main beam range because of fewer blockages to the main reflector.

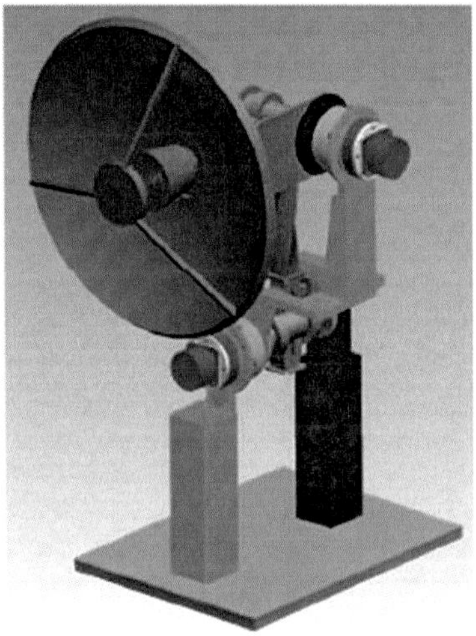

Fig. 3.29 Diagram of a ring-focus point beam double circularly polarized antenna

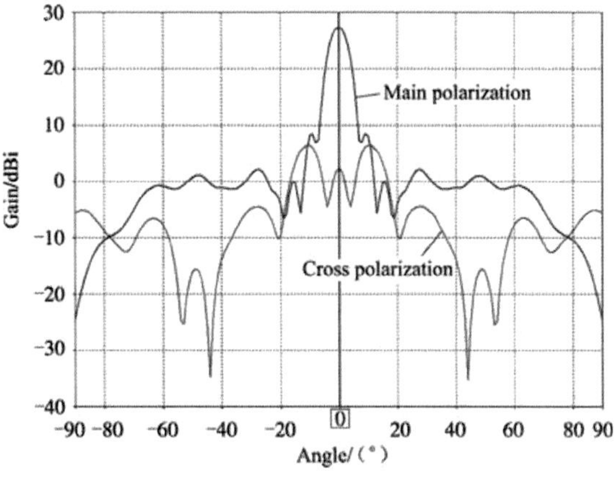

Fig. 3.30 Simulated diagram of the ring-focus double circularly polarized antenna (8.0 GHz)

References

1. Zhang Z, Ye Y (2004) Design of shaped reflector antenna for earth observation satellite. China Aerosp Sci Technol 1(2):7–11
2. Cheng Z, Lv S, Jiang X (2001) Integrated design of sum and difference of monopulse circular focus antenna. J Beijing Univ Aeronaut Astronaut 27(1):12–15

Chapter 4
Spacecraft GNSS Antenna

Abstract Global Navigation Satellite System (GNSS) is a space-based TT&C resource. GNSS is a powerful complement to the ground-based TT&C system. This chapter provides the design of single-frequency GNSS antenna, dual-band GNSS antenna for both low orbit (LEO) and high-orbit (GEO) spacecraft.

A relatively mature ground-based space TT&C network which can be networked with the world ground-based space TT&C stations has been established in China to meet the TT&C requirements of various types of satellites. Global Navigation Satellite System (GNSS) is a space-based TT&C resource, which locates the target spacecraft by multi-satellite one-way pseudo-code ranging and carrier ranging technology. The position of the target spacecraft can be determined by measuring pseudo-ranges of at least four GNSS satellites. GNSS is a powerful complement to the ground-based TT&C system. Compared with the ground-based TT&C system, GNSS has many advantages, such as around-the-clock, all-weather and global positioning information, wide coverage, long visible time, multi-target tracking ability and simple equipment, as well as high measurement accuracy.

At present, the most typical GNSSs include Global Positioning System (GPS) of the U.S., Galileo Satellite System of the European Union, Global Navigation Satellite System (GLONASS) of Russia and Beidou Satellite Navigation System (BDS) of China. Other navigation satellite systems include Japan's Quasi-Zenith Satellite System (QZSS) and India's Regional Navigation Satellite System (IRNSS), as well as augmentation systems of navigation systems, the Wide Area Augmentation System (WAAS) of the U.S. and the European Geostationary-Satellite Navigation Overlap Service (EGNOS). In the future, with the development of GNSS led by various countries, the satellite navigation system is about to enter a new era and will face the situation that nearly 100 navigation satellites coexist and are compatible with each other. Moreover, with abundant navigation information, the availability, accuracy, completeness and reliability of satellite navigation can be improved.

GNSS receiver processes all signals obtained from visible navigation satellites through the satellite-borne GNSS antenna. Although GNSS signals use different signal multiple access mechanisms (e.g., CDMA for GPS and FDMA for

R. You et al., *Technologies for Spacecraft Antenna Engineering Design*, Space Science and Technologies, https://doi.org/10.1007/978-981-15-5833-7_4

Table 4.1 Frequency and bandwidth of the GNSS system

System	Country	Operating frequency/MHz	Bandwidth/MHz
GPS	USA	L_1: 1 575.42	±10.23
		L_2: 1 227.6	±10.23
		L_3: 1 176.450	±10.23
GLONASS	Russia	L_1: 1602.000 (FDMA)	±0.511
		L_1:1575.42 (CDMA)	±5.11
		L_2: 1246.000 (FDMA)	±0.511
		L_2: 1242.000 (CDMA)	Under consideration
		L_3: 1202.025 (CDMA)	Under consideration
		L_5: 1176.450 (CDMA)	Under consideration
Galileo	EU	E_1: 1 575.42	±12.176
		E_6: 1 278.750	± 20.46
		E_5b: 1 207.140	± 10.23
		E_5: 1 191.795	± 25.575
		E_5a: 1 176.450	± 10.23
BDS	China	B_1: 1 559.052–1 591.788	± 2.046
		B_2: 1 162.220–1 217.370	± 12
		B_3: 1 250.618–1 286.423	± 12

GLONASS), they can be classified into at least three layers, e.g., carrier, pseudo-random noise (PRN) code (referred to as pseudo-code) and navigation message data code. The navigation message modulated in the signal includes satellite clock and orbit parameters. The GNSS receiver accurately calculates the space position of the satellite based on these parameters at the time of signal transmission.

Table 4.1 lists the frequency and bandwidth of GPS, GLONASS, Galileo and BDS (also known as BeiDou). Although the frequencies used in these satellite navigation systems are different, the frequency ranges are between 1150–1300 MHz and 1559–1611 MHz. Therefore, if a broadband navigation positioning receiving antenna capable of covering two bands is developed, it can be applied to different navigation systems and good versatility and compatibility can be achieved.

At present, the most common GNSS antenna is designed for L_1 of GPS constellation, which is called single-frequency GNSS antenna. GNSS antenna and receiver are compatible with different navigation systems, i.e., the frequency of GPS constellation L_1 is like that of Beidou navigation system B_1, and the GNSS antenna which covers both band of GPS constellation L_1 and band of Beidou navigation constellation B_1 is called single-frequency dual-mode GNSS antenna. As more GNSS are put into use, GNSS antennas compatible with different navigation systems will appear soon. However, antennas are required to have good reception performance in different frequency bands.

The performance of GNSS antenna depends on the operating mode of the navigation signal receiving system. Different operating modes have different requirements for positioning accuracy, velocity, operating environment and temperature. For GPS antennas required by the precise orbit determination satellite, the antenna is required to have the ability to operate in L_1 and L_2 bands, as well as good multipath suppression ability. The GNSS antenna of GEO orbit satellite receives side lobe signal of navigation constellation. The gain pattern of the antenna is different from that of the GNSS antenna of MEO and LEO satellite, and its working environment temperature changes more dramatically.

4.1 Main Technical Requirements of Spacecraft GNSS Antenna

GNSS signal receiving system can provide various functions in spacecraft applications, including satellite attitude determination (heading, pitch and rolling), time synchronization, orbit determination and absolute and relative positioning.

The utility of GNSS antenna is to convert the electromagnetic energy of radio signals sent by GPS, BDS, Galileo and other navigation satellites into the current that can be received by the receiver electronic devices. The size and shape of the antenna are critical, which determine the ability of the antenna to acquire weak GNSS signals. General antenna types include microstrip patch antenna, helical antenna, cross-dipole antenna, etc.

Each antenna has its own advantages and disadvantages:

1. Microstrip patch antenna has the advantages of low profile, simple structure, simple feeding mode and easily implemented circular polarization radiation. The weak points are narrow bandwidth and insufficient beam width;
2. The self-feeding quadrifilar helix GNSS antenna has the advantages of cardioids (heart-shaped) radiation pattern, wide-beam circular polarization radiation and high gain at low elevation, however, narrow bandwidth is a weakness
3. The advantages of cross-dipole antenna are low profile, simple structure and high reliability. Narrow bandwidth is still a drawback.

Generally, single-frequency GNSS antenna is used in spacecraft. The antenna only needs L_1 frequency receiving capability of GPS constellation (or compatible with B_1 frequency of Beidou navigation constellation) and hemispheric coverage pattern.

For high-precision positioning spacecraft, GNSS antennas generally operate in dual-band (receiving GPS constellations L_1 and L_2 at the same time, or compatible with Beidou navigation constellations B_1 and B_2). The pattern requires not only hemispheric coverage but also fast roll-down in horizontal position, higher polarization purity, multipath suppression ability and stable phase center.

In addition, the requirements of material, mechanical structure and thermal stability in space environment should also be satisfied with the spacecraft GNSS antenna.

The technical specifications of GNSS antenna vary from the different requirements of orbital altitude and positioning accuracy of spacecraft. Tables 4.2, 4.3, 4.4

Table 4.2 Technical requirements of single-frequency GNSS antennas for MEO and LEO satellites

Specifications		Requirements
Operating frequency/MHz		L_1: 1575.42 ± 10.23 B_1: 1561.098 ± 2.046
Gain/dBi	0°–45°	(Better than)-1
	45°–75°	(Better than)-3
	75°–85°	(Better than)-4
Polarization mode		RHCP
Standing wave ratio		<1.5
Electrical interface		SMA
Operating temperature/°C		−100 to +100

Table 4.3 Technical requirements of dual-band GNSS antennas for MEO and LEO satellites

Specifications		Requirements
Operating frequency/MHz		L_1: 1575.42 ± 10.23 L_2: 1227.60 ± 10.23 B_1: 1561.098 ± 2.046 B_2: 1207.14 ± 2.046
Gain/dBi	L_1 and B_1: 0°–70°	(Better than)-5.5 (on satellite)
	L_2 and B_2: 0°–65°	(Better than)-3.5 (on satellite)
Polarization mode		RHCP
Standing wave ratio		<1.5
Phase center change/mm	L_1 and B_1: 0°–70°	(Better than)7 (on satellite)
	L_2 and B_2: 0°–65°	(Better than)7 (on satellite)
Electrical interface		SMA
Operating temperature/°C		−100 to +100

and 4.5 show the technical requirements of GNSS antennas of single-frequency, dual-frequency, low-elevation zone high-gain, dual-frequency and high-orbit satellite, respectively.

4.2 Single-Frequency GNSS Antenna

Microstrip antenna, quadrifilar helical antenna and dipole antenna can be used for multiple mode navigation receiving antenna units working in a single frequency band (such as GPS satellite L_1 band and BDS satellite B_1 band).

Table 4.4 Technical requirements of high-gain dual-band GNSS antennas in low-elevation area of MEO and LEO satellites

Specifications		Requirements
Operating frequency/MHz		L_1: 1575.42 ± 10.23 L_2: 1227.60 ± 10.23 B_1: 1561.098 ± 2.046 B_2: 1207.14 ± 2.046
Gain/dBi	L_1 and B_1: 0°–80°	(Better than)-3.5 (on satellite)
	L_2 and B_2: 0°–65°	(Better than)-3.5 (on satellite)
Polarization mode		RHCP
Standing wave ratio		<1.5
Phase center change/mm	L_1 and B_1: 0°–75°	(Better than) 12 (on satellite)
	L_2 and B_2: 0°–75°	(Better than) 12 (on satellite)
Electrical interface		SMA
Operating temperature/°C		−100 to +100

Table 4.5 Technical requirements of GNSS antenna for high-orbit(GEO) satellite

Specifications		Requirements
Operating frequency/MHz		1559–1606
Gain/dBi	0°–30°	(Better than)7
	30°–37°	(Better than)5
Polarization mode		RHCP
Standing wave ratio		<1.5
Phase center change/mm		(Better than)0.5 (single unit)
Electrical interface		SMA
Operating temperature/°C		−172 to +112

4.2.1 Self-feeding Quadrifilar Helix GNSS Antenna

Self-feeding quadrifilar helix GNSS antenna, which is the most widely used antenna form adopted in Chinese spacecraft, forms a standard cardioid-shaped beam through a four-arm helix antenna. The laser welding process is applied to the stainless-steel antenna connection, and the electrical performance matching, machining and laser welding are completed at one time. The process has the advantages of no solder welding, reliable welding performance, excellent environmental adaptability, low-cost and easy implementation.

As described in 2.3, a resonant four-arm helical antenna can be considered to consist of two sets of bifilar helixes fed by a space orthogonal 90°, as shown in Fig. 4.1. The following is illustrated by an example of a bifilar helical antenna with

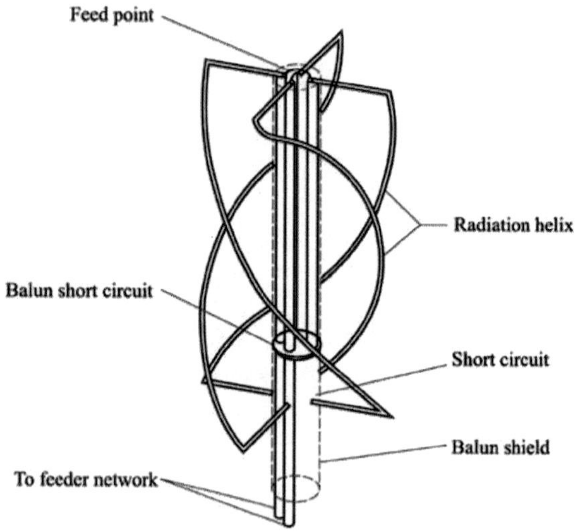

Fig. 4.1 Structure of four-axis helical antenna

a length of λ/2 and 1/2 turn.

The bifilar helical antenna can form a cardioid-shaped hemispheric beam with an amplitude pattern of $1 + \cos(\theta)$, and the maximum radiation is directed to the axial direction of the antenna, i.e., $\theta = 0°$. The two radiation components (E_θ, E_φ) of a four-arm helical antenna, which are orthogonal in space with equal amplitude and 90° of phase difference, can achieve circular polarization in the whole space and have uniform spatial phase distribution, so the four-arm helical antenna can be approximated as ideal point sources.

Generally, a four-arm helical antenna forming hemispheric beams requires a 3 dB/90° circularly polarized feeding network. In order to simplify the configuration of satellite-borne GNSS antenna, the feeding network can be realized by the input admittance of the antenna itself. The length of two bifilar orthogonal helix is so adjusted that one bifilum is shorter than λ/2, the other bifilum is longer than λ/2, and the two bifilar are, respectively, capacitive and inductive impedance, to meet the circular polarization feeding conditions. The antenna impedance matching can be realized by using λ/4 impedance transformation section. In this way, the circular polarization and impedance matching of the antenna can be realized without any peripheral components, and the antenna integration can also be improved.

Figure 4.2 shows the structure of the self-feeding four-arm helix antenna without the 90° phase-shifting network. The antenna has a compact structure and high integration. A radome is installed outside the helical antenna, which is made of quartz fiber composite material and is nearly transparent to electromagnetic waves. This radome will not cause pattern distortion and obvious loss. Figure 4.3 depicts the antenna pattern. The antenna is directly mounted on the satellite through the bottom flange, which has enough stiffness and strength. The antenna has been used on many satellites.

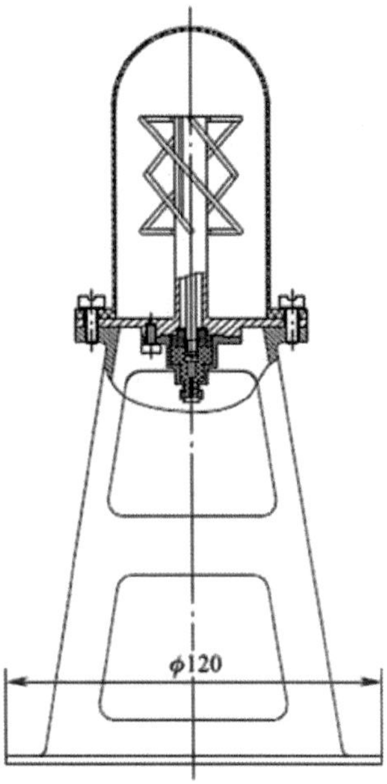

Fig. 4.2 Structure of self-feeding quadrifilar helix GNSS antenna

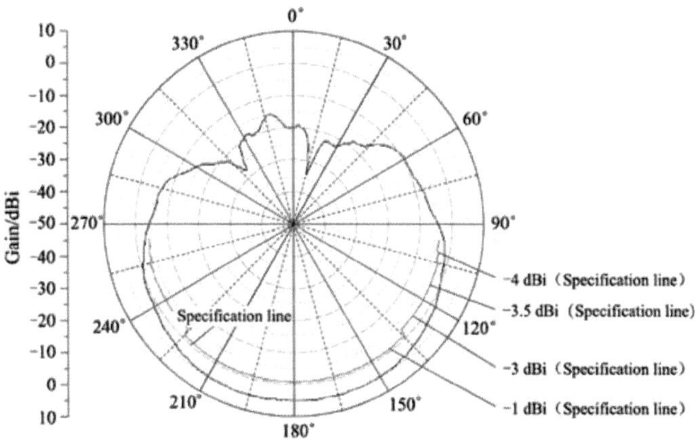

Fig. 4.3 Pattern of self-feeding quadrifilar helix GNSS (frequency = 1575 MHz)

4.2.2 Microstrip GNSS Antenna

The most common form of single-frequency GPS antenna working in L_1 band is microstrip antenna with a high dielectric constant and thick dielectric substrate. This type of antenna has many advantages, such as good coverage at low elevation, enough bandwidth, good axis ratio and low cost.

To analyze the radiation principle of microstrip antenna, a simple rectangular microstrip antenna is used as an example to explain its radiation mechanism with a transmission line model. If the length of the radiation element is L, the width is W, and the thickness of the dielectric substrate is h, the rectangular microstrip antenna can be equivalent to a microstrip transmission line with length L and an open circuit at both ends, as shown in Fig. 4.4. Because the thickness of dielectric substrate h is much smaller than the working wavelength, the electric field only varies along the length L direction, while the electric field distributes uniformly along the h direction. In the simplest case, the electric field along the width W direction is also approximated as not changing with position and the electric field distribution is shown in Fig. 4.5.

As shown in Fig. 4.5, the electric field can be divided into horizontal and vertical components at the two open circuit ends along the length direction. It can be seen that the horizontal component of the field has the same direction, and the fields generated in the far-field area overlap and add together, while the vertical component has the opposite direction, and the fields generated in the far-field area cancel out. Therefore, the rectangular microstrip antenna can be equivalent to two-element slot arrays separated by one-half wavelength with width h, length W and co-phase excitation.

GNSS antennas are all RHCP antennas. Taking the microstrip GNSS antenna shown in Fig. 4.6 as an example, its components are shown in Fig. 4.7.

The microstrip GNSS antenna consists of upper patch, lower patch, feed network, feed box, feed probe and SMA socket. The upper patch, lower patch and feeder network with flat structures are overlapped from top to bottom and fastened to a whole structure with feed box. The upper surface center of the upper patch and the

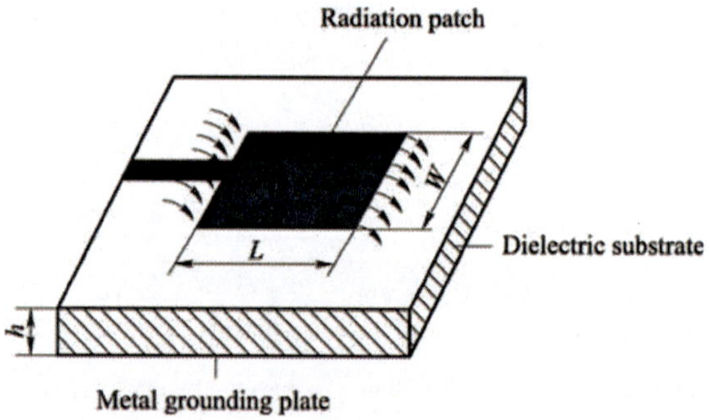

Fig. 4.4 Electric field distribution of rectangular microstrip patch antenna open circuit end

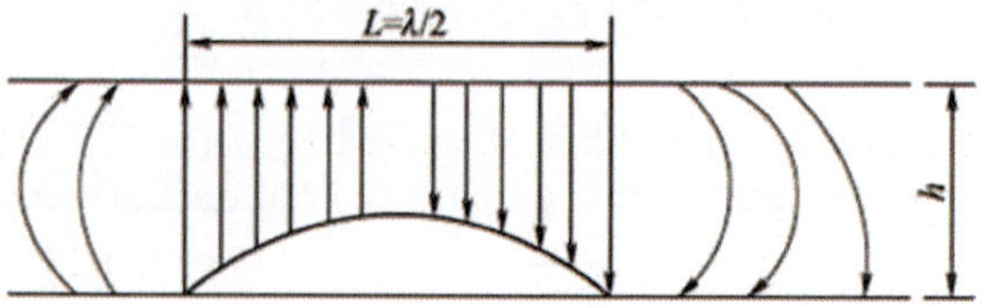

Fig. 4.5 Side view of field distribution

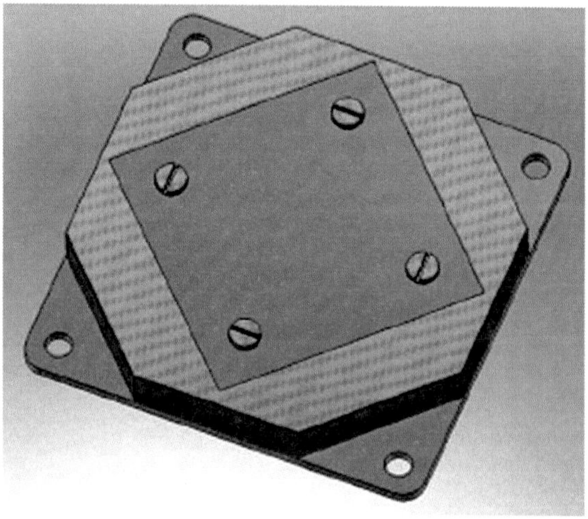

Fig. 4.6 Profile of a microstrip GNSS antenna

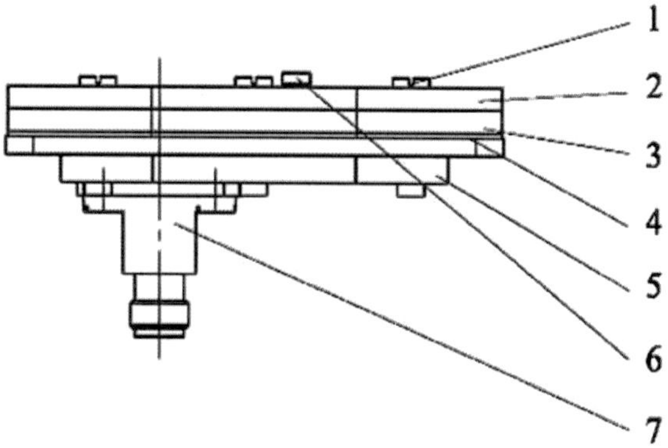

Fig. 4.7 Schematic diagram of microstrip GNSS antenna components. Legend: 1—positioning screw; 2—upper patch; 3—lower patch; 4—feeding network; 5—feeding box; 6—feeding probe; 7—SMA socket

lower patch is covered with a square copper coating layer, the lower surface of the lower patch also has a copper coating layer, but the lower surface of the upper patch is smooth. By adjusting the edge length of the square copper coating layer on the upper surface center of the upper patch and that of the lower patch, the upper patch and the lower patch can resonate on the desired GNSS receiving frequency band. The output end of the feed network and the upper patch can be electrically connected by feeding probe. The SMA socket is fastened on the feed box, and the inner conductor is electrically connected with the input terminal of the feed network.

The antenna feeds the signal into the feed network through SMA socket. The signal is divided into two sub-signals with equal amplitude and 90° phase difference after power dividing and phase shifting by the feed network and is output through two output ports. The output signal is fed to the surface of the upper patch through two feed probes to excite the radiation of the patch. The upper patch and the lower patch realize the radiation of the antenna, and the two patches have cascaded structures. The

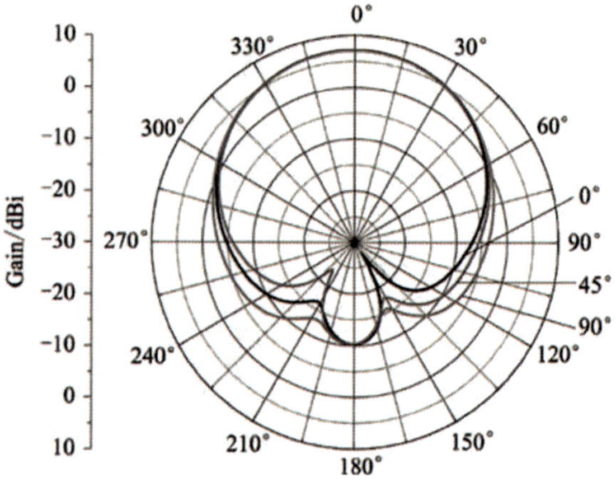

Fig. 4.8 Simulation results of microstrip GNSS antenna patterns

upper patch excites the lower patch to radiate through the coupling. The simulation results of antenna pattern are shown in Fig. 4.8.

4.3 Dual-Band GNSS Antenna

RHCP Dual-band GNSS antenna operates on L_1 and L_2 of GPS. At present, dual-band GNSS antennas on satellites mainly include microstrip antenna, quadrifilar helical antenna, air dielectric microstrip antenna, umbrella dipole antenna, etc. Most dual-band GNSS antennas used in satellites are microstrip antennas.

Dual-band GNSS antenna has the following characteristics:

1. Chokes are usually used to suppress the multipath effect when the installation dimension is allowed. Generally, in order to improve the receiving effect and suppress the multipath effect (the superimposed effect of reflected waves and direct waves caused by large scatters such as satellite or solar array), it is necessary to install chokes on the antenna. For example, some special satellite-borne GNSS antennas have chokes with a diameter of about 300 mm.
2. Weak directivity. Generally, GNSS antenna is a kind of hemispherical coverage antenna, which can receive visible satellite signals in the field of view around the antenna axis (±85°). Because of the wide main lobe range of the antenna, the ability of anti-interference and anti-multipath effect of the antenna is weak.
3. High stability of phase center. The phase center of GNSS antenna refers to the fixed position of the receiving point which is obtained after the receiver processes the GNSS constellation signal received by the antenna. In the application of precise positioning, the stability of the phase center of the antenna is required to be higher, generally within 10 mm.

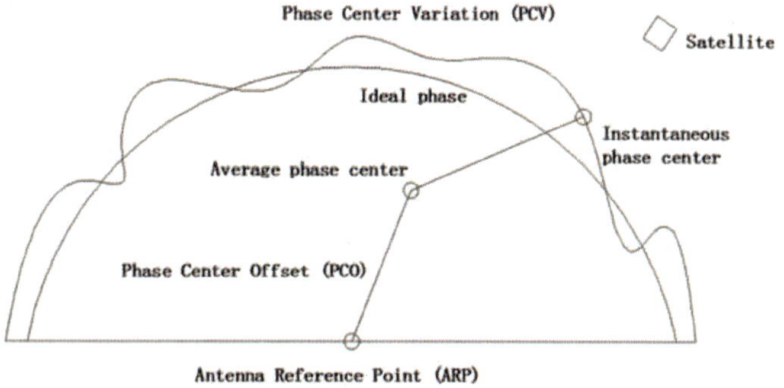

Fig. 4.9 Schematic diagram of phase center offset and phase center variation of the antenna

The phase center of GNSS antenna is generally described using the following two parameters: Phase Center Offset (PCO) and Phase Center Variation (PCV) [1]. Antenna phase center offset refers to the deviation between the average phase center of the antenna (the average of the instantaneous phase center of the antenna) and the antenna reference point (ARP); and the antenna phase center variant refers to the difference between the instantaneous phase center and the average phase center of the antenna, as shown in Fig. 4.9.

The phase center of the antenna varies with the direction of the transmitted signal. The phase center of the navigation satellite changes as it moves in the space. Therefore, the GNSS antenna should be designed as a minimized PCV.

For a certain antenna, PCO can be regarded as a fixed deviation. However, PCO is related to the operating frequency of the antenna, that is, the PCO of the antenna varies at different frequencies. PCV is related to the direction of the signal, and changes with the azimuth and zenith angle of the signal. For high-precision GNSS positioning, not only the PCO and PCV of antenna but also the influence of GNSS antenna installed on satellite surface on antenna phase center and the inconsistency of antenna phase center at different frequencies should be considered.

4.3.1 Dual-Band Microstrip GNSS Antenna

4.3.1.1 Antenna Unit Design

Dual-band microstrip GNSS antenna uses a probe-feeding excitation cross-slot microstrip antenna, as shown in Fig. 4.10.

Compared with general microstrip antennas, it has the following advantages:

1. The microstrip patch is excited by the cross-slot coupling form through the power-division phase-shifting feed network. The feed part of the antenna is isolated from the radiation part. The parasitic radiation generated by the feed line can be basically eliminated. Moreover, the feed mode of the antenna is not the probe

Fig. 4.10 Dual-band microstrip GNSS antenna unit

feed mode, so it is beneficial to reduce the heat input into the antenna feeding network through the probe, thus the working temperature environment of antenna feeding network is improved.
2. Reasonable design of cross-slot aperture can increase antenna bandwidth.
3. The symmetrical offset feeding mode has better axial ratio bandwidth.
4. The antenna structure is symmetrical and its electrical performance is axially symmetrical.

The main components of the antenna are shown in Fig. 4.11. The working principle of the antenna is described below in combination with the components of the antenna.

The components of the antenna from bottom to top are as follows:

SMA socket, antenna high-frequency signal interface, transmits the signal received by the antenna through the cable to the GNSS receiver.

Case cover, fixes SMA socket.

Feed network, power division phase shifting network, four signals with equal amplitude and 90° phase difference can be synthesized into one signal successively.

Dielectric substrates, alumina ceramic substrates, is favorable for reducing network size with a dielectric constant of about 9.6; of course, other substrates can also be adopted.

Cross-slot case can accommodate the feed network substrate, and the feed network excites the patch through the cross-slot on the case.

Patch substrate is made of polyimide material, used to support the patch, and has a high dielectric constant, which is advantageous for reducing the size of the patch and improving the low-elevation gain level of the antenna.

Patch is an antenna radiation element with square metal sheet, it converts electromagnetic energy of received GNSS satellite signal into high-frequency circuit signal.

Top cover is made of polyimide material, which is used to fasten the patch.

The circular polarization is realized by the four monopoles whose phase is 0°, 90°, 180° and 270°, it excites the cross-slot, respectively, and then half-space radiation is

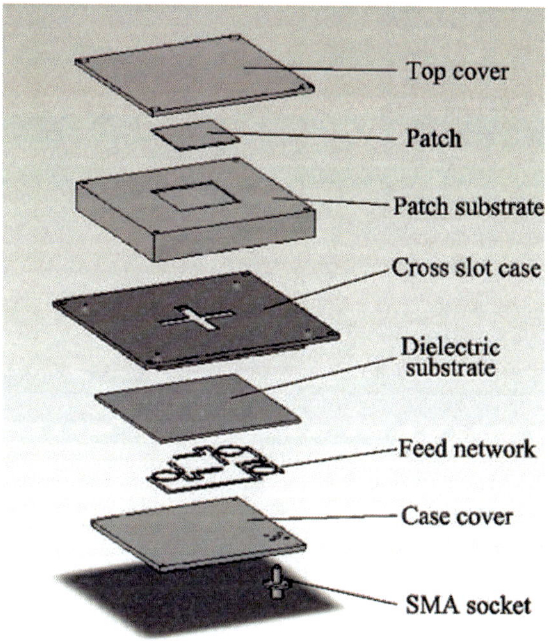

Fig. 4.11 Main components of GNSS antenna(CAST)

generated by using the patch as the director. Because of the strong coupling between the patch and the cross-slot, the size of the patch and the cross-slot should be designed synchronously. If the slot length increases, the back lobe of the antenna may be increased. Cross-slot and patch realize dual-band resonance, and then dual-band operation. The width of the slot also affects the coupling strength, which is less sensitive than the length of the slot.

The thickness and dielectric constant of the feed network substrate, the size of the patch and the coupling with the cross-slot are important parameters to determine the antenna bandwidth.

The design outcome of antenna VSWR is shown in Fig. 4.12.

4.3.1.2 Choke Design

When GNSS antenna is installed on the satellite surface, its radiation characteristics will be affected by multiple metal parts on the satellite surface, which will change the antenna pattern and influence the stability of the antenna phase center. In addition, the multipath signals generated by satellite solar arrays and satellite body are one of the main error sources affecting the measurement accuracy of the receiver. Therefore, in order to solve the above problems, choke is added outside the antenna unit, which can not only improve the stability of antenna performance but also effectively suppress the multipath signal coming from the back.

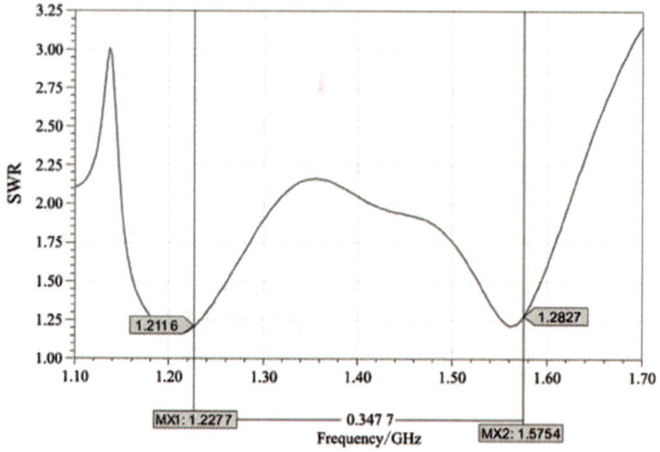

Fig. 4.12 Antenna VSWR

The dual-frequency GNSS antenna with a common choke is shown in Fig. 4.13, and its pattern and phase center stability are shown in Figs. 4.14, 4.15, 4.16 and 4.17, respectively. It can be seen that the change of L_1 phase center is 4.17 mm and that of L_2 phase center is 3.14 mm. Although the three chokes can improve the anti-multipath capability of the antenna, it will also greatly reduce the gain of the antenna at low elevation, thus affecting the ability of the system to receive satellite signals at low elevation. For satellites requiring continuous positioning on orbit, it is necessary to improve the gain in the low-elevation region.

Fig. 4.13 Dual-Band microstrip GNSS antenna with three chokes

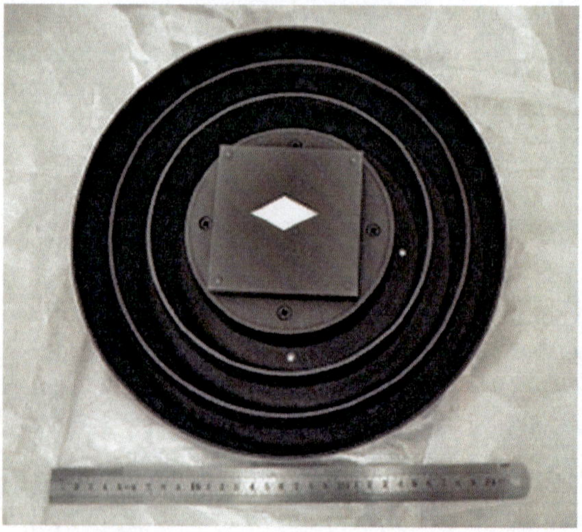

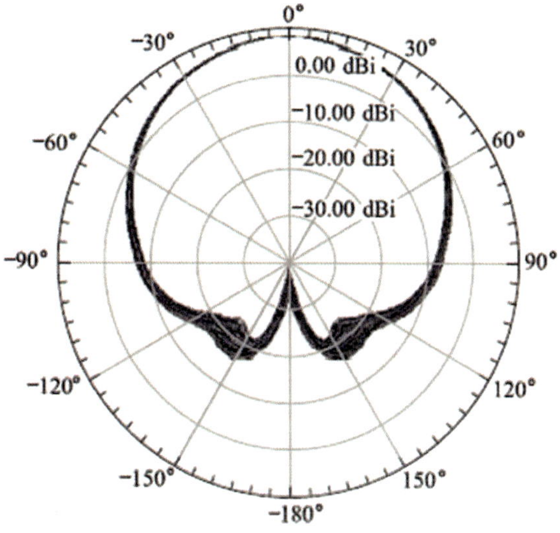

Fig. 4.14 Gain pattern (L_1, frequency = 1575 MHz)

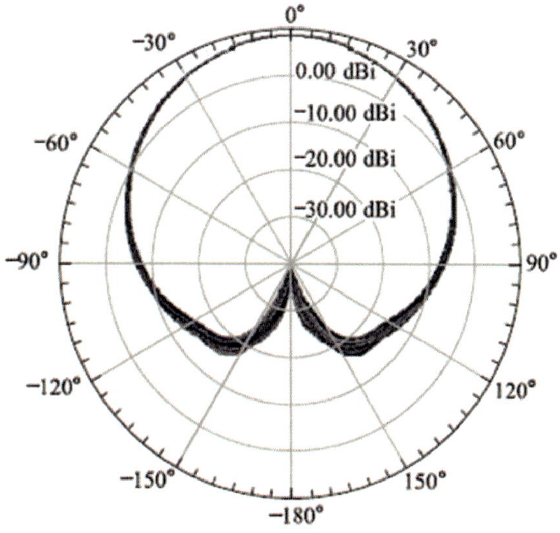

Fig. 4.15 Gain pattern (L_2, frequency = 1227 MHz)

With the addition of common choke, although the effect of satellite is reduced to some extent and the multipath effect is suppressed, the gain of antenna in low-elevation region is reduced, under which it may appear that the visible satellites on orbit is insufficient and the receiver is not positioning.

Therefore, three-dimensional choke structure (Fig. 4.18) is introduced, which can improve the antenna's low-elevation gain, reduce size and weight and make the antenna more suitable for installation on the satellite while suppressing the multipath effect.

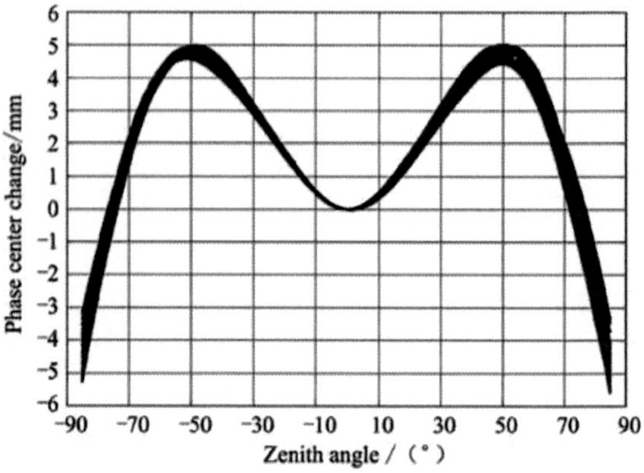

Fig. 4.16 Antenna phase center change (L_1)

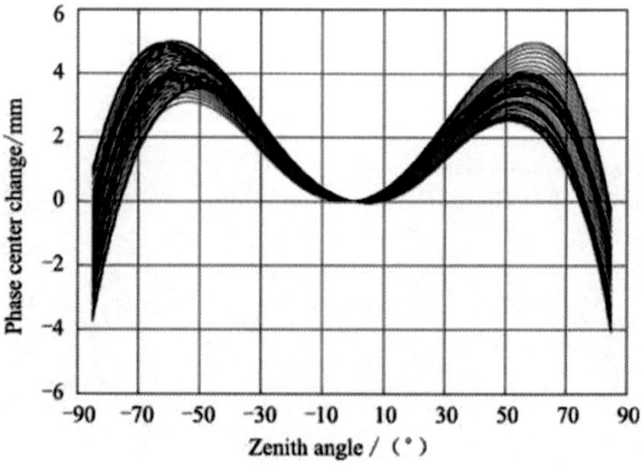

Fig. 4.17 Antenna phase center change (L_2)

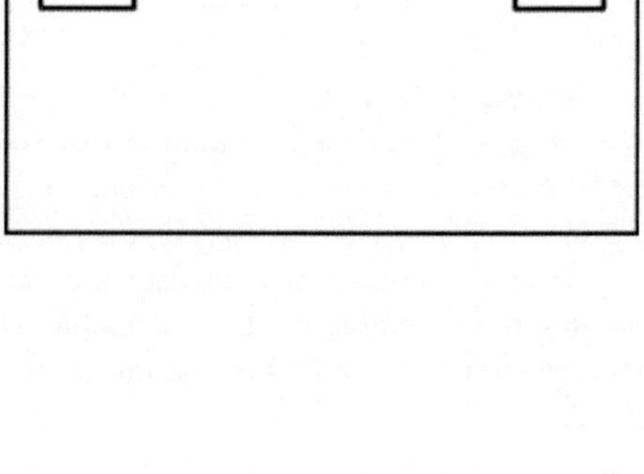

Fig. 4.18 Diagram of three-dimensional choke

A three-dimensional choke structure (Fig. 4.19) is developed according to the requirement of anti-multipath of antenna to improve the multipath suppression capability as much as possible without sacrificing the low-elevation gain. The stability of the pattern and phase center of this antenna is shown in Figs. 4.20, 4.21, 4.22 and 4.23. The change of L_1 phase center is 2.67 mm and that of L_2 phase center is 3.23 mm.

Fig. 4.19 Dual-Band microstrip GNSS antenna with three-dimensional choke

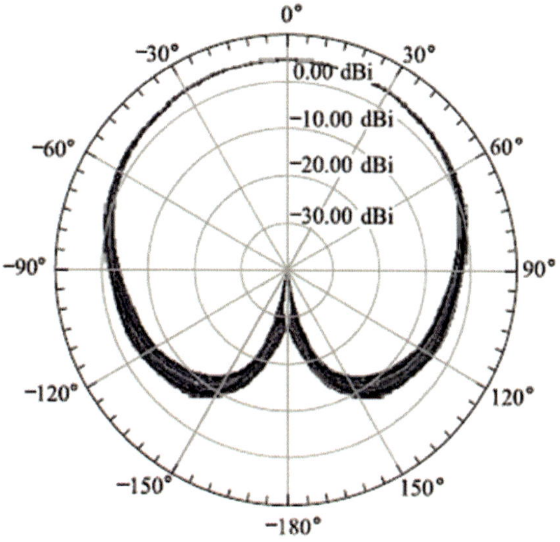

Fig. 4.20 Gain pattern (L_1, frequency = 1575 MHz)

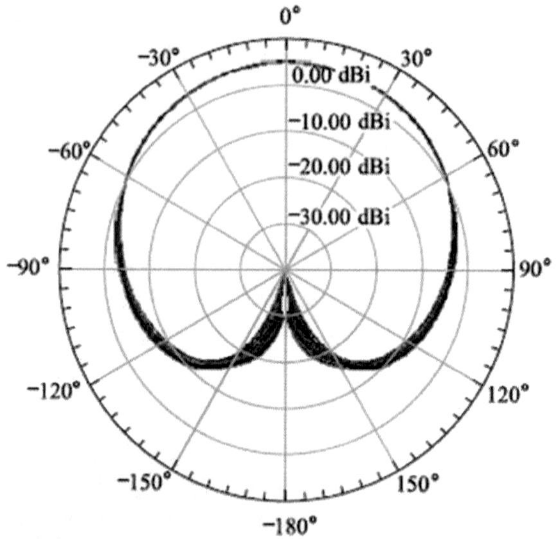

Fig. 4.21 Gain pattern (L_2, frequency = 1227 MHz)

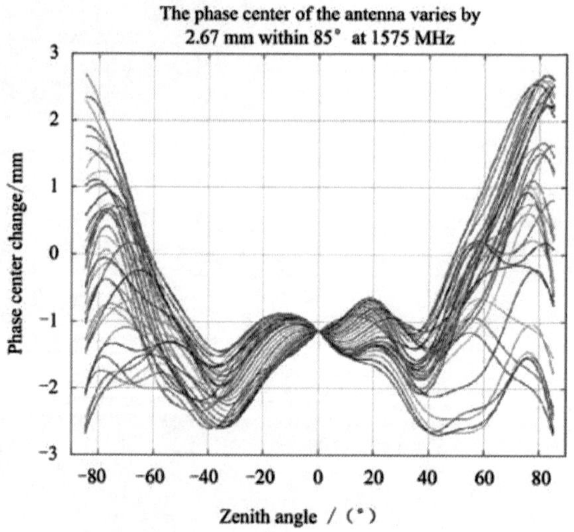

Fig. 4.22 Change of antenna phase center (L_1)

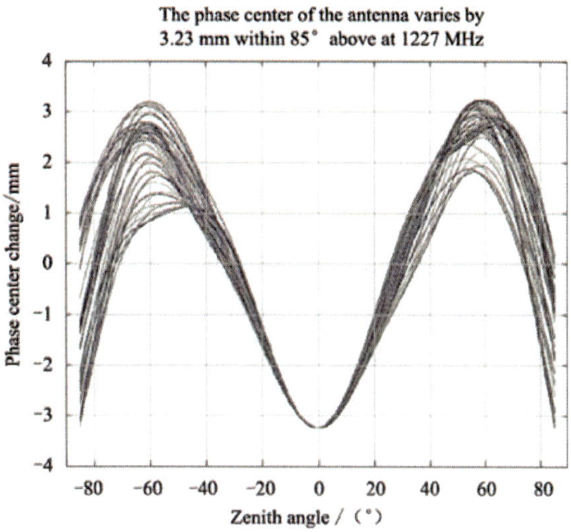

Fig. 4.23 Change of antenna phase center (L_2)

4.3.2 *Dual-Frequency PEC GNSS Antenna*

Swedish RUGA Aerospace Corporation has developed a series of different types of GNSS antennas [2–4], the most typical of which is the Patch Excited Cup (PEC) antenna used by SWARM satellite.

4.3.2.1 PEC Dual-Band GNSS Antenna Unit

The PEC antenna used in SWARM is a four-point feed antenna with a two-layer air medium, as shown in Fig. 4.24. There are two GNSS antennas on the satellite,

Fig. 4.24 PECGNSS antenna for SWARM [2]

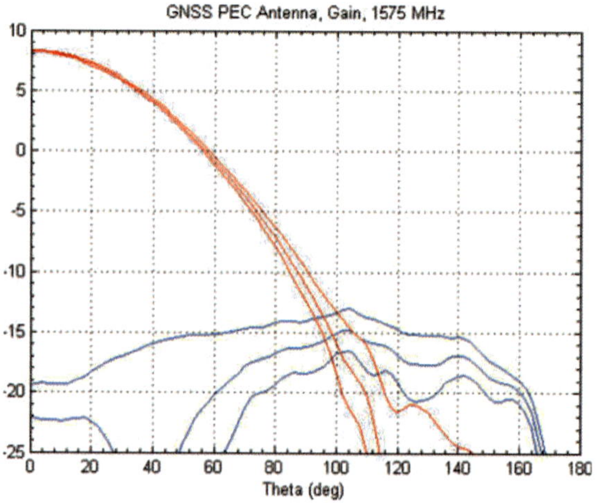

Fig. 4.25 GNSS PEC antenna typical patterns(co-polar red and cross-polar blue) [2]

which are mounted facing away from the Earth. The antenna has a height of 55 mm, a diameter of 160 mm and a weight of 325 g.

The PEC antenna is realized by two layers of patches placed in the reflector cup, in which the upper patch is excited by the lower patch coupling and the lower patch is fed by the strip line feeding network for four points. The feed network is independent and, respectively, excites the lower patch at different phases of 0°, 90°, 180° and 270° with the same amplitude to achieve circular polarization. Broadband matching is achieved by magnetic coupling between the feed point and the lower patch. Its typical pattern is shown in Fig. 4.25.

The antenna design is strictly constrained by the allowable size, so it has the advantages of lightweight and small size. However, the back lobe of the antenna is relatively large, which makes its radiation performance vulnerable to the interference of the satellite environment. Therefore, it has good performance when it is mounted on the edge of the satellite, while the performance is greatly affected by the satellite when the antenna is mounted on a large plane of the satellite.

4.3.2.2 PEC Dual-Frequency GNSS Antenna with Choke Rings

Considering the influence of the satellite environment, RUGA Aerospace Corporation has developed a PEC antenna with two chokes, which realizes the low back lobe of the antenna, improves the cross-polarization ability of the antenna, reduces the multipath effect of the satellite, and realizes the precise positioning of the satellite. The antenna with two choke rings [4] has a height of 87 mm, a diameter of 200 mm and a weight of 735 g. Its cross-polar pattern gain within range 0°–140° from −12 dB (no choke) decrease to −18 dB (with two choke rings) [3, 4].

4.4 High-Orbit GNSS Navigation Antenna

HEO spacecraft, especially geosynchronous orbit (GEO) satellites, have become an important orbital resource for Earth due to their unique high-orbit and geostationary characteristics. The application of navigation reception has broad prospects in TT&C system of HEO spacecraft.

In 2000, according to the operation requirement documents for GPS constellation issued in the U.S., the space user requirement was included for the first time. The service volume of satellite navigation was divided (Fig. 4.26) into the terrestrial service volume (TSV), which includes the airspace from the Earth's surface to an altitude of 3000 km, and the space service volume (SSV), which is the airspace from 3000 to 36,000 km. In 2008, NASA further divided the space service volume into two parts: MEO space service volume (3000–8000 km) and HEO/GEO space service volume (8000–36,000 km).

For TSV, the service objects include not only all users on Earth's surface (such as vehicles), users in the atmosphere (such as aircraft) but also Low Earth Orbit (LEO) satellites. Most navigation satellites are located above MEO orbit, so the number of GPS satellites observed by GPS receivers on LEO spacecraft and the level of received GPS signals are similar to those of ground GPS receivers, and the positioning accuracy can be comparable to that of ground users. At the altitude above 3000 km, the main beam of the signal transmitted by the GPS satellite antenna is blocked by the Earth, so it can only receive the side lobe or back lobe signal of the GPS satellite. Therefore, the antenna receiving the GPS signal should be specially designed. As shown in Fig. 4.26, when the orbital altitude of spacecraft is higher than that of GPS constellation (about 20,000 km), the receiving antenna is required to point to the center of the Earth, to receive GPS satellite signals from the opposite side of the Earth, and it is generally a side lobe signal of the GPS satellite.

In the development of GNSS antenna for a GEO orbital remote sensing satellite in China, PEC antenna scheme is selected referring to the experience of international spaceborne high-orbit GNSS antenna [3]. In this antenna, three patches are fed by

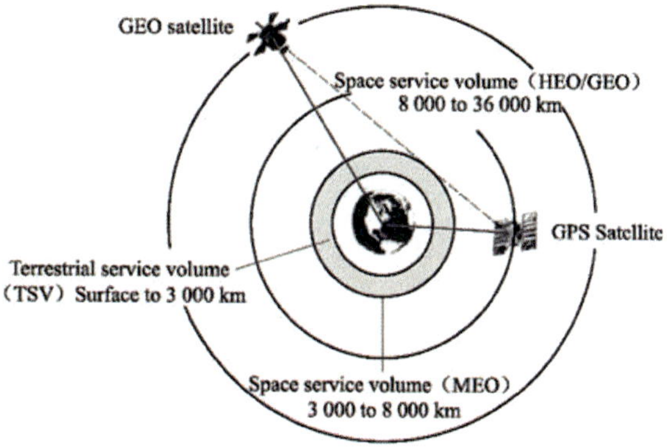

Fig. 4.26 Service volume description of different orbital altitudes

four-point circular polarization feeding, and the broadband matching of small antenna is realized by the reactance balance of the feeding structure itself. The antenna is designed as a flat plate structure to stabilize the phase center of the pattern. By using the reflector cup, the roll-off and front-to-rear ratio of the pattern is increased and the multipath effect is suppressed. Unlike the low-orbit GNSS antenna, a director is added to improve the main gain while ensuring circular polarization radiation. The antenna structure is shown in Fig. 4.27. The feed network of the antenna is composed of Wilkinson power dividers, which achieve the excitation of different phases of 0°, 90°, 180° and 270°. The whole network is enclosed in the antenna floor cavity, so it has good anti-electromagnetic interference and overall integration.

The antenna pattern design results are shown in Fig. 4.28. It can be seen that the side lobe of the antenna is less than −18 dB within the required frequency band and the gain of the antenna is more than 5.41 dBi within the range of ±37°.

The design advantages of this antenna scheme are as follows: cross-polarization is greatly improved compared with helical antenna; back lobe radiation is very small, which can effectively suppress the multipath effect.

Fig. 4.27 HEO GNSS antenna

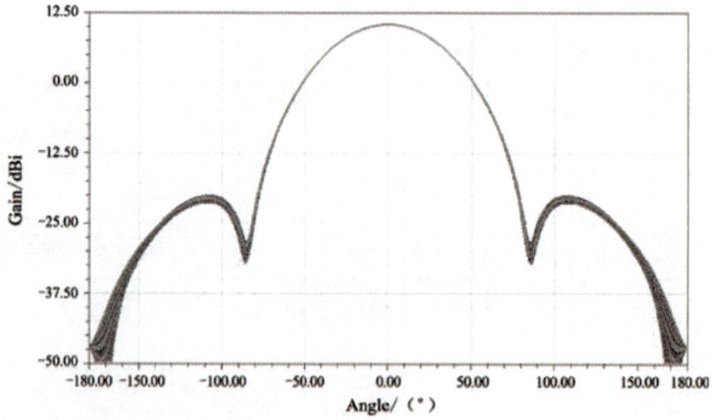

Fig. 4.28 Performance results of HEO GNSS antenna

References

1. Dilßner F, Seeber G, Schmitz M et al (2006) Characterization of GOCE SSTI antennas, computer science
2. Zackrisson J (2017) RUAG space activities in the TT&C, GNSS and data-downlink antenna field. In: 11th European conference on antenna and propagation (EUCAP)
3. Kurlberg N, Magnusson P (2009) A compact Rx/Tx dual circular polarization satellite antenna array element including diplexers. In: 3rd European conference on antennas and propagation, 2009, EuCAP
4. Zackrisson J, Öhgren M (2011) GNSS receive antennas on satellites for precision orbit determination. In: 25th annual AIAA/USU conference on small satellite, 2011

Chapter 5
Inter-satellite Link Antenna

Abstract Inter-Satellite Link (ISL) refers to the communication link between satellites, which can replace the ground relay station to do the autonomous operation of satellite constellation. This chapter provides two kinds of ISL antennas with beam scanning capability: the movable spot-beam reflector antenna and the phased array antenna.

Keywords Inter-satellite link · Reflector antenna · Phased array antenna · Bean scanning · Calibration

Inter-Satellite Link (ISL) refers to the communication link between satellites, which can replace the ground relay station to complete the autonomous operation of satellite constellation, and greatly reduced the number of ground injection stations (mainly for regular injection of orbiting data and clock calibration data). Autonomous operation of satellite constellation means that the satellite performs the functions or operations required for the mission by autonomously determining the constellation state and maintaining the constellation configuration on orbit without relying on ground facilities. An autonomous operation can greatly reduce constellation cost and system risk, which is an inevitable trend of development. Recognizing the shortcomings in operation and management of the traditional mode of ground-based TT&C, all countries are actively developing the autonomous operation technology of spacecraft.

The US GPS has added ISLs since the Block-IIR satellite, with which the accuracy of user positioning in 180-day forecast period was increased from 5000 to 6 m without the support of the ground master station, thus significantly improving the availability of GPS in war and other special environments. In GPS Block-IIR satellite, a ring array with ten helical antenna elements is installed on the toward-Earth side of the satellite, and the group delay uncertainty of the array antenna is controlled within 0.5 ns. Also, both the Crosslink Transponder Data Unit (CTDU) and the Crosslink Transponder Subsystem (CTS) are configured. CTDU is a time-division multiple access frequency-hopping spread-spectrum communication system, which uses UHF band, 5Mc/s pseudo-random code and 108 W radio frequency transmission power to provide precise ranging signals and exchange autonomous navigation data. GPS Block-IIR satellite implements dual-frequency one-way ranging. Within ranging frame, one Block-IIR satellite transmits ranging signals in the allocated time

R. You et al., *Technologies for Spacecraft Antenna Engineering Design*, Space Science and Technologies, https://doi.org/10.1007/978-981-15-5833-7_5

slot under ranging mode, and all other visible navigation satellites receive and process the ranging signals in UHF band, in which the plasma layer effect is eliminated by two different frequencies, thus completing pseudo-range measurement. Within the data frame, each satellite transmits data parameters related to itself in its allocated time slot under data transmission mode, mainly including the calculation result of the pseud-orange measurement, the estimated position and clock parameters of the satellite, and the corresponding estimated variance.

Milstar, DSC-III and LEASAT of USA all equipped with ISLs. Milstar represents the state-of-art of U.S. military satellites, which uses V-band antenna beam to form the inter-satellite crosslinks with six satellites. It has a strong ability of anti-interference and is less vulnerable. In addition, Ka-band ISLs technology, S-band ISLs technology and medium-low-orbit ISL technology is used in Iridium constellation, White Cloud Ocean surveillance satellite system, and Tracking and Data Relay Satellite System (TDRSS), respectively.

ESA's Artemis satellite has inter-satellite relay links that operate in Ka- and S-bands; the Russian Space Agency has added ISLs to its military and civilian relay satellite systems, working in L- and C-bands, respectively. Both the Galileo system of the European Union and the GLONASS system of Russia have started to do the research and equip ISLs. Japan also focuses on the ISLs system in tracking and relay satellite system and carries out Ka-band and S-band link technology research.

ISLs are equipped in global communication satellite constellations, reconnaissance formation flying systems and global navigation satellite constellations. Except for communication constellations, the latter two have inter-satellite communication and inter-satellite ranging functions, and GNSS constellations also have autonomous navigation functions. Beidou-3 satellite and some application satellites of China are also equipped with ISLs. Therefore, starting from the introduction of the characteristics of the ISLs system of navigation satellites, this chapter draws the functional requirements for the antenna of the ISL. Then, according to the requirements of the ISLs, two kinds of antennas with beam scanning capability—the movable spot-beam reflector antenna and the phased array antenna—are discussed in this chapter.

5.1 ISLs Characteristics of Navigation Satellites

In the global navigation satellite system, maintaining high accuracy of satellite orbit and clock difference determination is the key to ensure that the navigation receiving terminal obtaining the positioning or timing accuracy required by the system. Generally, in order to obtain precise orbit and clock difference parameters, many ground stations are required to be set up around the world for continuous observation. The precise orbit parameters can be obtained through a few monitoring stations after the construction of ISLs in the global navigation constellation and the ISLs can also increase the ephemeris injection frequency, both of which directly affect the navigation and positioning accuracy. When the uplink data cannot be injected into the satellite-borne navigation task processing unit, the autonomous operation ability of the constellation can be improved by ranging between the satellites and predicting the ephemeris and clock data. In addition, the ground station's TT&C and operational control information can be transmitted through ISLs, which can greatly improve the ground station's TT&C management capabilities for constellations.

Therefore, with ISLs, the navigation satellite constellation can support joint orbit determination and time synchronization of satellite and ground station, and real-time measurement and control management of global constellation, and implement autonomous navigation of navigation constellation, so as to improve the management ability of the ground system and enhance the autonomous operation ability of the space system.

The function of GPS ISLs is improved with the change of system tasks, from supporting information transmission of nuclear explosion detection to supporting inter-satellite ranging and information transmission of autonomous navigation, and then to fully supporting space integrated information network communication. Since 1978, the GPS constellation has developed to the third generation. The earliest Block-I had no ISLs, and Block-IIA had ISLs without ranging capabilities. In 1997, the Block-IIR series began to have autonomous navigation function in UHF ISLs, which used UHF helical array antenna. In the planning of next-generation GPS Block-III constellation, Ka- or V-band ISLs antenna will be used.

The research of ISLs in China started relatively late. First- and second-generation Beidou navigation satellites have no ISLs communication function. The function of ISLs has been added to third-generation Beidou satellite system. Two ISLs systems, namely reflector and phased array, have been tested and verified simultaneously in the test satellite. On this basis, the phased array ISL system was used in Beidou-3 navigation satellite.

The Global Navigation Satellite System (GNSS) of Beidou-3 project consists of 30 satellites, including 24 MEO+3 GEO+3 IGSO satellites, namely 24 Medium Earth Orbit (MEO), 3 Geostationary Orbit (GEO) satellites and 3 Inclined Geosynchronous Satellite Orbit (IGSO) satellites. The constellation orbit, as shown in Fig. 5.1, will provide global navigation services. Moreover, Ka-phased array antenna is mounted on the satellite to establish ISLs in Ka-band.

In terms of working mode, ISL antenna is mainly used to establish three types of link, namely static link, dynamic link and satellite-ground link. The static link is mainly used for ordinary inter-satellite communication and ranging, which continuously measure and control constellations to solve the problem of global satellite ground station distribution; the dynamic link is mainly used for inter-satellite ranging in special services, which makes the constellation run independently in a short time and maintain certain positioning accuracy; and the satellite-ground link is mainly used for on-orbit testing.

Static link can be established between MEO satellites that are permanently visible to each other. As for the dynamic link, the link topology is selected according to the optimal navigation and positioning accuracy requirements.

In terms of the object of link establishment, the ISLs antenna can be used to establish the low-low link between MEO satellites and the high-low link between MEO satellites and GEO/IGSO satellites.

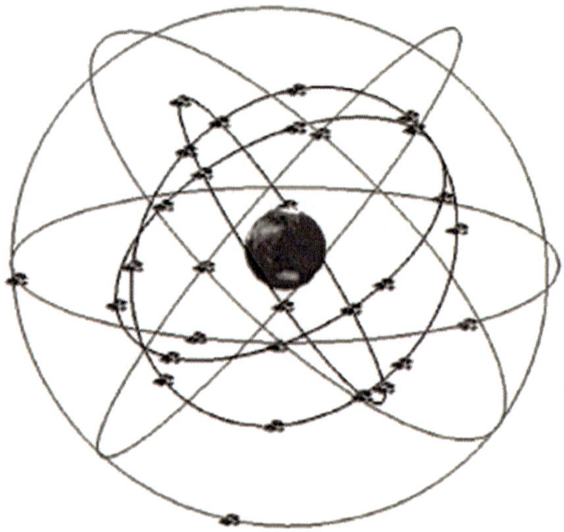

Fig. 5.1 Schematic diagram of the orbit of Beidou-3 constellation

5.1.1 Static Link

In order to guarantee the global TT&C function of navigation satellite constellation, it is necessary to establish static links, that is, ISLs should only be established between satellites whose distances are more than 5000 km and which are permanently visible to each other.

Firstly, the beam coverage of the low-low link of MEO satellite constellation is analyzed.

1. For the seven co-orbital satellites, except for one that is not visible due to the Earth occlusion, the maximum viewing angle of the other six satellites is $\theta = 68°$;
2. For the 16 off-orbital satellites, 4 of them have intermittent Earth occlusion and 2 of them have a motion range of θ more than 70°.

Therefore, the beam coverage of $\theta \in [0°, 70°]$ and $\varphi \in [0°, 360°]$ covers all visible co-orbital satellites and 10 off-orbital satellites, a total of 16 satellites.

The beam coverage of the high-low links is analyzed as follows:

For GEO/IGSO satellites, all MEO satellites are within ±46.3° cone angle, as shown in Fig. 5.2. Due to the higher orbit of the GEO/IGSO satellites, they are not permanently visible to the MEO satellites with limited beam coverage.

Therefore, the ISLs antenna beams have a beam coverage of $\theta \in [0°, 70°]$ and $\varphi \in [0°, 360°]$, which enables MEO satellites to cover 16 permanently visible MEO satellites and GEO/IGSO satellites to cover all permanently visible MEO satellites.

When Ka-reflector antenna establishes static links, each satellite can establish static links with specific multiple satellites and can switch the target according to the topology table. The static link of Ka-reflector antenna is a frequency division

Fig. 5.2 View angle of MEO/GEO/IGSO satellite

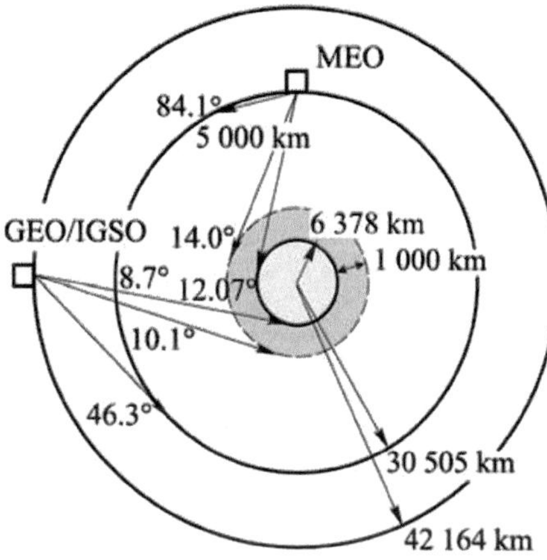

full-duplex system. Its advantage is to establish uninterrupted communication links, thus is suitable for inter-satellite communication.

5.1.2 Dynamic Link

Dynamic links are established between MEO, GEO and IGSO satellites that are more than 5000 km apart and visible to each other. According to the analysis of link pointing range, the scanning range of Ka-band ISLs antenna beams in the range of $\theta \in [0°, 60°]$ and $\varphi \in [0°, 360°]$ covers more than eight visible satellites, which can meet the requirements of establishing dynamic links between MEO/GEO/IGSO satellites.

Each satellite can establish dynamic links with all satellites in the antenna field of view by using Ka-phased array antenna to establish dynamic links. The dynamic link of Ka-phased array antenna is characterized by time-division same-frequency half-duplex. It can switch beam pointing quickly and has the function of on-orbit delay monitoring and calibration. The advantage of dynamic link includes that it has a better geometric distribution of spatial measurement and is suitable for inter-satellite ranging.

5.2 ISLs Reflector Antenna

The ISL reflector antenna is generally in the form of mechanical movable beam antenna, with which long-lasting ISLs between satellites with high gain and low side lobe can be established.

5.2.1 Demand and Task Analysis

According to the design of navigation satellite signal system, the ISLs reflector antenna with the function of establishing a persistent link works at two frequency bands in Ka-band by the frequency-division full-duplex mode. In order to keep the state of all satellites consistent, the Tx frequency and Rx frequency of the antenna can be switched on orbit through the duplex assembly.

ISLs reflector antenna is used not only for communication function but also for high-precision inter-satellite ranging so that the stability of delay and phase center should be guaranteed.

The ISLs reflector antenna is composed of radio frequency part, structure and mechanism part and thermal control part. The antenna pointing control is realized by an antenna servo controller. When the satellite launching, the antenna is in a folding state (Fig. 5.3) and locked on the satellite. After the satellite is on orbit, the locking/release mechanism is unlocked and the antenna is deployed in position and supported outside the satellite body under the drive of the deploying/locking mechanism. When operating on orbit, the two-axis driving mechanism drives the reflector antenna to rotate and perform beam pointing scanning. The radio frequency part is a fundamental element to realize the electrical performance of the antenna. This section focuses on the design of the radio frequency part of the antenna.

As shown in Fig. 5.4, the RF section of ISLs reflector antenna mainly includes reflector antenna assembly, rotating joint assembly, waveguide assembly, etc. The reflector antenna assembly includes reflector assembly, feed assembly and support rods, which are used to realize antenna radiation-related characteristics, such as gain, polarization, side lobe, phase center, etc. The antenna uses three identical rotating joint assemblies, which are, respectively, located on the deployment axis, the axis A and the axis B of the antenna, to realize the microwave signal transmission during the rotating process of the antenna. The waveguide assembly is used to connect the reflector antenna assembly and three rotating joint assemblies.

Considering the yaw of satellite, based on the constellation dynamic simulation analysis of the ISLs reflector antenna, the maximum speed of the two-axis pointing mechanism is about 0.42°/s, the cumulative rotational travel is not more than 9000 rotations and the cumulative revolutions number is not more than 25,000 turns during

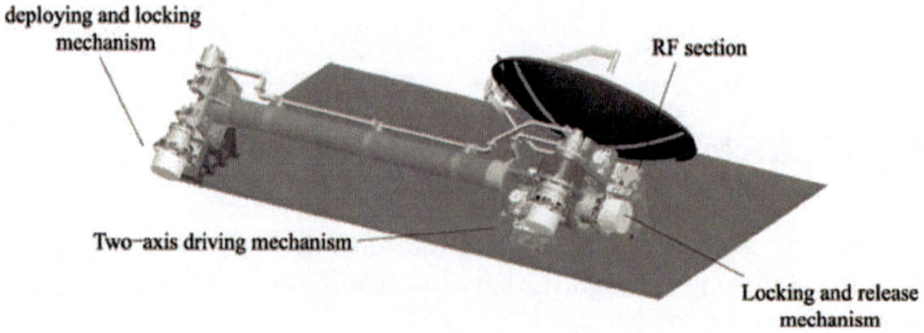

Fig. 5.3 ISLs reflector antenna profile (Excluding thermal control system)

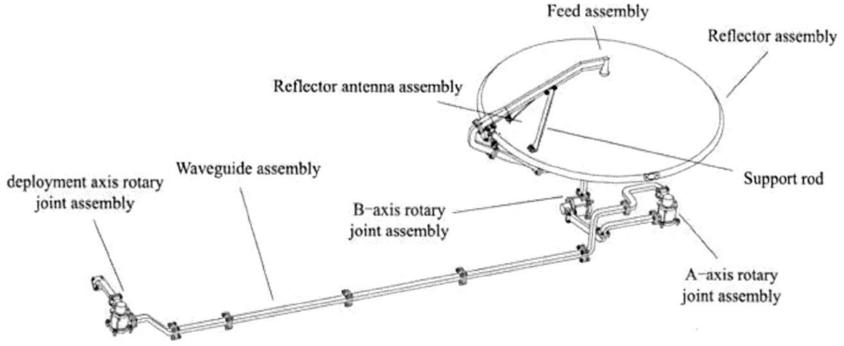

Fig. 5.4 RF section of ISLs reflector antenna

the lifetime. Here, one rotation refers to the cumulative rotation of 360°, and one turn means a complete reciprocating motion (including two changes in the direction of rotation).

5.2.2 Design of Reflector Antenna Assembly

The difficulty in the design of reflector antenna assembly is that it shall satisfy the constraints of high gain, low side lobe, envelope size, mechanical characteristics and thermal environment at the same time. Firstly, the antenna beam scanning range is a cone-shaped area with a half cone angle of 70°, so it is suitable for the focus-fed reflector configuration with symmetry structure, such as the focus-fed single reflector, the focus-fed Cassegrain double reflectors, and ring focus antenna. Secondly, the antenna is required to have a low side lobe characteristic of −28 dB. Due to the obstruction of the main reflector by the subreflector and the supporting structure, the side lobes are inevitably raised. Therefore, the focus-fed Cassegrain double reflectors and ring focus antenna with subreflector are basically excluded. Now, only the focus-fed single reflector antenna is available, and the general focus-fed single reflector is difficult to achieve the low side lobe characteristic of −28 dB. It is necessary to optimize the overall design of the feed and its supporting structure to minimize the blockage to the main reflector, and to use the feed waveguide with a low scattering profile to further reduce the side lobe.

In the general focus-fed single reflector antenna, the feed includes the connection between the feed waveguide and the circular polarizer, the connection between the circular polarizer and the horn, etc., and often occupies a larger size. The size of feed becomes reduced after optimization design by which the circular polarizer is fed by the rectangular waveguide directly from the side, and the transition size between the circular polarizer and the horn is shortened. In addition, the selection of circular polarizer is also very important. The size of the feed can be further reduced by using a more compact dentiform circular polarizer. The purpose of the circular polarizer is to form two electric field components in orthogonal directions, with equal amplitude and 90° phase difference.

For coaxially fed antennas, such as helical antennas and microstrip antennas, power divider and 90° phase shifter (or 3 dB bridge) can be used. While for waveguide-fed antennas, circular polarization can be formed more easily by utilizing the mode excitation and inherent characteristics of transmission in the waveguide. Three types of waveguide circular polarizer are shown in Fig. 5.5.

The circular and the square waveguides in Fig. 5.5 have an electric field in two orthogonal directions, with equal amplitude and same phase. If a number of protrusions, such as short pins, diaphragms and corrugates, are set on the inner wall of the waveguide in the vertical direction, the equivalent inner diameter in the vertical direction will be reduced, and the waveguide wavelength of the horizontal electric field will be increased. In case that the phase velocity of the horizontal electric field increases, the vertical electric field phase velocity remains unchanged. The phase difference of 90° will naturally occur after the transmission of the two electric fields for a certain distance, thus forming circular polarization. The circular polarizer is shown in Fig. 5.5a, b generates LHCP at the output port. Placing a dielectric plate in the horizontal direction of the circular waveguide is equivalent to slowing down the phase velocity of the electric field in the horizontal direction, while the phase velocity of the electric field in the vertical direction is unchanged. Therefore, the circular polarizer in Fig. 5.5c can produce RHCP at the output port.

Note that the horizontal and vertical electric fields in Fig. 5.5 are actually the components of the same waveguide mode electric field in the horizontal and vertical directions. When the electric field of the waveguide mode rotates 90°, all three circular polarizers in the figure output the corresponding reverse circular polarization.

The horn has a small aperture and can produce a wide beam, which is suitable for the irradiation of the focus-fed single reflector. The reduction of the overall size of the feed reduces not only the blockage of the reflector aperture but also its weight, which is conducive to the improvement of the overall fundamental frequency of the reflector antenna. Therefore, the constraints of electrical and mechanical properties are unified.

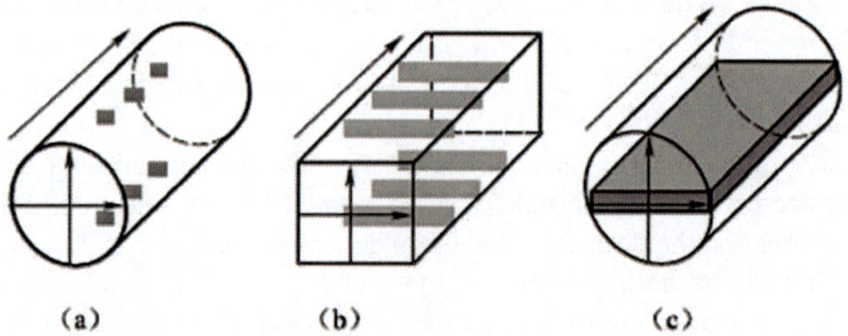

Fig. 5.5 Waveguide circular polarizer. **a** Pin type; **b** diaphragm or corrugated type; **c** dielectric sheet type

Reflector antenna assembly is in a severe thermal environment when working on orbit. The temperature range is −150 °C to +110 °C. In order to reduce the structural deformation caused by thermal environment, the reflector is made of aluminum honeycomb composite material with carbon fiber skin, and the feed and its supporting structure are made of low expansion alloy (commonly known as invar steel). The expansion coefficient of this material is about 0.8×10^{-6}/°C, which is close to that of carbon fiber braided fabric, so its thermal deformation can be neglected. The transverse (i.e., perpendicular to the focal axis) offset of the feed will affect the orientation of the antenna electric axis. When the antenna is working on orbit, the low-temperature condition is −150 °C, and the temperature difference from the normal temperature is the largest. Under this condition, the transverse shrinkage of the feed is 0.05 mm, and the corresponding electric axis tilt is negligible.

The feed waveguide of the horn will produce electromagnetic scattering, which will then result in higher side lobes. In order to reduce the side lobe, the irradiated surface of the feed waveguide (facing to the reflector) is designed as a wedge shape, which is beneficial to improve the mechanical stiffness of the feed waveguide. However, the feed cannot be supported solely by the feed waveguide because it cannot resist the force perpendicular to the direction of the feed waveguide. Therefore, two supporting rods are symmetrically provided on both sides of the feed waveguide. The connection position between the support rod and the feed waveguide is contradictory in electrical performance and mechanics, the electrical performance design requires that the connection point be far from the center of the reflector, while the mechanical design requires that the connection point be closer to the center of the reflector. Therefore, balance and compromise are required here.

The measured results of reflector antenna pattern after the joint optimization design of electrical performance and structure are shown in Fig. 5.6. Thus, the reflector antenna has achieved excellent design results, and all its performances meet the specifications.

Fig. 5.6 Contour pattern (dB is the unit of contour value)

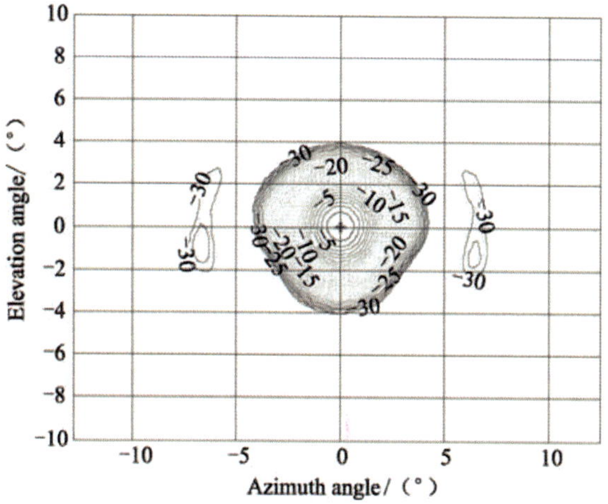

5.2.3 Design of Rotating Joint Assembly

The microwave rotating joint is an important part of the mechanical movable beam antenna, which realizes the transmission of microwave signal during the rotating of the antenna. Generally, such a function can also be implemented by flexible high-frequency cables. However, for satellite-borne applications, high-frequency cables face a higher risk of reliable operation and a series of problems such as delay stability, power capacity and insert loss after tens of thousands of bends in harsh thermal environments. Therefore, microwave rotating joint is used as a more reliable design scheme. For ISL reflector antenna, a microwave rotating joint with a large power capacity and good delay stability is required.

According to the structure, the microwave rotating joints can be divided into the coaxial, waveguide and coaxial hybrid types. According to the number of channels, the microwave rotating joints can be divided into a single-channel, double-channel and multi-channel.

The coaxial rotating joint can be seen as a coaxial line divided into two segments. According to the structure, they are divided into straight-through and right-angle elbow types, as shown in Fig. 5.7. Its advantages are simple principle and small volume when working in a lower frequency band, and its disadvantages are small power capacity, complex structure and many RF connectors.

The waveguide rotating joint is composed of rectangular waveguide and circular waveguide. The transmission of RF signal is implemented by waveguide mode conversion. According to the different methods of generating circular waveguide axisymmetric mode TM_{01}, it can be classified into short circuit mode, resonant mode and probe coupling mode, as shown in Fig. 5.8. Short-circuit mode suppresses TE_{11} mode in the circular waveguide by parallel short-circuit branch line; resonant mode cancels TE_{11} mode coherently and strengthens TM_{01} mode coherently by joint cavity length; probe-coupled mode transforms TE_{10} mode in rectangular waveguide to TM_{01} mode in circular waveguide through a coupling hole and probe. The probe-coupled mode requires the transition of TEM mode between TE_{10} mode (H_{10}) and TM_{01}

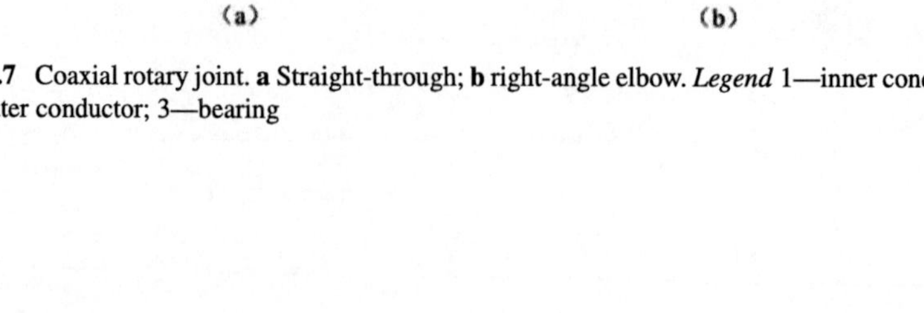

Fig. 5.7 Coaxial rotary joint. **a** Straight-through; **b** right-angle elbow. *Legend* 1—inner conductor; 2—outer conductor; 3—bearing

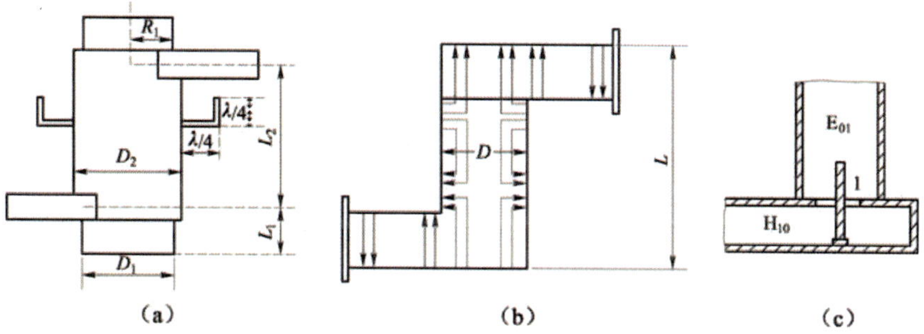

Fig. 5.8 Principle of waveguide rotating joint. **a** Short circuit type; **b** resonant type; **c** probe coupling type

mode (E_{01}), similar to the waveguide-coaxial hybrid mode, and also with the dielectric support coaxial part. High-power capacity is not allowed due to the limitation of the coaxial part.

The characteristic of waveguide-coaxial hybrid rotating joint is that it consists of two rectangular waveguides and they connect each other by the rotating part with coaxial structure, i.e., a pair of waveguide coaxial converters that can rotate. This is a common form of rotating joint. According to the different forms of coaxial waveguide conversion, waveguide coaxial hybrid rotating joint also falls into two types: as for type 1, the axis is the centerline of one of the waveguides, and as for type 2, the axis is perpendicular to the centerline of the two waveguides. The advantage of waveguide coaxial hybrid rotary joint is its simple principle, and its disadvantages are the complicated structure and the small power capacity.

After comparing various schemes, a waveguide-type rotating joint is chosen, whose waveguide cavity structure is used to realize the transformation from TE_{10} mode of rectangular waveguide to TM_{01} mode of circular waveguide. After simplifying choke slot, improving short-circuit circular waveguide, the compact volume and size are achieved, as shown in Fig. 5.9. Through the analysis and comparison of physical properties of materials, duralumin is selected, by which bright anodic

Fig. 5.9 Appearance of waveguide rotary joint

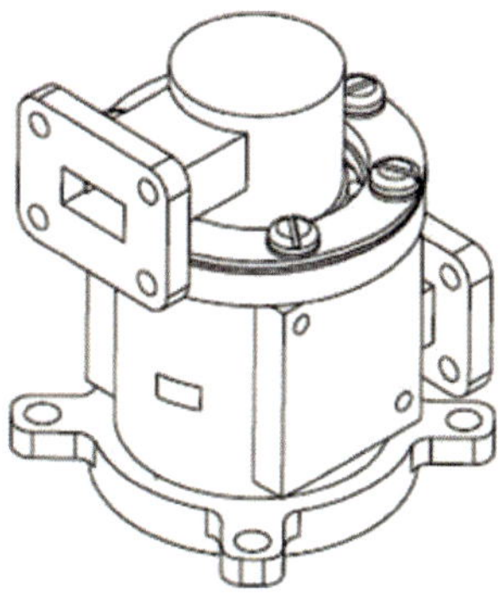

oxidation is adopted on the outer surface to meet the requirements of thermal control, chemical conductive oxidation is used on the inner wall to meet the requirements of low electrical loss. And then a pair of deep groove ball bearings is used.

For the waveguide rotary joint with on-orbit ambient temperature range of −170 °C to +100 °C, active temperature control measures are taken to ensure that its working temperature range is −25 °C to +80 °C (survival temperature). Because the thermal expansion coefficient of duralumin is larger and that of bearing steel is smaller, the bearing is tightened inside and loosened outside at high temperature and loosened inside and tightened outside at low temperature. Since the rotating joint is more likely to be under high-temperature condition, small clearance fit of outer ring and interference fit of inner ring can better adapt to the clearance change of external tightening and internal loosening. In addition, the bearing has a clearance of 0.02 mm in the axial direction, which can adapt to the axial thermal deformation.

Two M2 screw holes are provided on the side of the rotating joint for installing the wire-wound resistance heater. The external surface of the waveguide is treated by bright anodic oxidation. The analysis shows that the thermal control design of the rotating joint meets the requirement that the working temperature on orbit should not exceed −25 °C to +80 °C.

5.2.4 Design of Waveguide Assembly

In the design of the waveguide assembly, two factors in the space environment should be considered, one is exhaust; the other is the release of its own thermal stress. Generally, a pair of vent holes is set every 0.5 m. The vent holes are located at the broadside flange edge of the waveguide. The diameter of the vent holes is generally 1.0 mm, as shown in Figs. 5.10 and 5.11.

The analysis shows that the on-orbit temperature of the waveguide assembly is −95 °C to +55 °C, and the maximum temperature difference is 115 °C compared with the normal temperature. The temperature deformation of the waveguide assembly

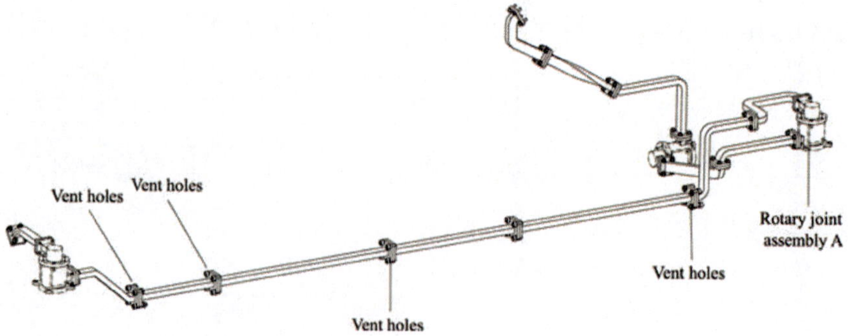

Fig. 5.10 Vent holes of waveguide assembly

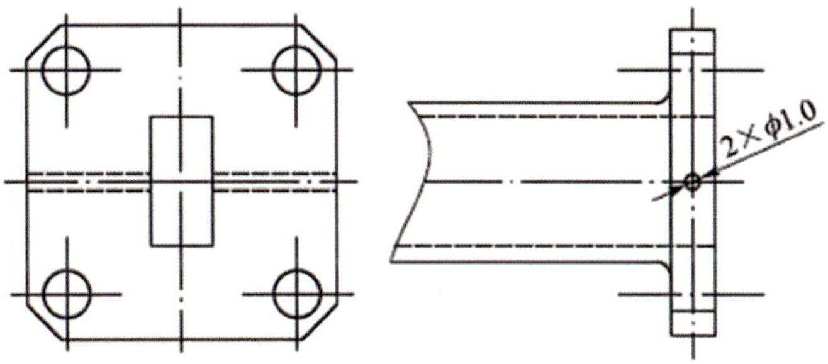

Fig. 5.11 Schematic diagram of waveguide vent hole

on the arm boom is different from that of the arm boom, so temperature stress will be occurred. The length of the longest boom is 780 mm, and the linear expansion coefficient is close to 0. The waveguide deformations are −1.8 to +0.1 mm, in which the shrinkage at low temperature is larger and the stress on the rotating joint is significant. One end of the waveguide assembly, which is connected to the rotating joint assembly of the deployment axis is fixed on the deployment locking assembly, and is not affected by the temperature stress; the other end of the connection with the rotating joint assembly A is not fixed, on which the stress effect is greater.

The temperature stress is eliminated by a soft waveguide. The specific measures are as follows:

1. Using soft waveguide: A 100-mm long soft waveguide is added to the position of waveguide 1, as shown in Fig. 5.12. The rest waveguide with inconsistent thermal deformation is 680 mm long, whose maximum deformation is about 1.5 mm. The available expansion of the soft waveguide is +2 mm, the waveguide deformation stress can be eliminated when the temperature changes.
2. Design of waveguide clips: two types of waveguide clips are used to fix the waveguide on the arm assembly.
   Fixed waveguide clips: a total of two, set at both ends;

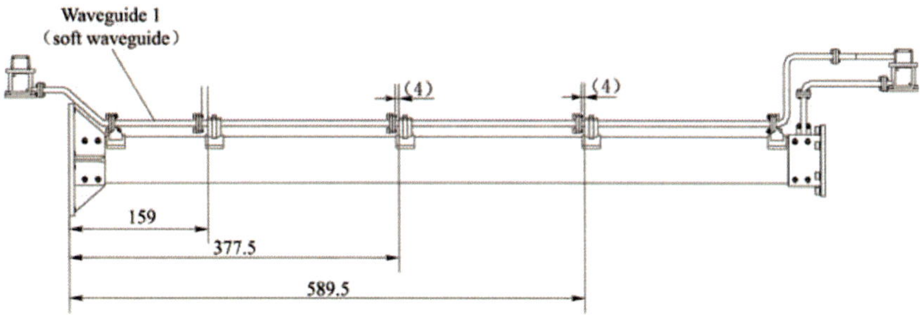

Fig. 5.12 Interface relation between waveguide module and arm bar

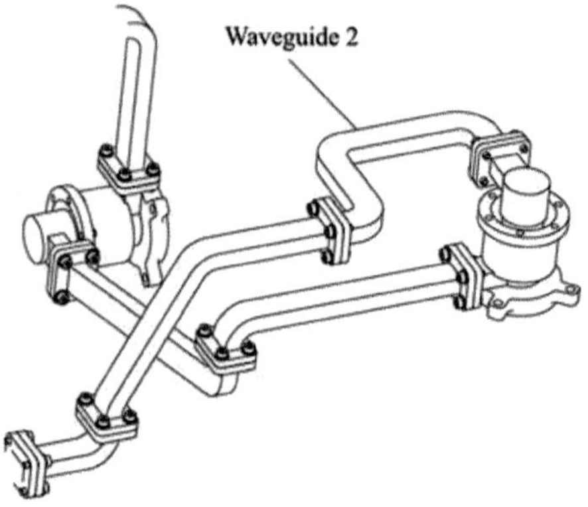

Fig. 5.13 Schematic diagram of waveguide 2 position

Sliding waveguide clips: a total of three, set in the middle. Silicone rubber is filled between the waveguide clip and the waveguide and is bonded on the arm boom with 420 glue.

3. A curved waveguide (waveguide 2) is adopted at the rotating joint of axis A, as shown in Fig. 5.13. The thermal deformation of the waveguide can be converted into a slight rotation, which compensates the thermal stress.

5.3 ISLs Phased Array Antenna

5.3.1 Demand and Task Analysis

Ka-band ISLs phased array antenna (Ka-phased array antenna for short) is a Tx/Rx community and time-sharing operation system. Similar to UHF ISLs, time slots are required to be divided when establishing ISL. It takes 30 time slots to establish the connection of all constellations. In terms of time, 1.5 min (90 s) is regarded as a period, and the phased array antenna completes the inter-satellite ranging task in each period. Each time slot is 3 s, which is divided into 1.5 s for transmitting and 1.5 s for receiving. Each of all 30 satellites carries the same inter-satellite equipment, and all ISLs have only one central frequency. For satellite-to-ground joint orbit determination and autonomous orbit determination, inter-satellite range is measured every 1.5 min.

Ka-phased array antenna has the advantages of fast beam switching and having no moving parts. However, there are also many constraints due to the characteristics of phased array antenna itself, such as small beam scanning angle, high side lobe, high power consumption, gain loss during beam scanning, phase center and delay stability, complex system and lower reliability. In the design of phased array antenna, measures should be taken to solve the aforementioned problems, and solutions that can be implemented in engineering should be obtained.

Firstly, the beam scanning range is closely related to the overall scheme of ISLs, which directly determines the number of visible satellites. The analysis shows that the number of visible satellites (more than eight satellites) is enough to ensure the required orbit determination accuracy for autonomous operation by using a pair of antennas with a flat-mounted layout and working mode with a scanning range of (±60°).

Secondly, the ISLs antenna adopts a system with co-frequency, time-sharing and Tx/Rx community from the viewpoints of lightweight, miniaturization, and low power consumption. This system not only simplifies the inter-satellite communication protocol but also integrates the transmitting and receiving antennas into one, thus reducing the weight and power consumption of the whole satellite. Also, the phased array antenna with the same phase center to transmit and receive improves the ranging accuracy. This system faces the difficulty of high integration of RF channels, but with the progress of technology, it can be effectively solved by using RF and control ASIC (Application Specific Integrated Circuit).

Thirdly, transmitting and receiving array should be optimized separately for the Tx/Rx community phased array antenna.

In order to reduce the antenna power consumption during the transmission, the density-weighted array is used by closing part of the transmission channel on the premise of ensuring enough EIRP value to meet the requirements. By simulating the amplitude weighting of the full array with the density weighting of the array elements, low side lobes can be achieved through fewer array elements.

In order to reduce the side lobe level during receiving, the array needs to be synthetically optimized to configure the amplitude of each receiving channel. By combining density weighting with amplitude weighting, a density weighting array is calculated with a triangular grid array as a reference, and the excitation amplitude of the array element is also weighted. After calculation, the side lobe level of this array is lower than before.

Fourthly, in order to meet the requirements of precise measurement of phase center and time delay, solutions such as ground and on-orbit amplitude-phase calibration, and channel consistency calibration under temperature environment can be adapted to the delay and phase center stability problems caused by on-orbit temperature environment changes.

Fifthly, for the problem of phased array antenna with high power consumption and difficult thermal control, the transmission power is reduced by reasonable designs of links and array size in close combination with the overall design of ISLs. Thermal control measures are taken to dissipate heat and control the temperature of phased array antenna.

5.3.2 *Overall Scheme of Phased Array Antenna*

According to the analysis of the preceding section, the Ka-phased array antenna should implement the following functions:

1. beam pointing and scanning function to achieve the coverage of antenna beams over the airspace of ±60°;
2. continuously tracking and pointing to a satellite according to the input beam pointing angle to establish ISLs;
3. rapidly switching beam direction according to the input beam pointing angle, that is, switching beam pointing from one satellite to another satellite to establish ISLs;
4. fast switching between transmission and receiving mode;
5. fast on-orbit delay monitoring and calibration;
6. fast on-orbit beam pointing monitoring and calibration, and;
7. suitable for inter-satellite ranging and communication.

In order to establish dynamic links with visible satellites as many as possible, ISLs phased array antennas should be installed on the toward-Earth side of satellites. In case that the antenna is on orbit, the central beam points to the geocentric position, which is defined as the zero starting point. By controlling the excitation phases of the antenna radiation elements, the antenna beam can be directed in any direction within the scanning range of $\theta \in [0°, 60°]$ and $\varphi \in [0°, 360°]$.

The layout design of phased array antenna determines the radiation characteristics, envelope size and mass of the antenna, which is one of the important steps of lightweight design.

Generally, antenna arrays can be divided into many types, such as rectangular arrays, ring arrays, and triangular grid arrays. The layout of planar array antennas is shown in Fig. 5.14. Two-dimensional array antenna is used if the antenna beam is to be scanned simultaneously in both azimuth φ and elevation θ directions. From the point of view of phased array antenna used on-orbit, triangular grid array is preferred for large scanning range.

Ka-phased array antenna includes radiation array, T/R assembly, feed network, control section, structure section, self-calibration section, power supply section, thermal control section and supporting ground equipment, etc. Each part consists of the following components:

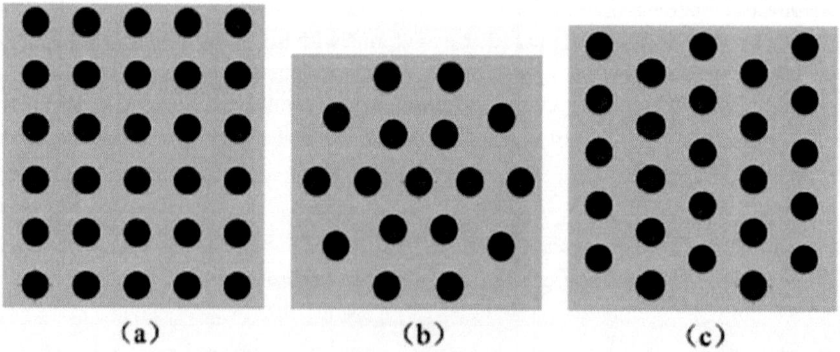

Fig. 5.14 Layouts of planar array antenna. **a** Rectangular array; **b** ring array; **c** triangular grid array

1. The radiation array consists of radiation element, feed cable, dielectric substrate and dummy element.
2. The T/R assembly consists of feed module (including phase shifter, attenuator, low-noise amplifier, power amplifier, control chip, encapsulation housing), power divider/combining network in the assembly, control and power supply circuit in the assembly, assembly housing, etc.
3. The feed network consists of power dividing/combining network, driving amplifier, microwave switch, etc.
4. The control unit consists of beam controller, sub-controller and control cable.
5. The structure section consists of antenna housing, assembly bracket and antenna mounting bracket.
6. The self-calibration section consists of Tx antenna, Rx antenna, calibration module and signal source.
7. The power supply section consists of DC/DC power supply module, power supply cable.
8. The thermal control section consists of heat pipe, condenser, thermistor and multi-layer thermal control assembly.
9. Ground equipment consists of ground controller, nitrogen-filled packing.

The block diagram of Ka-phased array antenna is shown in Fig. 5.15, in which the external RF interface, the external power supply interface, the external control

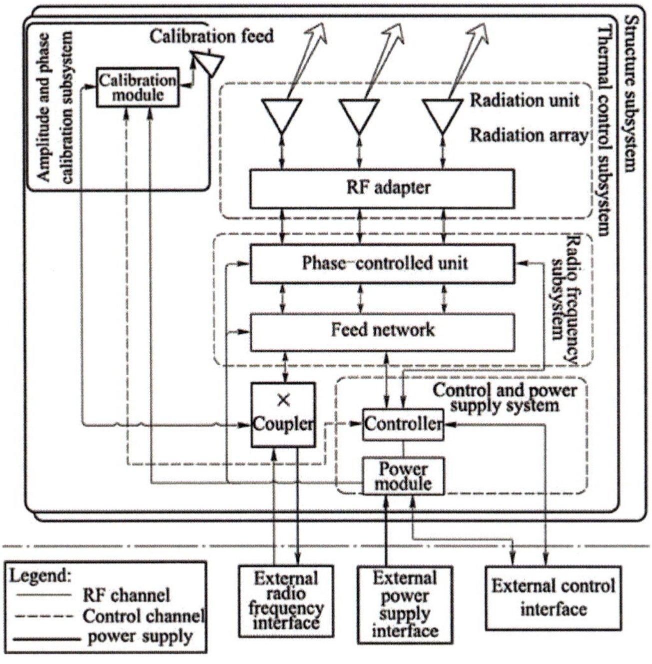

Fig. 5.15 Block diagram of the ISL phased array antenna

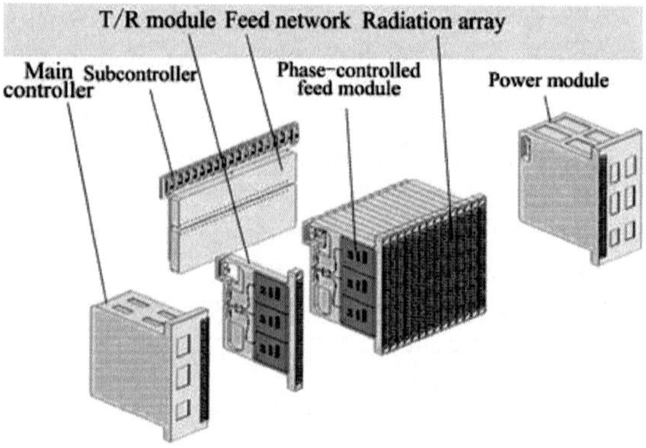

Fig. 5.16 Schematic diagram of ISL-phased array antenna composition modules

interfaces of the antenna, the internal RF channel, the power supply channel and the direction and relationship of the information flow of the control channel of the antenna are depicted.

The structure of Ka-phased array antenna is shown in Fig. 5.16. Each phased T/R assembly has the function of transmitting and receiving, and the radiation array is shared by transmitting and receiving. The antenna consists of 13 T/R assemblies. Three T/R feed modules are installed on both sides of each assembly, and each feed module has eight channels, thus there is a total of 624 channels, corresponding to 624 radiation array elements. Two rings of dumb elements are arranged around the radiation array to ensure the consistency of radiation pattern of array elements.

The internal power dividing/combining network of the T/R assembly is of a 1:3 scheme, and is connected to the T/R feed modules by a channeline (micro-coaxial cable) connection method. It is allocated by a 1:8 schemes in the feed module.

The power dividing and power combining network out of T/R assembly are designed separately. The power dividing network (including driving amplifier) adopts a 1:4 scheme, and connects to power dividing network with a 1:8 by channeline connection mode after driving amplifier; the power combining network adopts a 3:1 scheme, and connects to the network with 8:1 scheme by channeline connection mode.

By phased array antenna transmitting, the transmitted signal is divided into N RF channels through the external power dividing network and the internal power dividing network. Amplitude and phase of the transmitted signal are modulated by the amplitude and phase modulation chip, power of the signal is amplified by the multi-functional power amplifier chip, and the signal is finally transmitted by the radiation array. When transmitting, the receiving channel is turned off.

By phased array antenna receiving, M radiation elements corresponding to working channels receive the signal, which is firstly amplified by a low-noise amplifier chip, then modulated by an amplitude-phase modulation chip, and finally fed into the receiver by combining the power combining network inside the assembly and the power combining network outside the assembly.

5.3.3 Array Design of Phased Array Antenna

Array elements are usually arranged in rectangular or triangular grids. Rectangular grid has the advantages of regular arrangement, easy layout and producing of feed network. The triangular grid has the advantage of less mutual coupling effect among the elements, and fewer elements when the grating lobe suppression and the array gain are similar to that of rectangular grid.

To minimize the number of elements in the array, the ISL-phased array antennas are arranged in a triangular grid (Fig. 5.17). The total number of array elements of 624. The element spacing of 7.1 mm ensures that the grating lobes will not appear in real space when the beam is scanned to ±60° of any cut. Two circles of dumb elements are designed around the radiation array to ensure the consistency of radiation pattern of the array element.

For the Tx/Rx shared time-division phased array antenna, the transmitting and receiving arrays can be optimized separately.

The side lobes of regular triangular arrays with equal amplitude and phase excitation are usually −13 dB, which cannot meet the overall technical requirements. So the antenna array is optimized by array synthesis method. The optimization objective is to ensure that the EIRP value of the transmitting array antenna and the gain value

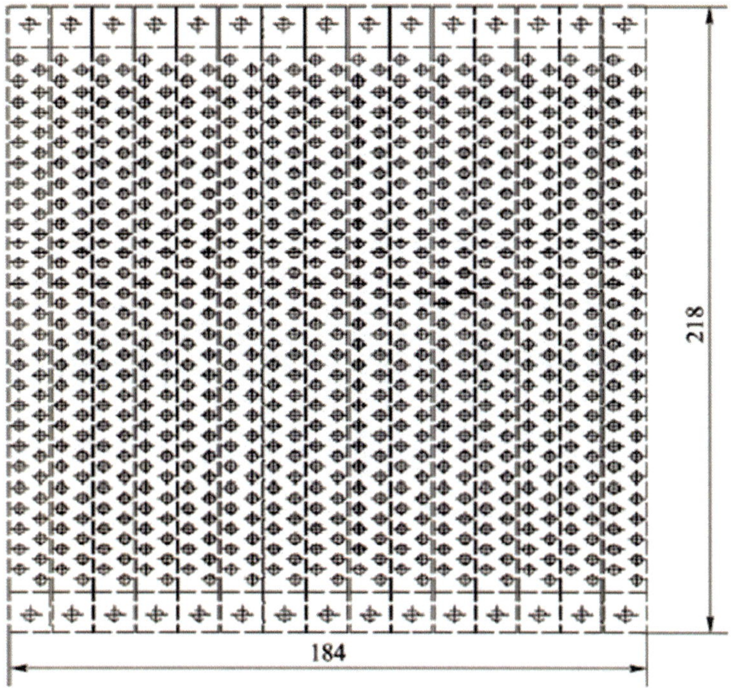

Fig. 5.17 Diagram of phased array antenna array

of the receiving array antenna satisfy the requirements of inter-satellite link, and the side lobe level of any cut is not greater than the required level. The fitness function is constructed base on this purpose, and the nonlinear least squares method and genetic algorithms are used to optimize the calculation. Through analysis, the side lobe level can be reduced by using a thinned array or thinned array plus amplitude weighting, and the side lobe level in an arbitrary angular domain is less than -15 dB.

5.3.3.1 Transmitting Array Synthesis Optimization

Using the density weighting of the array elements to simulate the amplitude weighting of the full array, the low side lobes can be realized with fewer array elements. For the uniform array shown in Fig. 5.17, a density-weighted array (thinned array) can be obtained by optimizing the decision to equidistant grid points with or without active array element.

Taking the full array in Fig. 5.17 as a sample, a density-weighted array with 420 active elements is obtained after optimization calculation, and its layout is shown in Fig. 5.18. The array layout can reduce side lobes, and the number of RF channels working at the same time is small. The pattern design results are shown in Fig. 5.19 and the related performance parameters are listed in Table 5.1. Beam scanning is controlled by a 5-bit phase shifter.

5.3.3.2 Synthetical Optimization of Receiving Array

In practical engineering design, the method of amplitude weighting has been proved to be an effective method for the design of low side lobe arrays. The combination of density weighting and amplitude weighting is used for analysis. Still taking the full array of Fig. 5.17 as a sample, the optimized calculation results in a density-weighted array with 544 active array elements, and the excitation amplitude of the array elements is also weighted (implemented by a 3-bit attenuator), its layout is shown in Fig. 5.20. The side lobe level of the array pattern is not more than −20 dB (Fig. 5.21), and the calculation results of the performance parameters are shown in Table 5.2. The array layout can reduce side lobes and gain loss.

5.3.4 Design of TR Assembly

Ka-band phased array TR assembly has a Micro-D connector for power supply and data line interface; each assembly has three feed modules on both sides, and each side has a blind plug-in RF coaxial connector (Sub-Miniature version P, SMP), which is used to connect with the system power dividing/combination network. The three-in-one power dividing/combination network and three-channel lines are used to allocate and transmit microwave signals. In order to effectively expand the bandwidth of the

0	0	0	0	0	1	0	1	1	1	0	0	0
0	0	0	0	0	1	0	1	1	1	1	1	1
0	0	0	0	0	1	0	0	1	1	1	1	1
0	0	0	0	0	1	0	0	1	1	0	0	0
0	0	0	0	0	1	1	1	1	1	1	1	1
0	1	0	0	0	1	1	1	1	1	1	1	1
0	1	1	1	1	0	1	0	1	1	1	1	1
0	1	1	1	1	0	1	0	1	1	1	1	1
1	0	1	1	1	1	1	1	1	1	1	1	1
0	0	1	1	1	1	1	1	1	1	1	1	1
0	1	1	1	1	1	1	1	1	1	1	1	1
1	0	1	1	1	1	1	1	1	1	1	1	1
1	0	1	1	1	1	1	1	1	1	1	1	1
1	1	1	1	1	1	1	1	1	1	1	1	1
1	0	1	1	1	1	1	1	1	1	1	1	1
1	0	1	1	1	1	1	1	1	1	1	1	1
0	1	1	1	1	0	1	0	1	1	1	1	1
0	1	1	1	1	0	1	0	1	1	1	1	1
0	1	0	0	0	1	1	1	1	1	1	1	1
0	0	0	0	0	1	1	1	1	1	1	1	1
0	0	0	0	0	1	0	0	1	1	0	0	0
0	0	0	0	0	1	0	0	1	1	1	1	1
0	0	0	0	0	1	0	1	1	1	1	1	1
0	0	0	0	0	1	0	1	1	1	0	0	0

Fig. 5.18 Array layout by transmitting (left half array, 1—equal amplitude excitation, 0—dummy element)

left-handed circularly polarized antenna, the dual-probe feeding method is adopted. The front side of the TR assembly is shown in Fig. 5.22, while the reverse side is completely symmetrically.

The block diagram of Ka-band phased array TR assembly is shown in Fig. 5.23. The TR module is the core of the whole TR assembly, and its principle block diagram is shown in Fig. 5.24. It is necessary to develop high-performance GaAs pHEMT multifunctional low-noise amplifier chip, high-performance GaAs pHEMT power amplifier chip, GaAs pHEMT amplitude-phase modulation chip with driving amplification function, CMOS SOS multifunctional digital chip integrating with temperature measurement, gain and phase temperature compensation and control, logic control, etc. Ka-band phased array TR module is developed through multifunctional low-noise amplifier chip, amplitude-phase modulation chip and multifunctional digital chip encapsulated by multi-layer Low Temperature Co-fired Ceramic (LTCC) metal. Each RF channel only needs 4 chips to complete the functions that used to require nearly 20 chips. The volume of each channel decreases greatly without the weakness

Fig. 5.19 Section view of array scanning characteristic by transmitting. **a** Array scanning characteristics for 0° section; **b** array scanning characteristics for 45° section, **c** array scanning characteristics for 90° section

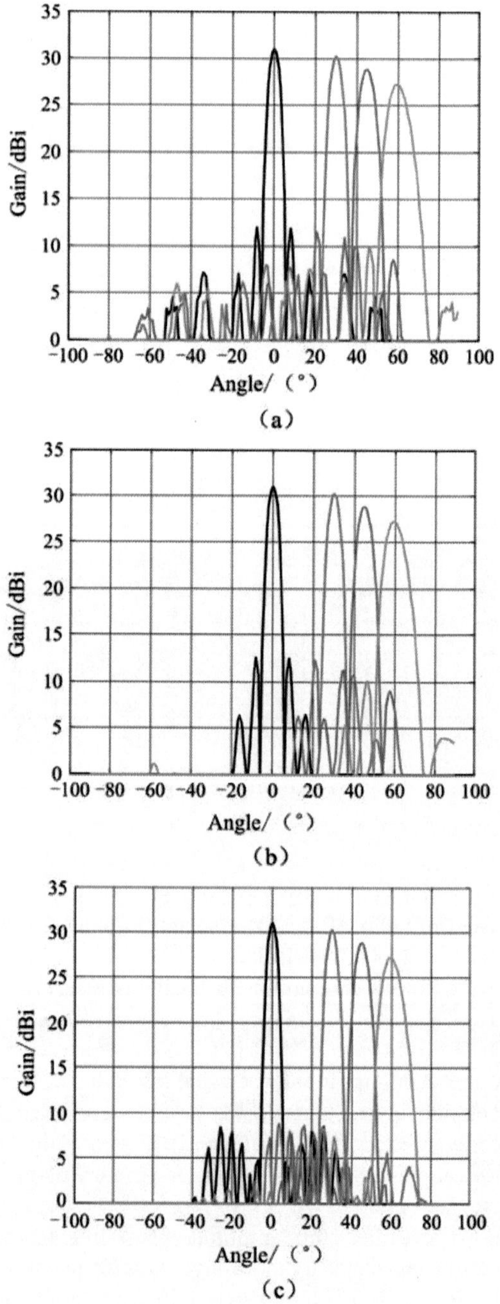

Table 5.1 Calculation results of performance parameters of transmitting array

Scanning angle /(°)	0		30		45		60	
	Horizontal	Vertical	Horizontal	Vertical	Horizontal	Vertical	Horizontal	Vertical
Peak gain G/dBi	30.5	30.5	29.6	29.6	28.3	28.3	26.6	26.6
±2° gain G/dBi	28.5	28.3	28.1	27.9	27.2	27.1	26.0	26
Max. side lobe level SLL_{max}/dB	−16.8	−20.5	−16.7	−20.2	−16.3	−19.2	−16.5	−17.4

−50	−5.25	−50	−5.25	−5.25	−50	−5.25	0	−5.25	−4.5	−5.25	−5.25	−5.25
−50	−5.25	−50	−5.25	−50	−4.5	−5.25	−2.25	−5.25	−3.75	−5.25	−2.25	−4.5
−50	−50	−50	−50	−50	−3.75	−5.25	−5.25	−5.25	−5.25	−1.5	−2.25	−1.5
−50	−50	−5.25	−5.25	−5.25	−5.25	−5.25	−5.25	−1.5	−4.5	0	−2.25	−1.5
−5.25	−5.25	−50	−5.25	−50	−4.5	−5.25	−3	−3	−3.75	−2.25	−3.75	−3.75
−5.25	−4.5	−5.25	−50	−5.25	−4.5	−5.25	−1.5	−5.25	−1.5	−1.5	−4.5	−1.5
−5.25	−5.25	−5.25	−3.75	−5.25	−5.25	−3	−5.25	−1.5	−2.25	0	−2.25	−0.75
0	−3	−3	−5.25	−2.25	−5.25	−3	−5.25	−3	−1.5	−0.75	0	−2.75
−4.5	0	−5.25	−3	−3.75	−5.25	−1.5	0	−2.25	0	−2.25	−0.75	−2.75
−5.25	−5.25	−1.5	−3	−1.5	−2.25	−0.75	−3	−2.25	−3.75	−0.75	−3	−0.75
−5.25	−3	−3.75	−3	−0.75	−2.25	−3	−2.25	−1.5	0	−3	0	−2.75
−3.75	−5.25	−3.75	−3.75	−2.25	−3.75	−0.75	−0.75	0	−3	−0.75	−0.75	−0.75
−3.75	−5.25	−3.75	−3.75	−2.25	−3.75	−0.75	−0.75	0	−3	−0.75	−0.75	−0.75
−5.25	−3	−3.75	−3	−0.75	−2.25	−3	−2.25	−1.5	0	−3	0	−2.75
−5.25	−5.25	−1.5	−3	−1.5	−2.25	−0.75	−3	−2.25	−3.75	−0.75	−3	−0.75
−4.5	0	−5.25	−3	−3.75	−5.25	−1.5	0	−2.25	0	−2.25	−0.75	−2.75
0	−3	−3	−5.25	−2.25	−5.25	−3	−5.25	−3	−1.5	−0.75	0	−2.25
−5.25	−5.25	−5.25	−3.75	−5.25	−5.25	−3	−5.25	−1.5	−2.25	0	−2.25	−0.75
−5.25	−4.5	−5.25	−50	−5.25	−4.5	−5.25	−1.5	−5.25	−1.5	−1.5	−4.5	−1.5
−5.25	−5.25	−50	−5.25	−50	−4.5	−5.25	−3	−3	−3.75	−2.25	−3.75	−3.75
−50	−50	−5.25	−5.25	−5.25	−5.25	−5.25	−5.25	−1.5	−4.5	0	−2.25	−1.5
−50	−50	−50	−50	−50	−3.75	−5.25	−5.25	−5.25	−5.25	−1.5	−2.25	−1.5
−50	−5.25	−50	−5.25	−50	−4.5	−5.25	−2.25	−5.25	−3.75	−5.25	−2.25	−4.5
−50	−5.25	−50	−5.25	−5.25	−50	−5.25	0	−5.25	−4.5	−5.25	−5.25	−5.25

Fig. 5.20 Array layout for receiving (left half array, element amplitude weighted, 64 elements cut out from four corners)

of RF performance. All the chips and elements used in the TR assembly have been homegrown, so the demand for mass production can be fully guaranteed.

5.3.5 Amplitude and Phase Calibration Design

When phased array antenna is on orbit, the relative amplitude and phase between individual RF channels will change due to temperature variation and device aging.

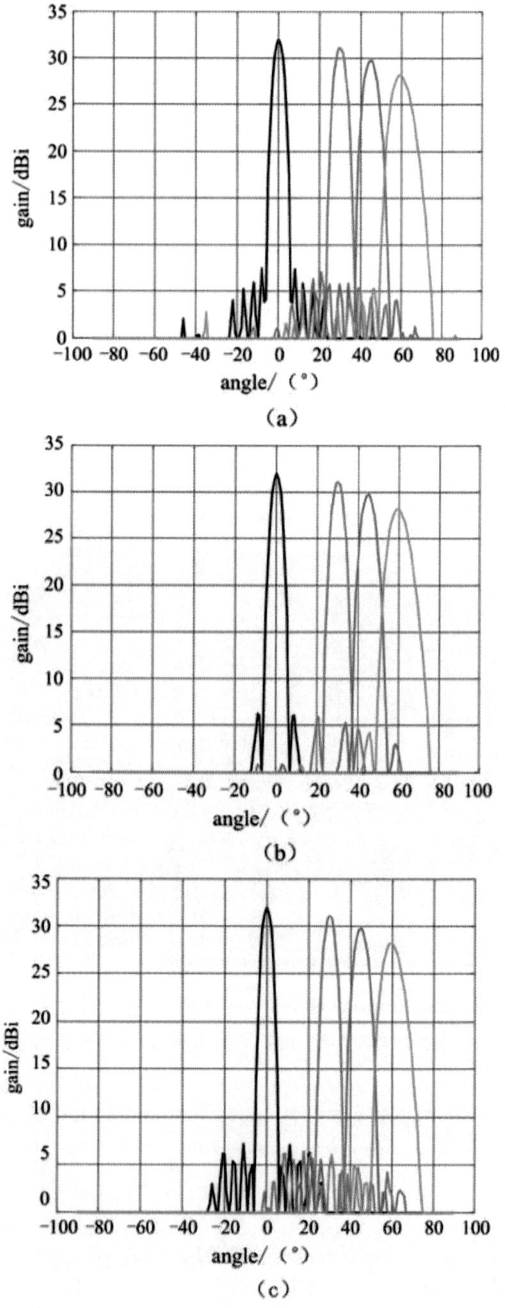

Fig. 5.21 Section view of array scanning characteristic by receiving. **a** 0° surface array scanning characteristics; **b** 45° surface array scanning characteristics, **c** 90° surface array scanning characteristics

Table 5.2 Test and calculation results of receiving array performance

Scanning angle /(°)	0		30		45		60	
	Horizontal	Vertical	Horizontal	Vertical	Horizontal	Vertical	Horizontal	Vertical
Peak gain G/dBi	31.5	31.5	30.6	30.6	29.3	29.3	27.5	27.5
±2° gain G/dBi	29.5	29.3	29.1	29.0	28.1	28.1	27.0	27
maximum side lobe level SLL_{max}/dB	−22.5	−22.6	−22.5	−23	−22.8	−22.4	−21.5	−20.1

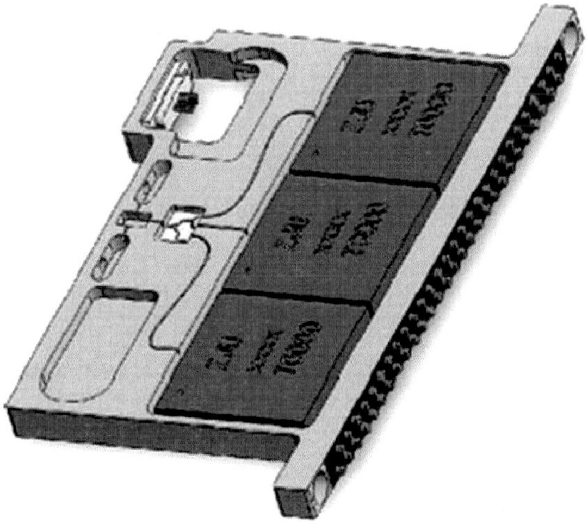

Fig. 5.22 Appearance of transceiver module

Especially when the temperature between the RF channels varies inconsistently, the relative amplitude and phase of each channel will change greatly. Therefore, in addition to the ground calibration before the phased array antenna is put into use, on-orbit calibration must be performed.

5.3.5.1 Calibration Method of Satellite-Borne Phased Array

Because the satellite has strict restrictions on payload weight and size, phased array calibration equipment should be as few as possible. Generally, ground calibration methods cannot meet the application of satellite-borne phased array. At present, there are three main calibration methods for the satellite-borne phased array:

1. Call amplitude and phase calibration table

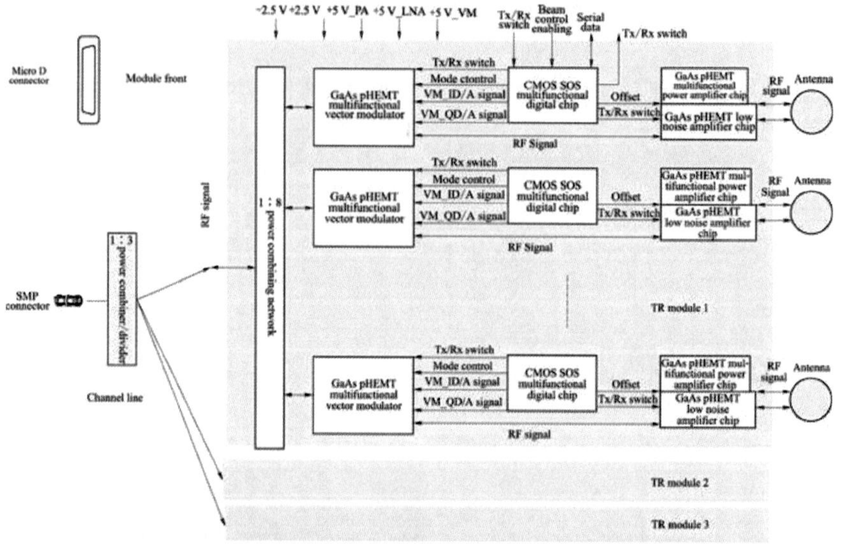

Fig. 5.23 Block diagram of transceiver components

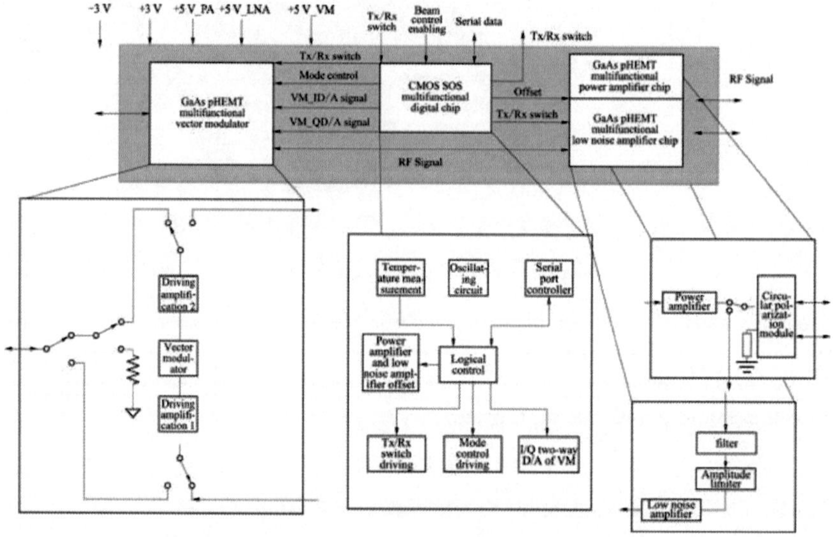

Fig. 5.24 Block diagram of transceiver module principle

The data of amplitude and phase of each phased array channel varying with temperature are measured under various ground temperature conditions, and then these data are stored in the phased array controller. When the phased array is on orbit, the working temperature of the phased array is measured, and then the stored amplitude and phase data are called from the memory according to the measured data of temperature. The amplitude and phase errors are compensated by phase shifter and attenuator of each channel. But this method needs a lot of measurements, and can only correct the amplitude and phase errors caused by temperature. This method has certain limitations since it cannot correct the amplitude and phase errors caused by device aging.

2. Unitary Transform Encoding (UTE) and Control Circuit Encoding (CCE)
   Unitary Transform Encoding (UTE) and Control Circuit Encoding (CCE) are two algorithms for long-distance calibration of transmitting/receiving active phased array systems using time-division multiplexing orthogonal coded signals. UTE method with high accuracy can effectively correct phased array systems based on Digital Beam Forming (DBF), because the amplitude and phase of each element required for DBF are digitally controlled in the baseband. Baseband digital signals are firstly converted into analog signals and then modulated to the desired RF carrier frequency. UTE method can also be applied to calibrate analog beamforming systems, but in analog applications, an additional hardware should be added after beamforming phase shifter, so as to provide accurate signal encoding. If the additional hardware fails, the calibration system will not run smoothly.
   In CCE method, the phase array system of analog beamforming is coded orthogonally by controlling the conduction of each delay circuit of phase shifter. CCE method can be accurately coded without additional hardware circuits, so it is one of the choices of long-range calibration analog beamforming phased array.
   Both UTE and CCE methods can correct the amplitude and phase errors caused by the temperature, inconsistency of device characteristics and aging. The advantages of these two methods are that the number of measurements is less in the process of calibration. The disadvantages of these two methods are that the amplitude and phase must be measured, and the measurement of phase is usually more complex, which will lead to the complexity of the equipment. In order to obtain the complex gain in the calibration data, coherent detection must be performed at the receiver using one or more phase invariant signals transmitted from the transmitter. For the calibration of satellite-borne phased array antenna, the coherent detection system must be specially designed to offset the Doppler shift caused by satellite movement and the phase noise caused by the clock asynchronization of satellite and ground receiving station.
3. Rotating Element Electric-Field Vector Method (REV)

Rotating Element Electric-Field Vector Method (REV) only needs to measure the relative amplitude change of the signal, but does not need phase measurement, and uses less calibration equipment. Its disadvantage is that it needs more measurement times to achieve calibration. On-orbit REV calibration can be divided into satellite-ground path calibration and on-satellite path calibration. In the satellite-ground path calibration system, the calibration antenna and signal source are placed on the ground, and used far-field REV calibration method can simplify the on-satellite equipment. The disadvantages are that the ground system is involved and the coordination is complex. For short-range calibration of on-satellite path, a Tx/Rx calibration antenna with certain height and corresponding signal processing device should be equipped on the satellite. Its advantages are that there is no interaction with the ground system and short-range can better guarantee signal quality.

The REV method controls the phase shifter at the front end of the phased array element by software. As the phase-shift value of a single element changes, the electric field intensity change of the synthetic electric field vector is measured. The test results are processed mathematically to obtain the amplitude and phase values of the element. Because only the magnitude of the synthetic electric field vector is required to be measured, the measurement accuracy and implementability are better than the phase measurement. Therefore, REV method is very suitable for calibration of satellite-borne phased array.

5.3.5.2 REV On-Orbit Calibration Process

As shown in Fig. 5.25, the synthetic electric field vector of an array is the superposition of the electric field vectors of each element. The element electric field vector is proportional to the product of the excitation amplitude and phase of the radiated electric field in this direction. When the phase shift of the phase shifter of an element changes, the electric vector of the element rotates with the radius of a circle and the position of the endpoint of the synthetic electric field vector falls on this circle. The

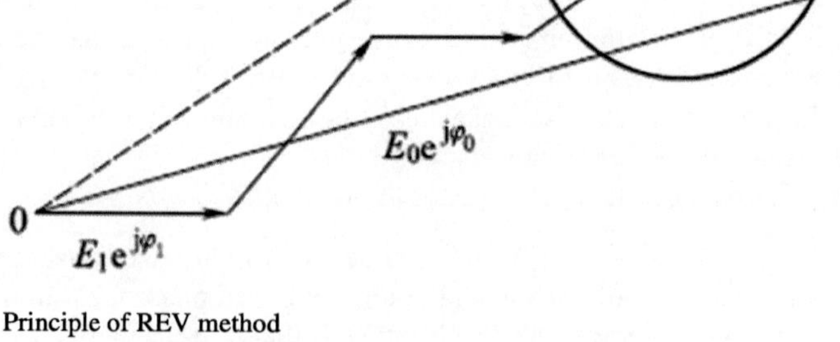

Fig. 5.25 Principle of REV method

principle of REV method is to measure the amplitude fluctuation of the synthetic electric field vector and then calculate the amplitude and phase of the rotating element by mathematical calculation. The reasonable random phase distribution of each element in the array is to ensure that the effect of a single element on the variation of the synthetic electric field is large enough.

The REV on-orbit calibration system consists of a calibration signal source, a calibration antenna, a calibration module, a controller, a control computer and a phased array. The calibration process is that the calibration signal source generates continuous wave signal; the phase shift of the corrected element phase shifter is changed by the controller; the field strength data are measured by the calibration antenna and the calibration module, and then the amplitude and phase errors of each channel are calculated by the calibration program of the controller based on the field strength data, then the amplitude and phase errors are compensated by the corresponding state of attenuator and phase shifter changed by the controller.

Figure 5.26 shows the calibration process of on-satellite path REV method. First, the on-satellite path calibration is carried out by the initial far-field REV method and the satellite-borne calibration antenna with initial REV method on the ground, resulting in two sets of amplitude and phase error data (A_0, φ_0) and (A_1, φ_1). These two sets of data are stored in the control computer as standard data. By on-orbit calibration, the relative amplitude and phase errors (A_2, φ_2) are obtained by REV method with satellite-borne calibration antenna, and then the amplitude and phase errors are compensated by $[-A_0 + (A_2 - A_1)]$ (dB) and $[-\varphi_0 + (\varphi_2 - \varphi_1)]$ according to standard data (A_0, φ_0) and (A_1, φ_1).

Figure 5.27 depicts the schematic diagram of the orbital amplitude and phase calibration system. The calibration antenna needs a special mounting bracket, usually at the edge of the phased array, to avoid affecting the normal communication. The key hardware of the calibration system is signal source and calibration module. The signal source provides a continuous wave signal with adjustable frequency to correct the phased array at different frequencies. The calibration module is used to receive the signal strength received by the phased array in the calibration process, and the

Fig. 5.26 REV amplitude and phase calibration process

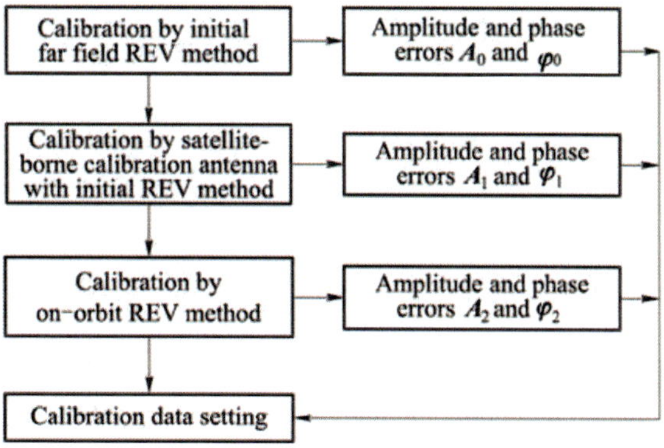

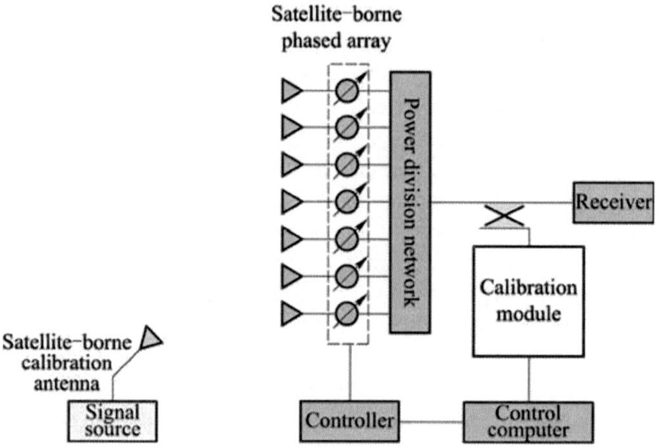

Fig. 5.27 Schematic diagram of on-orbit amplitude and phase calibration

signal is sent to the control computer after amplification, detection and analog-to-digital conversion. The calibration is performed by the controller according to the calculated calibration data.

5.3.5.3 Large-Scale Array Calibration Method

Based on the international REV method, the grouped REV method is proposed [1], which can be used in the calibration of satellite-borne phased array. The grouped REV method can improve the signal-to-noise ratio of measurement and increase the dynamic range of signal for signal detection. REV method requires less hardware and no forerunner sources and can correct the transmitting antenna pattern.

As shown in Fig. 5.28, there are two complex signals $C_1\exp(\mathrm{j}\varphi_1)$ and $C_2\exp(\mathrm{j}\varphi_2)$, where C_1, C_2, φ_1 and φ_2 is the amplitude and phase of signal 1 and signal 2, respectively. Let C be the amplitude of the synthesized signal, according to the

Fig. 5.28 Principle of REV method

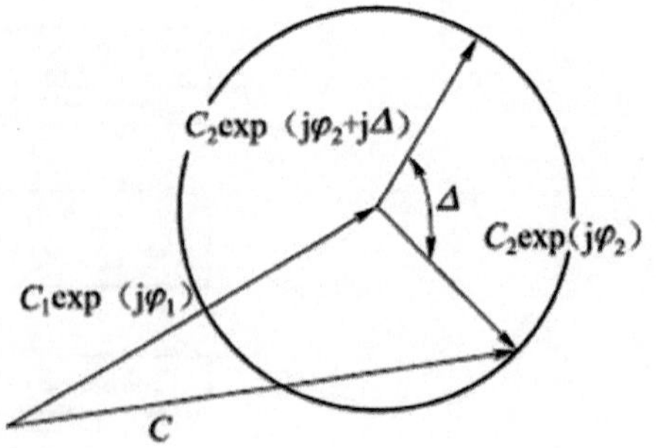

parallelogram rule, the following formula is obtained:

$$C^2 = C_1^2 + C_2^2 + 2C_1C_2\cos(\varphi_2 - \varphi_1 + \Delta). \tag{5.1}$$

In case that the phase shifter is n-bit discrete phase shifter, the phase state that the phase shifter can provide is $M = 2^n$. So Formula (5.1) can be modified as.

$$C^2(i) = C_1^2 + C_2^2 + 2C_1C_2\cos(\varphi_2 - \varphi_1 + i\delta). \tag{5.2}$$

This expression is in the form of cosine function, and in the process of i changing from 0 to $(M - 1)$, the cosine function completes a periodic change. Fourier analysis can be used to derive $\left(C_1^2 + C_2^2\right)$, C_1C_2 and $(\varphi_2 - \varphi_1)$ from the sequence $\{C^2(i)\}$.

When the grouped REV method is used, the phases of multiple array elements are simultaneously rotated. Suppose the phased array has N antenna elements and taking the mth observation point as an example when the observation point is located in the far-field of the phased array (Fig. 5.29), the array pattern is simplified as.

$$E = P(\hat{r}_{m,1})\sum_{i=1}^{N} E_i e^{j\varphi_i} e^{-jk\hat{r}_{m,1}\cdot\mathbf{d}_i}, \tag{5.3}$$

where $\mathbf{d}_i = x_i\hat{x} + y_i\hat{y} + z_i\hat{z}$ is the position vector of the i-th array element relative to the first array element, and is obviously $\mathbf{d}_1 = 0$; $\hat{r}_{m,1}$ is the unit vector of the position vector $\mathbf{r}_{m,1}$ of the first array element to the m-th observation point. Since it is concerned with the relative amplitude and phase correlation rather than the absolute value of that of the array element, the term irrelevant to i can be omitted from the above formula, thereby obtaining.

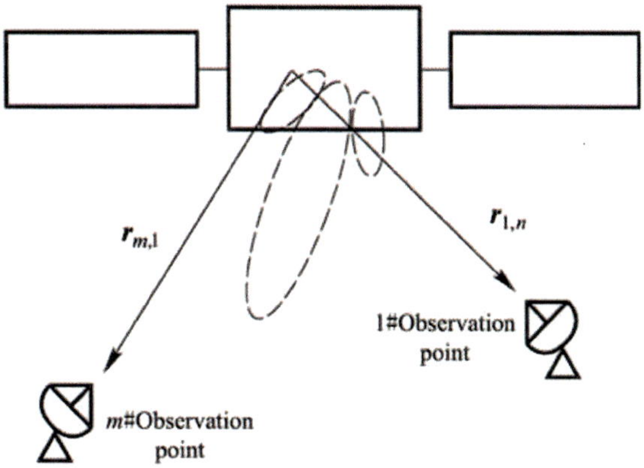

Fig. 5.29 Phase and position of multipoint observation calibration array element

$$E = \sum_{i=1}^{N} E_i e^{j\varphi_i} e^{-jk\hat{r}_{m,1}\cdot \mathbf{d}_i}. \tag{5.4}$$

This expression is a very typical expression of the synthetic pattern of phased array.

According to associative law of addition, Formula (5.4) can be written in an addition form of two sets of complex signals:

$$E = \sum_{i=1}^{I} E_i e^{j\varphi_i} e^{-jk\hat{r}_{m,1}\cdot \mathbf{d}_i} + \sum_{j=1}^{J} E_j e^{j\varphi_j} e^{-jk\hat{r}_{m,1}\cdot \mathbf{d}_j}, \tag{5.5}$$

where $I + J = N$.

The two sets of complex signals can be determined by the REV method, denoted as Z_1, Z_2 (Fig. 5.30), and the equations for the complex signals of the array elements are obtained.

$$\begin{cases} \sum_{i=1}^{I} E_i e^{j\varphi_i} e^{-jk\hat{r}_{m,1}\cdot \mathbf{d}_i} = Z_1 \\ \sum_{j=1}^{J} E_j e^{j\varphi_j} e^{-jk\hat{r}_{m,1}\cdot \mathbf{d}_j} = Z_2 \end{cases}. \tag{5.6}$$

Here, the phase shifter of each antenna element constituting Z_2 component should be synchronously adjusted with the rotation of Z_2 component, that is, when Z_2 component rotates, the phase shifter of each antenna element in the group increases the same phase. The complex signals Z_1 and Z_2 can be solved by REV method, and two equations are obtained. By changing the scheme of grouping, enough equations can be obtained to solve the complex signals of each array element. The calibration

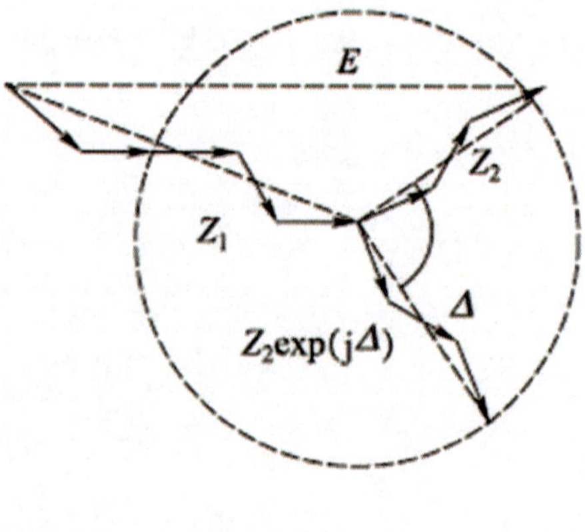

Fig. 5.30 Principle of transmitted signal synthesis of phased array antenna

accuracy can be further improved by the comprehensive processing of calibration data from multiple observation points.

When the amplitude and phase of the transmitting array are corrected, the calibration signal source provides the continuous wave signal, which is transmitted through the phased array by the coupler. Moreover, phased array controller controls the phase state of the phase shifter. After the calibration module completes data acquisition, the measured amplitude change data is sent to the controller. The phase state of the phase shifter is adjusted accordingly after the calculation of the amplitude and phase calibration data by the controller.

Reference

1. Minggang L, Zhenghe F (2007) Combined rotating-element electric-field vector method for calibration of large-scale phased array antenna. Chin J Radio Sci 3:380–384

Chapter 6
Communication Satellite Payload Antenna

Abstract Based on the analysis of the characteristics of the payload antenna of communication satellite, this chapter goes deep into the functions design of some typical antennas, such as shaped reflector antenna, multi-beam antenna and adaptive nulling antenna. Finally, this chapter briefly introduces the technology of deployable large-scale mesh reflector and its development trend.

Communication satellite, a kind of man-made Earth satellite on orbit, receives, amplifies or processes communication signals from ground transmitting stations, and then sends them back to one or more ground receiving stations. The first commercial satellite communications were deployed in the mid-1960s. After more than 50 years of development, it has developed from an alternative technology to a core transmission technology and penetrated into all the infrastructure of global communications. At present, communication satellites can be widely used in data, voice and video transmission applications, and provide services to fixed, broadcasting, mobile, personal communications and dedicated network users.

Communication satellite has become one of the most important application satellites. Its payload (including communication antenna and transponder) undertakes the task of receiving, processing and transmitting. In commercial and military communications, satellite services mainly fall into the following types [1]:

1. **Fixed Satellite Service (FSS)** covers parts of the country or intercontinental areas in C, Ku and Ka bands using shaped regional beams.
2. **Broadcast Satellite Service (BSS)** provides downlink beam coverage for territorial regions (such as mainland China, mainland America) and has local enhancement capability to compensate for rain attenuation.
3. **Personal Communication Service (PCS)** achieves K/Ka-band multi-beam coverage by using multi-aperture reflector antenna. Each reflector uses multiple feeds to form corresponding beams. Personal communication and user-to-user data transmission are realized by satellite. Generally, the forward link is established through the ground station-satellite-user link, while the backward link is the user-satellite-ground station link.
4. **Mobile Satellite Service (MSS)** provides communication services to mobile users via satellites. General mobile communication satellites work in lower

R. You et al., *Technologies for Spacecraft Antenna Engineering Design*, Space Science and Technologies, https://doi.org/10.1007/978-981-15-5833-7_6

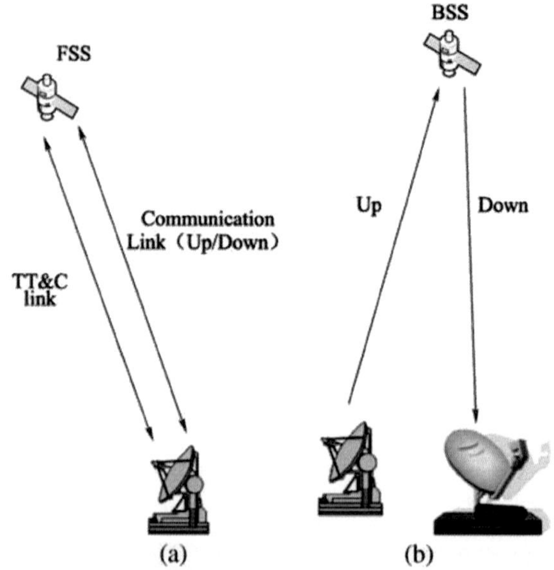

Fig. 6.1 Schematic diagram of typical satellite-ground links. **a** FSS; **b** BSS

frequency bands such as UHF, L, and S. Large deployable mesh antenna is required to achieve higher antenna gain, and large-scale feed array is used to realize overlapping beams on the ground.

5. **Inter Satellite Service (ISS)** provides communication links and data transmission between satellites, and uses satellite constellation to provide global services by using mechanical steerable dual-reflector antenna with automatic tracking capability. However, some systems also are still using phased array antennas.

Figure 6.1 depicts satellite link of FSS and BSS. In FSS, it is required to provide shaped beams for uplink and downlink in domestic area, while in BSS, shaped beams are only provided for downlink to meet the requirement of receiving broadcasting signals from domestic terminals. The uplink only provides the spot beam covering the ground station to reduce the requirement of transmitting EIRP of the ground station and improve the anti-interference capability of the satellite uplink.

Communication satellite payload usually requires multiple reflector antennas and transponders to support composite payload function and to provide multiple services for business area. This requires that the satellite platform be able to provide the layout space for multiple large aperture reflectors, which can be folded to meet the stringent envelope requirements of the rocket fairing when launching, and deployed on orbit. Common deployable reflectors are mounted on the east and west panels of satellite platforms. In a few cases, they are mounted on the satellite-to-ground panel. Some small reflectors without deployment mechanism, TT&C antennas, and beacon antennas will also be installed on the satellite-to-ground panel. Because the feed assembly handles high power and need to have the special Passive Inter-Modulation (PIM) performance, the relative installation position of the feed and the

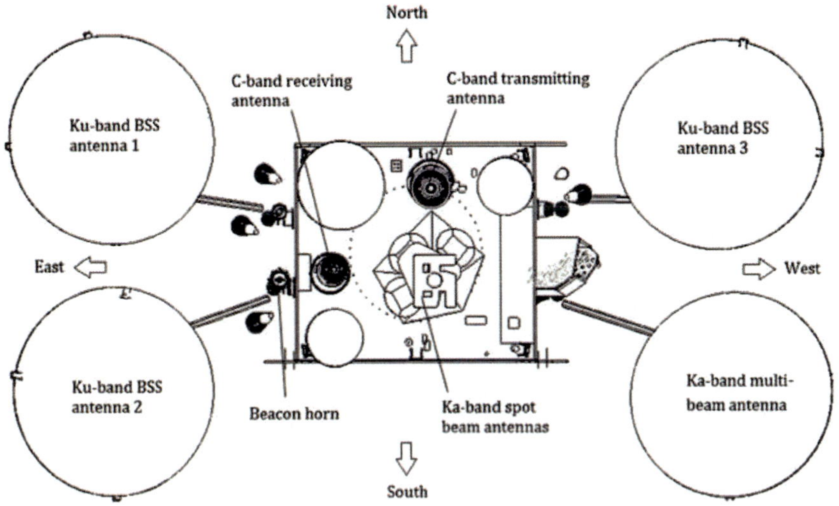

Fig. 6.2 Layout of typical deployable antenna

reflector should be considered comprehensively. Therefore, the feeds of the antennas on the satellite east and west panel are usually installed in the east point and west point of the platform and on the to-ground panel. Figure 6.2 shows a layout of typical unfolded antenna on satellite, in which the solar panels are deployed along north–south direction and tracks toward the sun.

The reflector antenna commonly used in communication satellite will adopt different reflector types according to the requirements of satellite payload and platform layout. The common reflector antennas include single offset reflector, Gregory double reflector (including offset and symmetrical configuration), linear polarization multiplexing double grating antenna, forward-fed Cassegrain antenna, offset Cassegrain antenna, large deployed mesh antenna, single reflector image antenna, wide-angle scanning double-reflector antenna (e.g. bias-fed offset Cassegrain antenna, forward-fed offset Cassegrain antenna) and confocal reflector antenna. Other antennas commonly used in communication satellites include lens antenna (including dielectric lens and waveguide lens) and phased array antenna. Phased array antenna can provide reconfigurable beam function, which is more suitable for military communication applications, although it is limited by bandwidth, solid-state amplifier efficiency, power consumption, weight, cost and other factors. All in all, the above antennas have been put into use on Chinese satellites.

At present, reflector antennas are widely used in communication satellites. Based on the analysis of the characteristics of the payload antenna of communication satellite, this chapter goes deep into the functional design of some representative antennas, such as shaped reflector antenna, multi-beam antenna and adaptive nulling antenna. Finally, this chapter briefly introduces the technology of large-scale mesh reflector and its development trend.

6.1 Communication Satellite Payload Antenna Characteristics

Communication satellite antenna technology has made great progress in the past decades. In addition to the traditional FSS, the development of BSS, mobile communication satellite and high-capacity personal communication satellite service has played a significant role in promoting the development of satellite-borne communication antenna technology. Each satellite service requires different types of communication antennas to meet different performance requirements.

6.1.1 Development and Characteristics of International Spaceborne Communication Antennas

Back to the 1960s and 1970s, reflector antennas were used to provide high-gain service to the coverage area in communication satellites such as Echo1, Telstar, Syncom-III and Early Bird. Among them, Telstar satellite verifies the C-band transparent forwarding technology for the first time.

In the 1970s and 1980s, space-borne communication antenna and ground station antenna had been developed greatly. In reflector antenna, the standard parabolic reflector is often illuminated by the feed cluster to form shaped beams, so the problems of coupling between feed arrays and complex Beam-Forming Network (BFN) are required to be solved. A C-band and Ku-band shaped reflector antenna, in which only a single feed is used to illuminate the reflector had been developed in the UK. In this antenna, shaped beam coverage of the corresponding ground service area is obtained through the shape of the reflector, which overcomes the shortcomings of the former reflector antenna, greatly accelerates the development progress and reduces the cost of the antenna. In addition, single reflector or double reflector is often used to realize the application of double linear polarization or double circular polarization. The United States and Canada had successfully developed a double-grid antenna and applied it to several satellites. Double-grid antenna technology uses two reflectors to reflect two orthogonal linear polarized electromagnetic signals, which can save layout space by multiplexing the physical space of two reflectors.

Since the end of the twentieth century, multi-beam antenna has been greatly developed, which can achieve frequency reuse and provide high-gain beam coverage for the service area. Typical satellite applications include M-Sat, Thuraya, ACeS, Globalstar, Anik-F, WGS, MUOS and AEHF. In multi-beam antenna technology, spectrum reuse technique can increase the effective bandwidth to 4–30 times of the original bandwidth. From 2000 to 2010s, BSS also increased local channel live broadcasting through multi-beam applications. For example, several satellites operated by EchoStar and DirecTV can provide live broadcasting services and local channel live broadcasting services.

Recently, several high bit rate personal communication satellites have been launched, providing hundreds of spot beams, greatly improving the frequency reuse rate and obtaining large bandwidth capacity communications. High bit rate personal communication satellites require satellite platforms to provide high-power output to support 100 Gb/s satellite communication capacity. The development of high-power satellite platform can carry complex, high-power, multi-band satellite payloads and realize the integration of military applications and commercial services on a satellite.

In the 1980s and 1990s, Harris and AstroMesh (now Northrop Grumman) developed a large deployable mesh antenna for mobile satellite applications in L and S bands, which provides a 5–12 m aperture reflector antenna that can be folded into a small volume envelope to meet the launch requirements. After the satellite is injected into orbit, the antenna is deployed. Recently, the development and flight of 18–22 m or larger aperture commercial and military antennas have been completed in the aerospace industry, and the Ka-band and higher frequency deployable antenna technology is being studied.

6.1.2 Development and Characteristics of Chinese Satellite Communication Antenna

Although the development and application of communication satellite antenna technology in China are lagging, it has basically experienced the same process as that in other countries. The earliest reflector antenna of communication satellite had been successfully applied to DongFangHong-2 test satellite, shaped reflector antenna technology in SinoSat-2 (2006), SinoSat-3 (2007), ChinaSat-10 (2011), ChinaSat-11 (2013), Nigcomsat-1R (2011), PakSat-1R (2011) and other commercial satellites, and double-grid antenna technology in VeneSat-l (2008) and SinoSat-6 (2010). In the application of multi-beam antenna, the multi-beam antenna with four-color reuse system has been first applied in ChinaSat-16, which is the first personal communication satellite with high bit rate and large capacity in China, and the satellite communication capacity supporting 100 Gb/s level has been realized. In the application of mobile communication satellite, China has realized S-band mobile communication service in Tiantong satellite (2016) by using large deployable mesh antenna and multi-beam antenna technology. The high-precision deployable antenna technology in China has also been developed and applied in the Tianlian relay satellite.

Several payload antennas of Chinese communication satellites are shown in Fig. 6.3.

In summary, space-borne communication antenna can be divided into spot beam antenna, shaped beam antenna, multi-beam antenna, adaptive nulling antenna, etc. The corresponding satellite services are shown in Table 6.1.

Spot beam antenna is an antenna that provides pencil-beam and form circular or elliptical coverage area in the field of view of the satellite-to-Earth. It can carry out

Fig. 6.3 Several types of payload antennas for Chinese communication satellites. **a** Shaped reflector antenna; **b** double-grid antenna; **c** large deployable mesh antenna; **d** high-precision deployable antenna

satellite-ground communication or data transmission. In order to improve the flexibility of communication satellite coverage area, the spot beam antenna is generally equipped with driving mechanism, so it is called the mechanical steerable spot beam antenna. It can make the beam move flexibly in the field of view of the satellite-to-Earth, and realize not only the communication between satellite and ground fixed station, but also the communication between satellite and related mobile station communication equipment such as vehicle carried, ship-borne and airborne device. For mechanical steerable spot beam antenna technology, see the introduction in other chapters.

Table 6.1 Relations between antenna technology and communication satellite service

Service type	Communication satellite antenna technology	Technical implementation characteristics
Fixed satellite	Spot beam and shaped beam antenna technology	Improving beam coverage efficiency in service area
Direct broadcast satellite	Spot beam, shaped beam antenna technology	Improving downlink beam coverage efficiency and anti-interference ability on the uplink
Personal communication	Multi-beam antenna technology	Improving minimum gain in multiple beams and enhancing system *C/I* capability
Mobile satellite	Multi-beam antenna technology	Improving minimum gain in multiple beams and enhancing system *C/I* capability
Inter-satellite link	Beam tracking reflector antenna technology and phased array antenna technology	Improving beam tracking accuracy and meeting multi-access communication capabilities as much as possible

Note C/I is the ratio of carrier power to interference power

6.1.3 Key Technology of Communication Satellite Antenna Engineering Application

6.1.3.1 Antenna Performance Design

In general, the communication satellite payload consists of antenna and transponder, in which the antenna provides specific polarization radiation pattern for the coverage area. Antenna gain required for closed-loop communication link, transceiver duplex, high cross-polarization isolation for polarization reusing system, as well as area isolation and side lobe shaped function. The key parameters of communication link are gain noise temperature ratio (G/T) and Equivalent Isotropically Radiated Power (EIRP). In the design, these two parameters are related to the antenna gain in the transceiver band, and the quality of the communication link is determined by the Signal-to-Noise Ratio (SNR), which directly depends on the antenna gain. In general, antenna designers spend a lot of time on antenna coverage gain optimization, so that the antenna gain can approach the theoretical value. Other technical challenges in antenna design include reducing antenna front-end loss, high-power tolerance of feed assembly, low Passive Inter-Modulation (PIM) and improving polarization isolation.

6.1.3.2 Feed Assembly Design

The design of communication satellite antenna should consider the applicability of space environment, including space thermal alternation, high-power handling, Passive Intermodulation (PIM), electrostatic discharge, vibration, impact, and noise, etc. Generally, antenna hardware consists of reflector assembly and feed assembly. The reflector assembly achieves coverage of the ground with a specific gain and beam shape, and the feed assembly is a key component in the satellite antenna system that determines the polarization, bandwidth, power withstanding capability, PIM level, electrostatic discharge, etc. of the overall antenna system. Feed assembly can separate different frequency band signals with sufficient isolation, and provide good free space matching to realize the irradiation of reflector or be used as radiation element of direct array. Typical feed assembly is composed of horn, orthogonal mode transducer, filter or diplexer, polarizer, adapter, power division/combination network and RF interface. At present, the feed assemblies of communication satellites mainly work in bands of UHF, L, S, C, X, Ku, K, Ka, EHF or V.

In engineering application, the payload system of communication satellite requires that the feed has the characteristics of lightweight, low loss, high power and high reliability. In the design of feed structure, the environmental adaptability of satellite launch and space environment should be satisfied, and the size optimization, weight reduction and structure integration should be carried out. In practice, the focus on individual performance indicators often leads to the degradation of the overall performance of the feed, resulting in redesigns and affecting product development. The order of PIM and handling power level directly affects the manufacture method of the feed products. Generally, the 19th order and above of PIM can be ignored, but measures must be taken to reduce the lower order of PIM level, such as reducing flange connection, increasing flange pressure, using choke flange. Furthermore, considering the expansion and contraction of the feed structure at the extreme temperature, it is necessary to design the protection bandwidth outside the working frequency band to cope with the change of the feed performance with temperature.

6.1.3.3 Reflector Assembly Design

Reflector assembly generally uses metal surface, composite surface or metal mesh surface. In general, the metal surface is used for the reflector with aperture less than 1 m, the carbon fiber composite reflector for less than 3 m, and the deployable metal mesh antenna for the reflector with larger aperture. The shape surface accuracy of the reflector has a significant influence on the beam performance of the whole antenna, and the curvature realization ability of the shaped surface is also an important factor affecting the material selection of the reflector and the realization of the antenna mold. In addition, the thermal stability of the reflector is very important to the on-orbit performance of the communication antenna. The Thermal Deformation (TED) characteristics of the reflector should be paid attention to in engineering design, and the influence of the thermal deformation on the antenna beam performance should be evaluated. The ohmic loss of the reflector is also different due to its surface treatment and material characteristics, and there are obvious differences in the loss of different frequency bands. In the application of Ka-band multi-beam antenna, the lightweight and modularization of reflector are of great significance to the engineering application of multi-aperture reflector antenna system. Furthermore, it is necessary to carry out special design on the selection of the honeycomb core and the reflector back structure.

Carbon fibers can be used as surface materials of antenna reflectors due to its conductive property and good radio frequency characteristics in the frequency band below 30 GHz (can be extended to 60 GHz under certain conditions). The reflector of space-borne reflector antenna usually adopts sandwich structures, whose upper and lower layers are thinner carbon fiber composites, and the middle layer is the honeycomb core supporting the reflector, as shown in Fig. 6.4. The composite panels of the upper layer are formed by the reflector mold to meet the electrical performance requirements. The honeycomb layer and the composite panels of the lower layer ensure a good structural stability of the reflector and satisfy the stiffness performance requirements of satellite applications. The upper and lower layers are divided into three types, i.e., unidirectional strip layer, braided fiber layer and tridirectional fabric layer. The honeycomb core in the middle can be divided into three honeycomb structures: aramid fiber, carbon fiber and aluminum honeycomb.

Carbon fiber composites are a kind of composites with certain conductivity, which are manufactured by adding different carbon fibers into the resin matrix and mixing them with certain methods. Carbon fibers have directional conductivity, so the morphology, structure and distribution of carbon fibers in composites determine the properties of materials. The electric, magnetic, thermal conductivity, specific heat, thermal expansion coefficient and mechanical properties of carbon fibers along the fiber axis and vertical to the fiber axis are different and anisotropic. The cross-polarization performance of reflector antenna is deteriorated due to the depolarization effect of the reflector by the direction difference of the first layer. With the

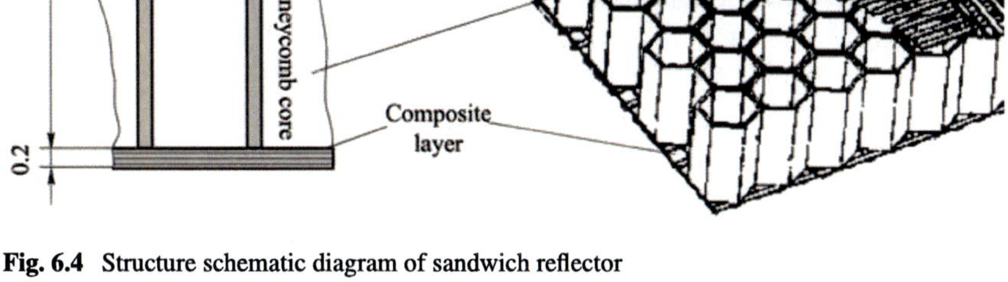

Fig. 6.4 Structure schematic diagram of sandwich reflector

increase of frequency, the depolarization characteristics of carbon fiber composite layers are enhanced, and the depolarization effect of layers at 30 GHz increased cross-polarization level by more than 10 dB compared with that at 10 GHz.

Frequency-selective surface and double-grid antenna (polarization-selective surface) are often applied in the reflector. By laying (or embedding) frequency-selective structure or metal strips on the reflector surface, the separation of electromagnetic waves with different frequencies or polarizations can be achieved in space, and finally realizing the function of multi-frequency/multi-polarization antenna.

6.1.3.4 Two-Axis Driving Mechanism

The two-axis driving mechanism, which is the core of the steerable antenna, enables the antenna to rotate around two orthogonal axes. In the mechanical steerable beam antenna, the driving mechanism is generally *X–Y* type, which consists of two-axis rotating units and related brackets. The main components and functions are as follows:

1. ***X*-bracket** is fixed to an unfolding mechanism or a supporting structure to determine the *x*-axis direction of the mechanism.
2. **Intermediate bracket** can rotate around *X*-axis and determine *Y*-axis direction.
3. ***Y*-bracket** is fixed to the antenna and is rotatable around the *Y*-axis.
4. ***X*-axis rotation unit** can rotate the intermediate bracket around X-axis under the drive of servo controller, and feedback the actual rotation angle signal of the intermediate bracket to the controller.
5. ***Y*-axis rotation unit** can rotate *Y*-bracket around *Y*-axis under the drive of servo controller, and feedback the actual angle signal of *Y*-bracket to the controller.
6. **Low-frequency cable** is a control cable of antenna pointing mechanism.

At present, the common antenna *X–Y* driving mechanism configurations include double-arm bracket type, two-axis coplanar type and single-arm bracket type. The typical configurations are shown in Figs. 6.5, 6.6 and 6.7, respectively. Table 6.2 gives the configuration characteristics of the two-axis driving mechanism and its application suggestions.

There are active and passive thermal control modes of two-axis driving mechanism. Active thermal control is generally a unidirectional heating type, which can solve the problem of too low temperature of the mechanism. It has stronger low-temperature thermal control function, but no high-temperature thermal control function. The high-temperature thermal control of the mechanism depends on passive thermal control mode, but its thermal control function is relatively weak. Therefore, special attention should be paid to the high-temperature condition of pointing mechanism in thermal analysis and thermal control design. The common thermal control measures for two-axis driving mechanism are shown in Table 6.3.

In the engineering application of communication satellite antenna, antenna assemblage and calibration technology, environmental test technology and antenna system

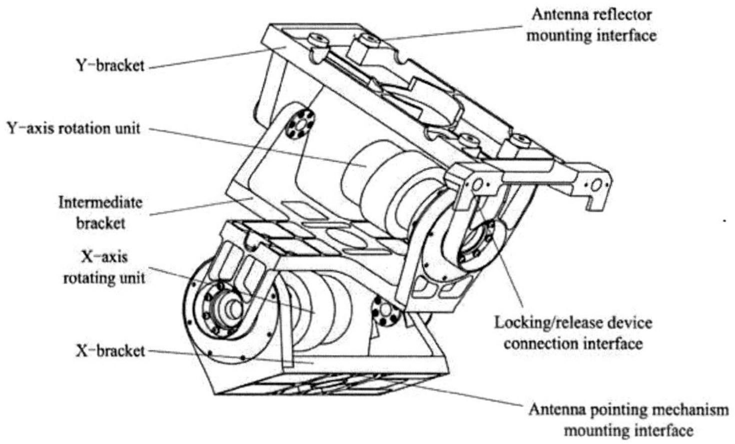

Fig. 6.5 Schematic diagram of two-axis driving mechanism with double-arm bracket type

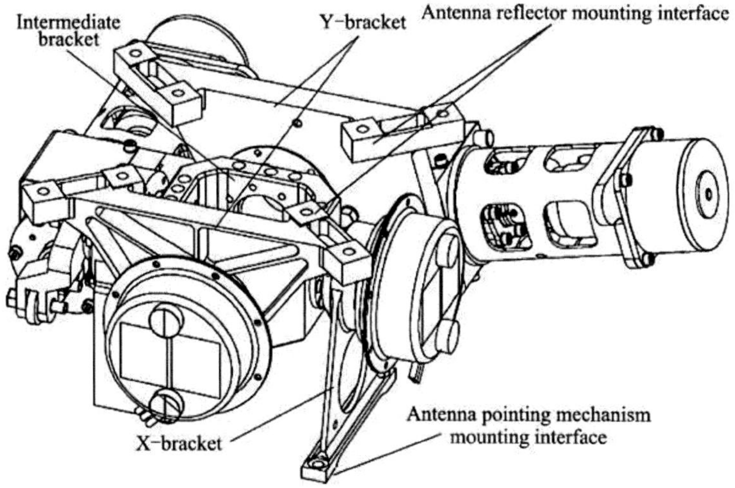

Fig. 6.6 Schematic diagram of two-axis driving mechanism two-axis coplanar type

measurement technology are also involved. In addition to physical appearance inspection and traditional antenna measurement methods, it is necessary to carry out special test scheme design and support of hardware and software technical capability for optical measurement of structural orientation and reflector shaped surface, environmental adaptability verification (mechanical test, thermal vacuum test, etc.) and antenna PIM measurement as well as multipaction measurement. The verifications of the final performance, structure and material property of the antenna on orbit are all finished successfully on the ground.

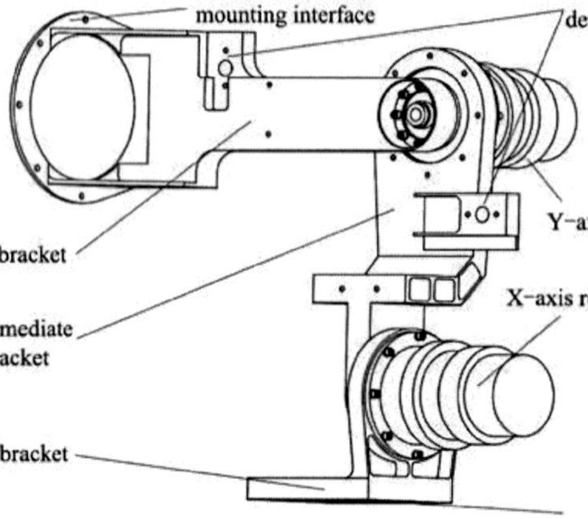

Fig. 6.7 Schematic diagram of two-axis driving mechanism with single-arm bracket type

Table 6.2 Characteristics of two-axis driving mechanism and application suggestions

No	Items	Double-arm bracket type	Single-arm bracket type	Two-axis coplanar type
1	Size	Large	Small	Medium (but not high)
2	Weight	Large	Small	Medium
3	Rotation range	Large	Large	Medium
4	Carrying capacity	Medium	Low	High
5	Stiffness	Medium	Low	High
6	Applicable RF transmission mode	1. Waveguide + rotating joint 2. Flexible RF cable	1. Waveguide + rotating joint 2. Flexible RF cable	Flexible RF cable
7	Manufacturing cost	High	Low	Medium
8	Recommended application range	For mechanically steerable beam antenna with high-performance requirements, complex RF channel, wide range of rotation, high requirements for mechanical strength and stiffness	For mechanical steerable beam antenna with small size and weight of the antenna assembly, wide range of rotation, and large restriction on the size and weight of the pointing mechanism	For mechanical steerable beam antenna with big size and weight of the antenna assembly, small range of rotation, and lower operating frequency

Table 6.3 Common thermal control measures for two-axis driving mechanism

Thermal control modes		Thermal control measures	Applications
Active	Heating with a heater	Electrical heating of the mechanism components using a film heater, armored heater or electric heating wire, temperature control by the thermistor and corresponding controller	To heat the parts that are too cold
Passive	Coating thermal control multilayer	Blockade radiant heat exchange between the mechanism and the outside by thermal control multilayer	1. To prevent sudden changes in the temperature of the mechanism during on-orbit operation 2. To prevent the mechanism from heating up too much when it is heated by solar radiation on the sun-facing side 3. To prevent the mechanism from too cool caused by the cooling radiation in the back sun side
	Change surface state	Control the temperature of the part by spraying a temperature-controlled paint, pasting a thermal control film or electric plating a specific coating on the surface of the part to change the emissivity and solar absorption rate of the part	To control the temperature rise and fall range of the mechanism caused by radiation heat exchange
	Change the heat conduction state	1. Add insulation material on the installation interface to block the conduction heat exchange between the mechanism and the outside; 2. Strengthen the conduction heat exchange between the mechanism and the outside by applying a heat transfer medium and increasing the mounting contact surface	To control the temperature rise and fall range of the mechanism caused by the heat transfer between the interface of the pointing mechanism and the outside

6.2 Shaped Reflector Antenna

With the development of satellite communication, the antenna beam coverage is required to match the shape of the service area in order to improve the Effective Isotropically Radiated Power (EIRP) and G/T value. And the beam formed is called shaped beam. Shaped beam allocates the radiation energy spatially according to the link demand mostly based on the budget of the system link, and the coverage of the regions is performed with unequal gain according to different service areas.

The shaped beam antenna is widely used in payloads of FSS and BSS service. By widening the standard high-gain spot beam into medium-gain beam, the pattern is synthesized for the coverage area to obtain the shaped beam of the uplink, downlink or uplink and downlink. As shown in Fig. 6.8, two methods are often used in the shaped beam:

1. **Multi-feed-shaped beam antenna** is a standard paraboloid antenna with feed array and complex beamforming network.
2. **Shaped reflector antenna** is a shaped reflector with a single feed. The reflector surface has a particular deformed configuration.

Multi-feed shaped beam antenna is an application of multi-beam antenna, which consists of a group of feeds associated with one or more reflectors (such as Gregorain double reflectors) to form multiple beams corresponding to the feeds, and the multiple beams are superimposed in space to form shaped beams. In general, a feed (including horn, duplexer, orthogonal mode transducer, filter etc.) corresponds to a beam projected as a circle. The feed array is arranged near the focus of the reflector antenna in a manner similar to the shape of the coverage of the service area, and uses the beamforming network to realize amplitude and phase adjustment for each beam. All the beams are weighted in the space by the amplitude and phase to form a shaped beam that satisfies the coverage requirement. The large size of the feed

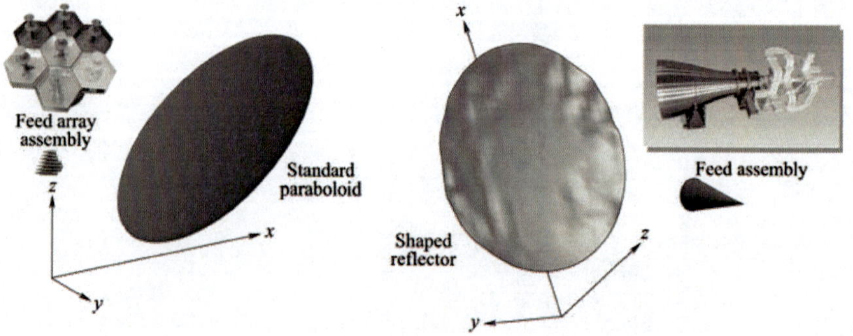

Fig. 6.8 Schematic diagram of shaped beam antenna. **a** Multi-feed-shaped beam antenna; **b** shaped reflector antenna

array and the complexity of beamforming network (especially the design of broadband phase shifter and attenuator) will lead to additional network losses, especially at lower operating frequencies (such as L-, S- or C-band), thus limiting factors to be applied on satellites. However, the antenna on orbit can adjust beam shape flexibly by changing the amplitude and phase of each beam. Therefore, with the development of digital beamforming technology, the multi-feed-shaped beam antenna will be one of the smart satellite antenna technologies.

Shaped reflector antenna is a form of shaped beam antenna developed after multi-feed-shaped beam antenna. Shaped reflector antenna usually has only one feed assembly (including horn, duplexer, orthogonal mode transducer, filter, etc.). By shaping the reflector surface, the amplitude and phase distribution of the radiation aperture are changed to form a radiation beam with special shape. Shaped reflector antenna has fewer components. Because the reflector itself has a wider working frequency band, it can realize broadband (multi-band)-shaped reflector antenna through the design of a group of complex broadband (multi-band) feed assembly. Thus, the arrangement of the feed assembly can be achieved with only a limited space on the satellite.

Compared with multi-feed-shaped beam antenna, shaped reflector antenna has the following advantages: (1) simple structure, lightweight and mature technology; (2) no amplitude and phase errors caused by beamforming network, no unit mutual coupling of feeds, and accurate and reliable design of beam coverage; (3) negligible leakage loss, and the directionality of beam coverage is high; however, the leakage loss of multi-feed-shaped beam antenna is about 1.2 dB; (4) without beamforming network, and the ohmic loss is small. The beam edge gain of the shaped reflector antenna is 0.5–2.0 dB higher than that of the multi-feed reflector antenna (depending on the service area, polarization and space limitation of the antenna). The processing of shaped reflector antenna surface is very difficult, especially for high-frequency reflectors (such as Ku and Ka bands), which need high-surface manufacturing accuracy to ensure that beam performances (such as antenna gain, suppression isolation and polarization discrimination) meet the requirements. At the same time, the adjustable flexible reflector is immature, so the shaped reflector antenna does not have on-orbit variable performance.

The shaped reflector antenna has been widely used in broadcasting and communication satellite antenna based on the advantages of technology, performance, cost and schedule. This section mainly introduces the shaped reflector antenna.

6.2.1 Design of Shaped Reflector Antenna—Representation Method of Reflector Surface

Shaping a reflector is usually based on a standard paraboloid, that is, the redistribution of the amplitude and phase of the radiation field of the antenna is achieved by changing the shape of any part of the reflective surface to form a shaped beam. Since the actual surface of shaped reflector is continuous and smooth, the surface is composed of an ideal paraboloid and a deformed surface, and the latter is generally quadratic derivable. Therefore, a shaped reflector is usually represented as a composite reflector of a basic surface and a deformed surface. Among them, the basic surface is usually parabolic, while hyperboloid or ellipsoid can also be used according to demand, and even an already shaped surface can be used as a reference reflector surface; the deformed surface is usually represented by a set of orthogonal basis functions in order to ensure its continuity and derivability. At present, Zernike polynomial and B-Spline polynomial are two general basis functions for representing deformed surface of reflector.

1. Zernike polynomial.

Zernike polynomial expansion is a very effective way to describe the surface of shaped reflector antenna with moderate deformation. Zernike polynomial is a complex function, however, only the real part is enough for describing the surface of reflector antenna. The Zernike polynomial is used to represent the reflector as follows:

$$z' = \sum_{m=0}^{M} \sum_{n=m}^{N(m)} Z_n^m(x, y) = \sum_{m=0}^{M} \sum_{n=m}^{N(m)} a_n^m R_n^m(t) \cos\big(m(\varphi - \varphi_n^m)\big), \tag{6.1}$$

where a_n^m is the amplitude coefficient, φ_n^m is the reference direction value of the reflector, t and φ are the parameters of the reflector in the aperture projection coordinate system, and $R_n^m(t)$ is the Zernike polynomial function of position, including item $t^m, t^{m+2}, \ldots, t^n$.

The definition of $R_n^m(t)$ and the orthogonality of Zernike polynomials are as follows:

$$R_n^m(t) = \begin{cases} \sum\limits_{k=0}^{(n-m)/2} (-1)^k \frac{(n-k)!}{(\frac{n+m-2k}{2})!(\frac{n-m-2k}{2})!} \cdot \frac{t^{n-2k}}{k!} & \begin{cases} t \leq 1 \\ n > m \geq 0 \\ n - m = \text{even} \end{cases} \\ 0 & \text{other} \end{cases} \tag{6.2}$$

Zernike polynomial is orthogonal over the unit circle, so it is necessary to project the real aperture of reflector into the unit circle for calculation. Considering that the polynomial may vary greatly at the edge of the circle, the actual reflector aperture should be less than 1 when projected to the unit circle. Since each mode of the Zernike polynomial $R_n^m(t)$ corresponds to a deformation of the whole unit circle (Fig. 6.9), the Eq. (6.2) denotes that the deformed surface of the reflector is the weighted sum of $m \times n$ kinds of deformation of the unit circle, which can be adjusted by the amplitude coefficient α_n^m; this characteristic also shows that the reflective surface represented by Zernike is a global representation method. Any change of the coefficient will bring about the change of the whole shape, and the shape optimized by the polynomial is also relatively smooth; the values of m and n are independent of the size of the

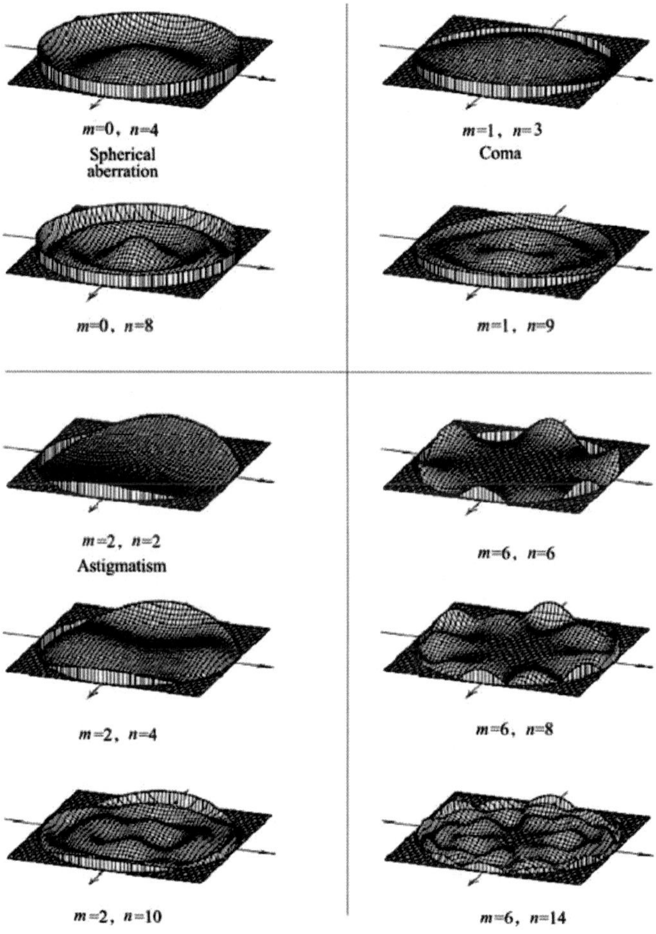

Fig. 6.9 An example of Zernike mode characterizing reflector surface deformation

reflector, which only indicates the shape-forming ability of the reflector. The larger the value of m and n, the more the shape-forming modes are to be provided, the stronger the ability of the reflecting surface is to be shaped, and the more severe deformation of the reflecting surface. In addition, it is known from the characteristics of the Zernike polynomial that each mode represents a shape under a rotationally symmetric structure when $m = 0$ and n is even, so the deformation control can be realized by limiting the m and n parameters when the reflective surface requires deformation of a rotationally symmetric structure.

2. B-spline basis function

When the reflector is represented by B-Spline polynomial of B-spline basis function, the surface equation of the reflector is as follows:

$$z' = \sum_{i=1}^{NX}\sum_{j=1}^{NY} c_{ij}B_{ij}(x', y') \quad \begin{cases} x'_{\min} \le x' \le x'_{\max} \\ y'_{\min} \le x' \le y'_{\max} \end{cases}, \tag{6.3}$$

where c_{ij} is the coefficient of spline function, B_{ij} is a B-spline polynomial, and each spline function is nonzero only in some part of the reflector antenna, which combines to form an effective global reflector.

$$B_{ij}(x, y) = B_i(x)B_j(y), \tag{6.4}$$

where B_i is a cubic spline polynomial, which is related to a series of so-called nodes. $t_1, t_2, \ldots, t_M$ constitute a sequence, i.e., $-\infty < t_1 < t_2 < \cdots < t_M < \infty$, so B_i is a continuous third-order polynomial within $(M-1)$ intervals of these points and has continuous first and second derivatives for x at each node. Cubic spline polynomials have three characteristics:

1. The cubic spline polynomial $B_k(x)$ is uniquely determined by five points: t_{k-2}, t_{k-1}, t_k, t_{k+1} and t_{k+2}.
2. $B_k(x) = 0$, when $x \le t_{k-2}$ or $x \ge t_{k+2}$.
3. $B_k(x) > 0$, when $t_{k-2} \le x \le t_{k+2}$.

The definitions of cubic spline polynomials are as follows:

$$B_k(x) = (t_{k+2} - t_{k-2}) \sum_{m=k-2}^{k+2} \frac{(x - x_m)_+^3}{w'(t_m)}, \quad t_1 \le x \le t_M, \tag{6.5}$$

where

$$(x - x_m)_+^3 = \begin{cases} 0 & x < t_m \\ (x - x_m)^3 & x \ge t_m \end{cases} \tag{6.6}$$

$$w_k(t_m) = \prod_{l=k-2}^{k+2} (x - t_l) \tag{6.7}$$

Unlike the Zernike polynomial, each term of the cubic B-spline reflector basis function represents only the deformation at five points (except the boundary) near its location, and the other ranges are independent of that point, as shown in Fig. 6.10. The deformation density of the whole reflector is determined by the number of m and n. The more the number of m and n, the denser the reflector mesh divided, the smaller the deformation unit of the reflector, and the more severe the deformation of the reflector. The mesh density in the direction of m and n can be different. The reflector represented by B-spline is a local representation. Therefore, the shaped deformation is more severe than that of Zernike coefficient.

To sum up, the shaped surface of reflector can be represented by the global basis Zernike function and the local basis B-spline function, respectively, and different

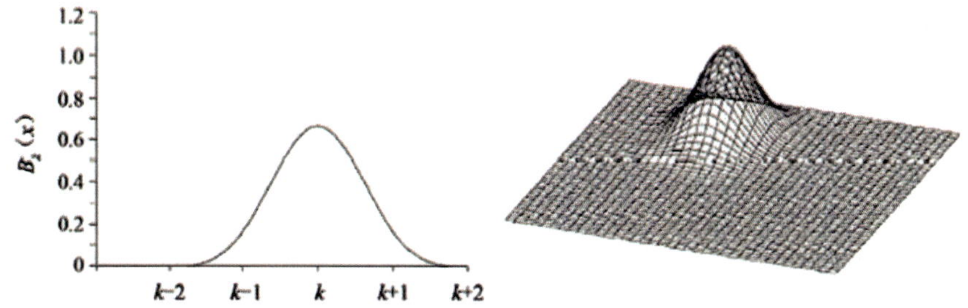

Fig. 6.10 Examples of unit basis functions

functions can be used for different types of deformed surfaces. The antenna aperture is smaller relative wavelength, the weaker the ability to be shaped, and the smaller the choice of the basis function coefficient will be. On the contrary, the larger aperture relative wavelength, the stronger the shaped ability and the larger the choice of the basis function coefficient is. The basis functions expressed by these two kinds of polynomial can cover most cases. For a shaped reflector with special requirements, such as a mesh reflector that requires a node to control the shape, some special basis functions are needed.

6.2.2 Optimizing Reflector Shape

Physical Optics (PO) theory is used to calculate the far field of reflector antenna (see Annex A2), so as to achieve accurate analysis and control of its performance.

Currently, there are many algorithms to optimize the design of shaped reflector antenna, among which Minimax algorithm is the most practical one in engineering applications. This algorithm can set the objective function flexibly according to the actual needs. If the design parameters are appropriate, the optimization results can be ensured to meet the design requirements even at the worst evaluation point.

Historically, the first Minimax problem was raised by French mathematician Jame Waldegrave in a study of a two-player card game in 1713, but the term of Minimax was not used by him at that time. Because of the particularity of the objective function, in recent years, with the development of computer technology, Minimax algorithm has been widely used in engineering design, electronic circuit planning, game theory and other fields, and has attracted much attention.

In the optimization design of shaped reflector, the mathematical model of Minimax problem is shown in Eq. (6.8):

$$\min_{x \in X} F(x)$$
$$F(x) = \max_{1 \le j \le J} f_j(x), \tag{6.8}$$

Fig. 6.11 Expected pattern (dBi is the unit of contour value in the figure)

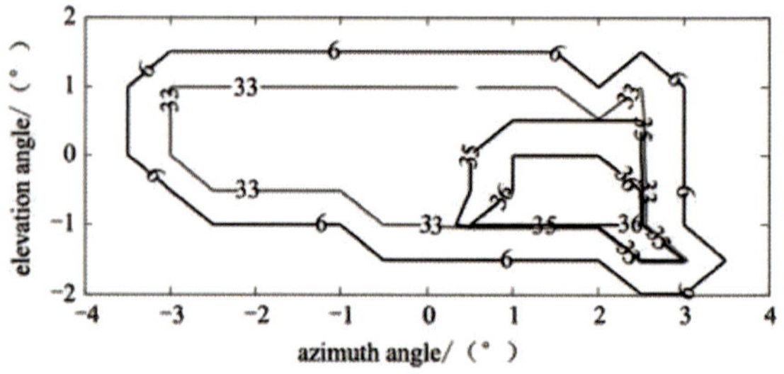

where $\boldsymbol{x} = (x_1, x_2, \ldots, x_I)^{\mathrm{T}} \in \boldsymbol{R}^I$ is a vector containing I elements, which is composed of parameters representing the shape of the reflector; $f_j(x)$ is a function of x, $j = 1, 2, \ldots J$, where J denotes the number of observation points selected in the service area, and $f_j(x)$ denotes the difference between the antenna performance calculation result and the design index of the jth observation point. The specific expression is shown in Formula (6.9):

$$f_j(x) = w_j\left(D_j - G_j\right), \tag{6.9}$$

where D_j is the design index requirement of the jth observation point, G_j is the calculation result of the jth observation point under the design parameter x, and w_j is the weight coefficient.

The expression $F(x) = \max\limits_{1 \le j \le J} f_j(x)$ means that the value of function $F(x)$ is the maximum value of J functions $F_j(x)$ ($j = 1, 2, \ldots, J$) when the variable x is determined to be a specific value. If $j = M$, the difference between the gain of the M_{th} observation point and the design index requirement is the greatest. The purpose of the optimization is to find the minimum value of the function $F(x)$ within a certain range X of x, that is, the gain of the M_{th} observatory is increased by adjusting the optimization variable x, the difference between it and the design index is reduced, and finally the $F(x)$ approaches zero.

Taking the coverage of the U.S. map as an example, the shape pattern of the U.S. map is obtained by optimizing the shape-surface expansion coefficient of the reflector under single feed irradiation using the shaped reflector antenna technology. The desired pattern is shown in Fig. 6.11, the optimized contour pattern is shown in Fig. 6.12, and the deformation of the surface shape of the reflector relative to the standard paraboloid is shown in Fig. 6.13.

6.2.3 *Structure Realization*

In general, a shaped reflector antenna has different structural configurations corresponding to the satellite layout. Taking the DFH Fourth Platform in China as an example (Fig. 6.14), the shaped reflector antenna (antenna 3) on the to-Earth side

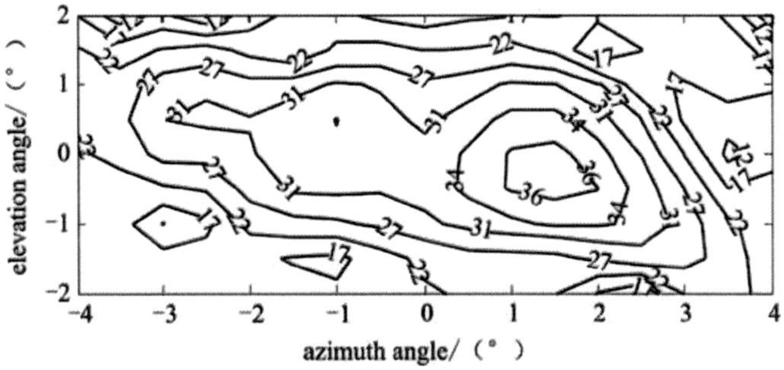

Fig. 6.12 Optimized contour pattern of the U.S. territory (dBi is the unit of contour value in the figure)

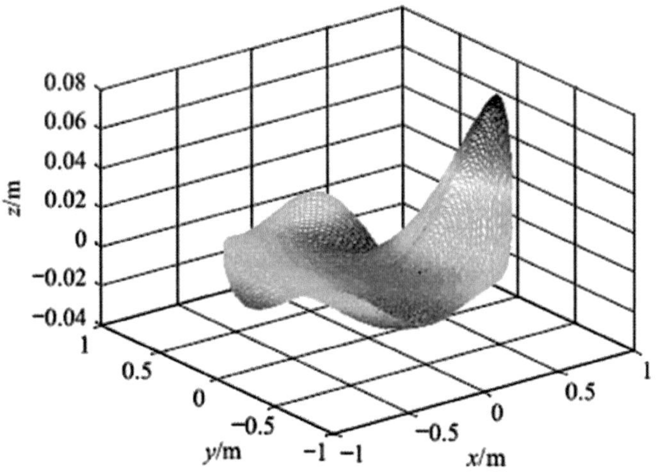

Fig. 6.13 Surface deformation of reflector

of satellite has relatively simple structure, which consists of feed assembly, main reflector and sub-reflector as well as their supporting towers. The reflector of the antenna located on the east–west direction deck of the satellite platform is required to be folded in the launching state, and deployed after the satellite on orbit, so the antenna is more complex, generally composed of feed assembly, shaped main reflector, sub-reflector, reflector deployment arm, reflector locking/releasing device, feed and sub-reflector support tower, etc.

In order to meet the needs of users, the feed assembly generally made of aluminum alloy has various structures. The surfaces of horn and mounting bracket are usually anodized. Other parts of the feed assembly can be processed by special technologies such as silver plating and white paint coating. The carbon fiber skin and aluminum honeycomb sandwich structure are often used to reduce the mass of the shaped

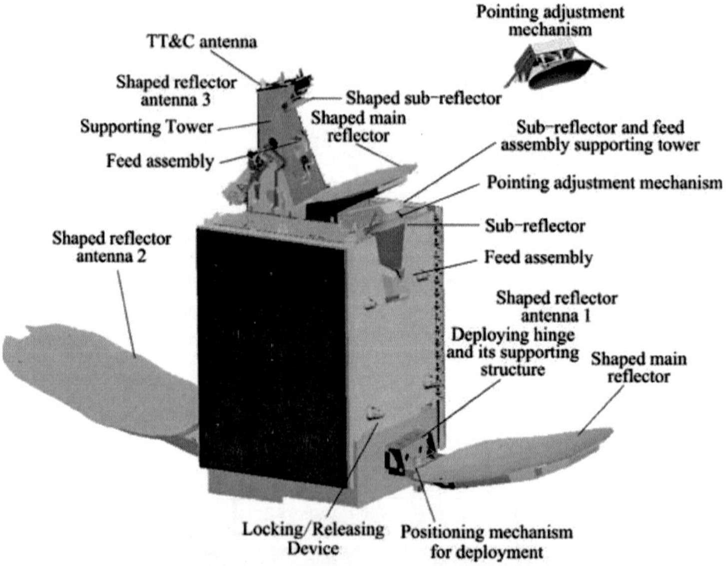

Fig. 6.14 Shaped reflector antenna composition and satellite layout

reflector. The reinforcements on the back of the shaped reflector are provided to control the shape accuracy of the reflector and bear the load when launching; the front of the reflector is coated with anti-static white paint to reduce the temperature gradient caused by the shielding; and the back and stiffener are covered with multi-layer insulation components. In order to improve the pointing accuracy of antenna, Adjustable Pointing Mechanism (APM) is sometimes used to increase the pointing adjustment ability of beams on orbit. APM mounted on the main reflector or sub-reflector is a spatial mechanism that implements micro-adjustment of the antenna pointing ($\pm 1.5°$). The antenna beam pointing can be fine-tuned by changing the reflector's pointing through APM.

6.3 Multi-beam Antenna

Multi-beam antenna can form multiple independent spot beams at the same time. It has the following significant advantages:

1. the increased capacity of satellite communication system. Since multi-beam antenna can form multiple independent spot beams at the same time, spatial isolation and polarization isolation can be used to achieve multiple frequency reuse, which greatly increases the applicable bandwidth and improves the communication capacity of the system;

2. simple ground receiving equipment to guarantee personal mobile communication. Because the multi-beam antenna covers the service area with multiple narrow beams, the gain of the antenna is greatly improved compared to the shaped beams, and the EIRP of satellite transmitting is greatly improved. In this way, users can receive satellite signals using small-scale antenna, thus reducing the ground receiving equipment, providing technical support and guarantee for personal mobile communication global roaming service; and
3. higher system flexibility. Multi-beam antenna has the ability of beam scanning and beam reconstruction, by which the system has great advantages in multi-purpose, anti-interference and enhancing the on-orbit satellite's survivability.

The multi-beam antenna can be classified into three basic types according to the type of antenna, namely, reflector, lens and direct radiating phased array. Due to the mass and volume constraints of satellite antenna, lens antenna is rarely used in satellite multi-beam antenna, and only a small number is used in military communication satellite.

Direct radiating phased array antenna can achieve beam scanning and side lobe suppression by controlling the phase of array elements as well as coverage and scanning in larger angular domain. However, the huge number (hundreds) of space-borne array elements are needed to meet the requirements of high gain of satellite-ground communication links. Under the same gain requirement, the number of phase-shifting control assemblies is increased by times the number of phased array beams, and the difficulty of realizing power division network is further increased (especially in Ku band). As a result, the cost, volume, weight, power consumption and heat consumption of the entire phased array antenna are relatively large. At present, military and civil communication satellite systems require a wide range of coverage, and the number of beams needed is at least 20, which makes it almost impossible to use a direct radiating phased array antenna on a satellite; at the same time, phased array beams cannot achieve real-time communication between different beam positions, the inter-beam frequency reuse is difficult, and the system capacity cannot be effectively expanded. Therefore, the space-borne direct-radiating phased array antenna is suitable for flexible coverage of service area with a small number of beams (up to 8 beams in X-band, up to 4 beams in K-band) and medium gain. However, it is limited in the application of large-capacity and wide-coverage satellite communication.

At present, reflector multi-beam antennas are most widely used in geostationary orbit communication satellites, in which parabolic reflector (or other shaped reflector) is illuminated through a set of lateral offset-focus feeds (known as feed arrays, in most cases consisting of horns) to form multi-beam, so it is also called multi-beam antenna with reflector and feed array.

There are three types of reflector multi-beam antennas: single-aperture single-feed per-beamforming, single-aperture multi-feed per-beamforming and multi-aperture single-feed per-beamforming multi-beam antenna. As the name implies, single-aperture single-feed per-beamforming multi-beam antenna uses one aperture to form multiple beams by each feed corresponds to a beam; single-aperture multi-feed per-beamforming multi-beam antenna uses one aperture to form multiple beams by several feeds corresponds to a beam, and the amplitude and phase excitation coefficient of the feed array are optimized; multi-aperture single-feed per-beam antenna forms the multiple beams by several apertures together, and the beams formed by the different apertures are arranged at intervals.

6.3.1 Application and Characteristics of Multi-beam Antenna

In the early 1980s, single-aperture single-feed multi-beam antenna technology was developed, and has been widely used on orbit. The U.S. Spaceway-3 launched in August 2007 is equipped with Ka-band phased array transmitting antenna and multi-beam receiving antenna [2]. Spaceway-3 forms corresponding spot beams dynamically according to the user's needs through downlink phased array single-beam transmitting antenna, which provides or enhances the satellite capacity and bandwidth in a certain area according to users' actual needs to realize the flexible dynamic management of system capacity and coverage. While, in order to achieve the normal coverage to the service area, a single-aperture single-feed multi-beam antenna is adopted as the receiving antenna, which can form 112 beams to cover the service area normally.

Ka multi-beam antenna developed by ASTRIUM GmbH under the "MEDUSA" program is a typical single-aperture multi-feed multi-beam antenna. The technology is used in Express AMU1 satellite (also known as Eutelsat 36C satellite) developed by Russian Satellite Communications Corporation (RSCC) and Eutelsat, which was launched in December 2015 and verified on orbit. In program MEDUSA, only one transmitting antenna and one receiving antenna are used to provide multiple spot beams for the satellite broadband interactive system, and the full coverage of Ka-band multi-beams in the service area is realized by "four-color" frequency polarization reuse technology. The service areas covered are shown in Fig. 6.15. The feed array forms 18 beams through an analog Beam-Forming Network (BFN), covering Syria, the Arabian Peninsula, Egypt and parts of Libya.

The Ka-band multi-beam transmitting antenna in program MEDUSA is an offset single reflector antenna, whose feed array consists of 74 horns and the core component is a complex beamforming network. The beamforming network achieves low return loss, high isolation and high-precision power allocation in the performance and compact and miniaturized design in structure through a large number of high-performance couplers and phase shifters. The calculation results of downlink coverage performance of "MEDUSA" Ka-band multi-beam antenna are shown in Figs. 6.16 and 6.17. As can be seen from the figures, the antenna gain is over 41 dBi and the *C*/*I* over 16 dB in the whole service area of the multi-beam antenna, which realizes the task of covering the wide service area. A physical picture of the feed array is shown in Fig. 6.18.

Ka-SAT with a high-power Eurostar E3000 satellite platform was launched in the third quarter of 2010, which has a launch weight of 5.8 t, a total power of 15 kW, a payload power of 11 kW, and an on-orbit design life of 15 years [4].

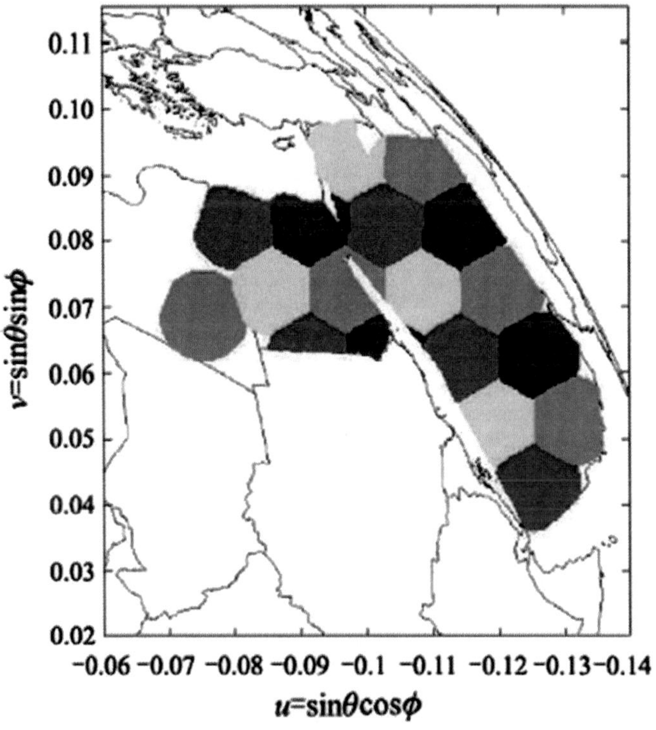

Fig. 6.15 Schematic diagram of multi-beam antenna service areas [3]

Ka-band antenna of Ka-SAT broadband satellite system is a four-aperture multi-beam Tx/Rx community antenna developed by Canadian MDA Company, and four antennas are, respectively, arranged on the satellite east and west wall panels (two of each). The antenna uses a long-focus offset single reflector with an aperture of 2.6 m. The Tx/Rx community and highly integrated feed array have 86 elements, which are divided into four groups. Each group illuminates one of the four reflectors, respectively, and each feed forms a beam. A total of 86 Ka-band spot beams are designed to cover continental Europe, North Africa and the Persian Gulf [5]. Moreover, the satellite uses more advanced Ka-band spot beam design. The total capacity of the satellite exceeds 70 Gb/s, which can meet the application requirements of 1 million home terminals. In order to cooperate with the operation of Ka-SAT broadband satellite system, Eutelsat will establish a ground network equipped with eight gateways to implement the management of all broadband services of Ka-SAT and end-user services, and build a complete new infrastructure of satellite broadband communications.

The ViaSat-1 was successfully launched in October 2011. It is the first large-scale Ka-band broadband communication satellite launched worldwide with a total data throughput of over 100 Gb/s. It is designed and built by Laura Space Systems (SS/L)

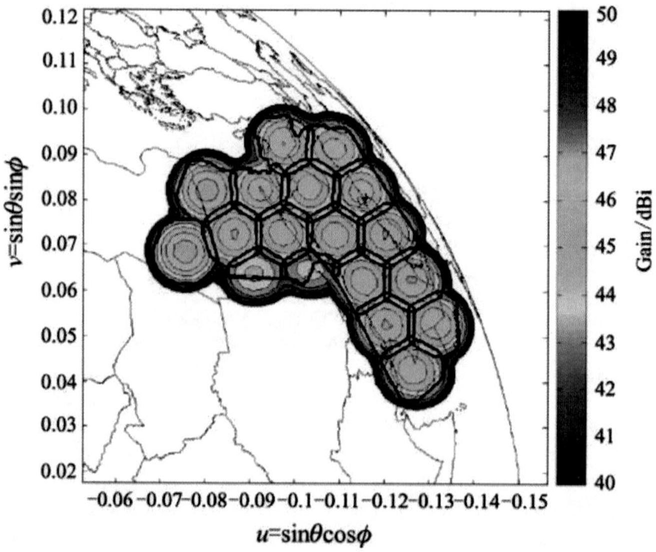

Fig. 6.16 Multi-beam antenna service gain budget (frequency = 18,200 MHz) [3]

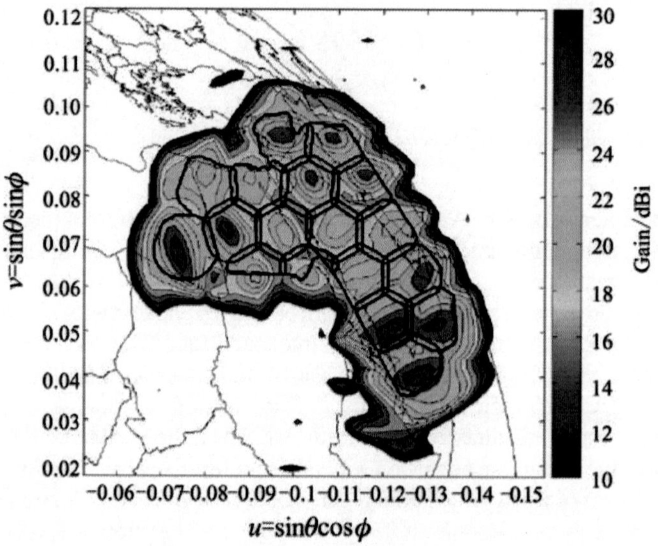

Fig. 6.17 Multi-beam antenna service *C*/*I* budget (frequency = 18,200 MHz) [3]

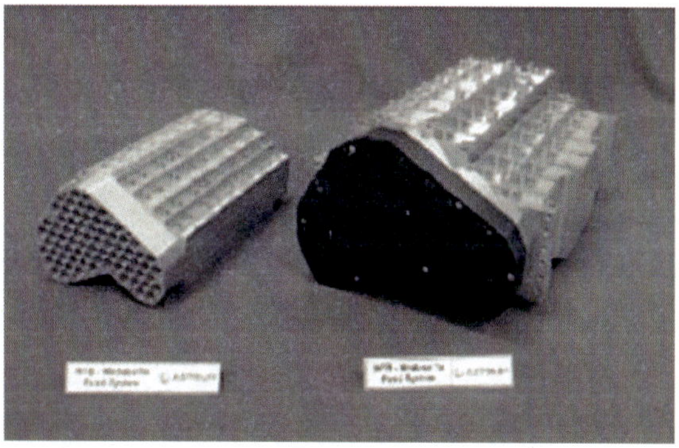

Fig. 6.18 Picture of feed array [3]

of the United States. The satellite with a Ka-band payload and an on-orbit design life of 15 years uses a highly reliable 1300 Type satellite platform, and covers 75% of the continents of the United States, Alaska, Hawaii and Canada. ViaSat-1 has the following characteristics:

1. ViaSat-1 is the heaviest Ka-band communication satellite ever launched in the world.
2. ViaSat-1 is the highest capacity broadband communications satellite in the world (up to 140 Gb/s).
3. ViaSat-1 has 72 spot beams for broadband services in the United States and Canada.
4. ViaSat-1 broadband satellite system includes SurfBeam 2 ground system, also known as ViaSat SurfBeam 2 broadband satellite network system.

The Ka-band antenna of broadband satellite ViaSat-1 uses four-aperture multi-beam Tx/Rx community antenna. Four antennas with a dual-reflector with an aperture of 2.4 m are, respectively, arranged on the satellite east–west wall panel (two of each). The Tx/Rx community and highly integrated feed array have 72 elements, which are divided into four groups. Each group illuminates one of four reflectors, respectively [6].

The Ka-band broadband satellite communication system of ChinaSat16, which has independent intellectual property rights by China, has a four-aperture multi-beam antenna (Fig. 6.19) with 26 user beams, 3 feed beams and total system capacity of 20 Gb/s. The satellite launched in April 2017 with a design life of 15 years can cover China and offshore areas. It is used in remote education, medical treatment, internet access, airborne and ship communications, emergency communications and other fields.

The Ka-band four-aperture reflector multi-beam antenna of ChinaSat16 covers the central and eastern regions and coastal areas of China with 26 beams (beam width 0.8°), providing Ka-band broadband communication services for small ground terminals.

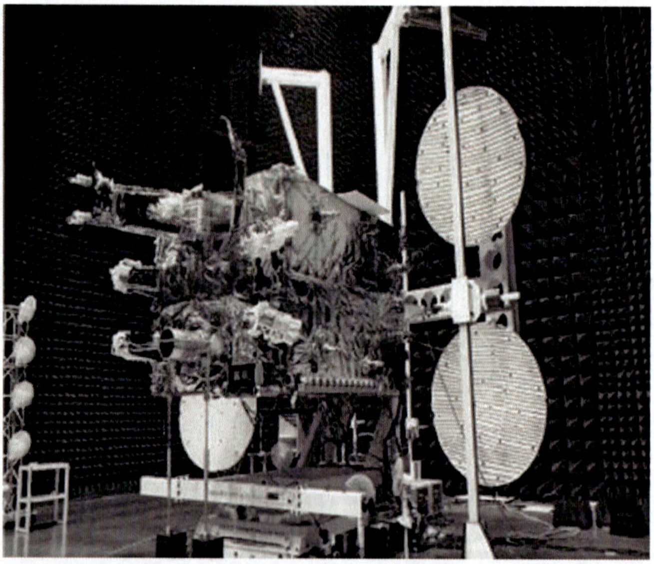

Fig. 6.19 Schematic diagram of the ChinaSat16 multi-beam antenna

6.3.2 Design of Reflector Multi-beam Antenna

Reflector multi-beam antennas can use smaller feed array to achieve high gain multiple beams, which reduces the complexity of the system. At present, there are two main forming methods of reflector multi-beam antenna, one is basic beamforming method, and the other is enhanced beamforming method, which is more complex. The basic beamforming method shown in Fig. 6.20 is characterized by one beam formed by each horn the reflector. The enhanced beamforming method shown in Fig. 6.21 forms multiple beams by several equivalent feeds illuminating the reflector, which is synthesized through the combination of multiple horns. For example, beam 2 is the superposition of per-beams formed by horn 1, 2, 3, 7, 8, 9, 19 at different amplitude and phase coefficients.

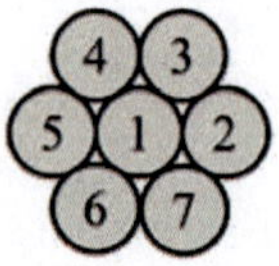

Beam number	Horn number
1	1
2	2
3	3
4	4
5	5
6	6
7	7

Fig. 6.20 Schematic diagram of basic beamforming method

Fig. 6.21 Schematic diagram of enhanced beamforming method

Beam number	Horn number
1	1,2,3,4,5,6,7
2	1,2,3,7,8,9,19
3	1,2,3,4,9,10,11
4	1,3,4,5,11,12,13
5	1,4,5,6,13,14,15
6	1,5,6,7,15,16,17
7	1,2,6,7,17,18,19

Reflector multi-beam antennas can be divided into single-aperture basic beamforming, single-aperture enhanced beamforming and multi-aperture basic beamforming multi-beam antenna.

6.3.2.1 Single-Aperture Basic Beamforming Multi-beam Antenna

Single-aperture basic beamforming multi-beam antenna (also known as single-aperture single-feed multi-beam antenna) forms independent spot beams by illuminating one reflector using a plurality of feeds located at different positions on the focal plane, achieving multi-beam coverage over the service area. Feed array is usually arranged in a hexagonal honeycomb with very small spacing (Fig. 6.22), which minimizes the gap between the spot beams to better cover the service area. Because the feed array is located on the focal plane, all the feeds except the feed at the center have a transverse offset relative to the focus, and the direction and offset are different, so that a group of independent and non-coinciding spot beams with approximately equal width and uniform distribution in the far-field region will be formed by the reflector, as shown in Fig. 6.23. The size of each spot beam is mainly determined by the aperture size of the antenna, while the spacing between the spot beams is determined by the arrangement of the feeds.

In order to realize the overlapping and seamless coverage of adjacent beams, the taper of the feed with general aperture from 0.8 to 1.0 wavelength is generally selected from −3 to −5 dB in the single-aperture single-feed multi-beam antenna. The side lobe of antenna beams is high (about −15 dB) due to the scattering and diffraction

Fig. 6.22 Arrangement of feed arrays

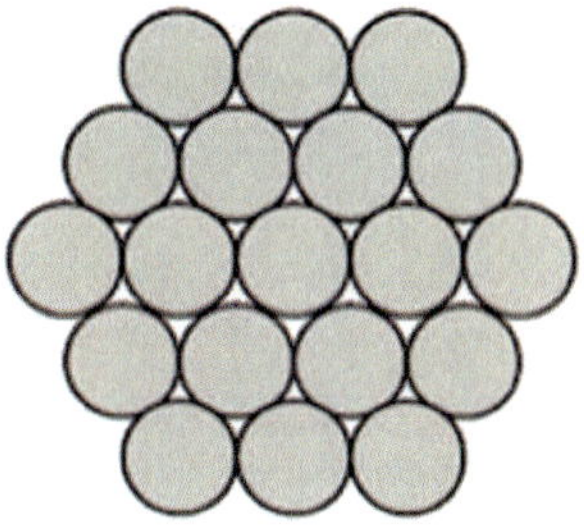

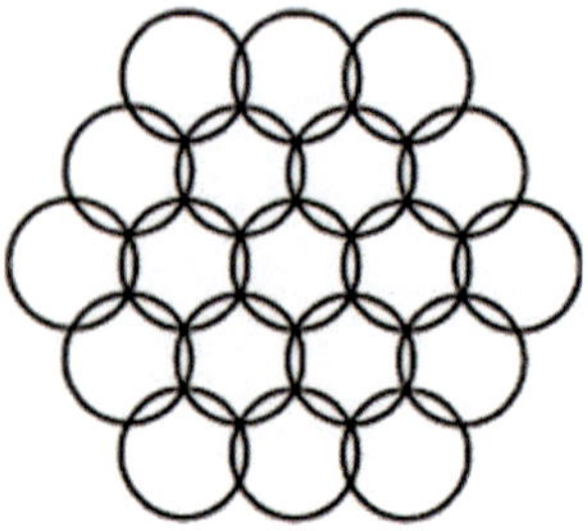

Fig. 6.23 Spatial distribution of beam

effect of the reflector edge, resulting in more serious leaking radiation of the feed after defocusing, so this type of multi-beam antenna has lower aperture efficiency, and lower gain, which is suitable for multi-beam systems with sparse arrangement and less frequency reuse.

6.3.2.2 Single-Aperture Enhanced Beamforming Multi-beam Antenna

Single-aperture enhanced beamforming multibeam antenna (also known as single-aperture multi-feed multi-beam antenna) uses the concept of Multi Feed Per Beam (MFPB) and combines multi-feed to obtain a number of equivalent feeds to irradiate reflector, and ultimately form multiple beams. Several commonly used beamforming methods are shown in Fig. 6.24.

The technology of single-aperture multi-feed multi-beam antenna can effectively improve the equivalent aperture of equivalent and the beam overlapping level by beamforming with multi-feed and sharing part of the feeds, while the spacing of beams remains unchanged. At the same time, the side lobe level of the beam in other same frequency and co-polarization beam regions is effectively controlled, and the *C*/*I* performance in the coverage area is improved by optimizing the amplitude and

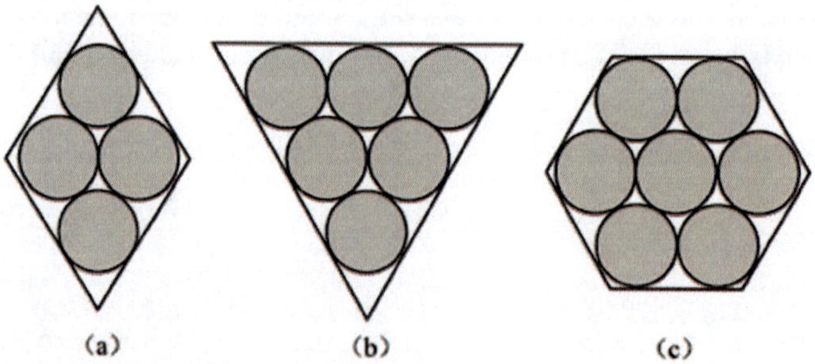

Fig. 6.24 Schematic diagram of beamforming method. **a** Four-feed beamforming; **b** six-feed beamforming; **c** seven-feed beamforming

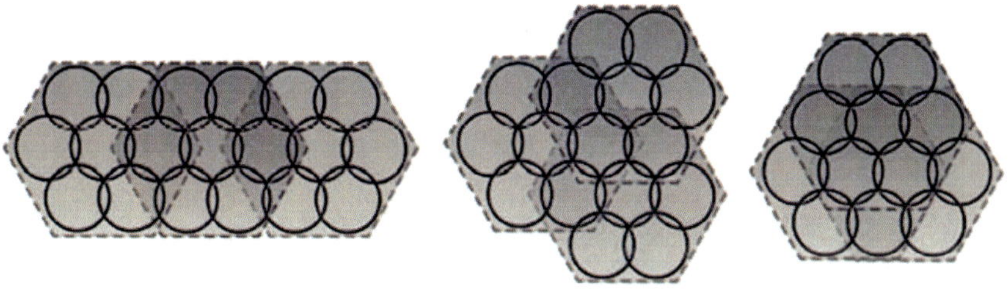

(a) (b) (c)

Fig. 6.25 Schematic diagram of feed sharing modes. **a** Single-feed sharing; **b** dual-feed sharing; **c** four-feed sharing

phase excitation of each feed. Figure 6.25 illustrates the schematic diagram of several feed sharing modes.

The feed array of a single-aperture multi-feed multi-beam antenna is the core component of the antenna. Its function is to form multiple beams in a limited area, and improve the directivity of the beams by focusing energy on the reflector. When designing the parameters of antenna, it is necessary to compromise the feed array and the reflector to meet the system requirements. On the one hand, the feed aperture should be large enough to obtain the required edge taper and gain, and reduce lacking loss, suppress side lobes of the pattern to achieve the design target of *C*/*I*; on the other hand, the feed array should be compact enough to minimize the insertion loss of the feed system and improve energy utilization through miniaturization design, thus satisfying the expected design beam density.

6.3.2.3 Multi-aperture Basic Beamforming Multi-beam Antenna

Multi-aperture basic beamforming multi-beam antenna (also known as multi-aperture single-feed multi-beam antenna) arranges the beam formed by feed arrays corresponding to different apertures at intervals. Because the distance between the spot beams formed by the same aperture antenna is increased, larger aperture feed is allowed, which reduces the energy leaking loss of the feed and the side lobe level of the antenna, and improves the gain of the antenna and the reusability of the same frequency beam. In addition, each feed in reflector corresponds to a beam to avoid the complex feed network of single-aperture multi-feed antenna. Multi-aperture single-feed multi-beam antenna has become a common form of broadband high-throughput satellite antenna because of wide coverage, high gain, low side lobe and high *C*/*I*.

Figure 6.26 shows the position relationship between the four-aperture antenna and the corresponding multi-spot beam. Among them, Beam *A* is formed by antenna aperture *A*, beam *B* by antenna aperture *B*, beam *C* by antenna aperture *C*, beam *D* by antenna aperture *D* and the beams formed by different antenna apertures are arranged alternately. The side lobe of each beam is low, thus meeting the requirements of system *C*/*I*.

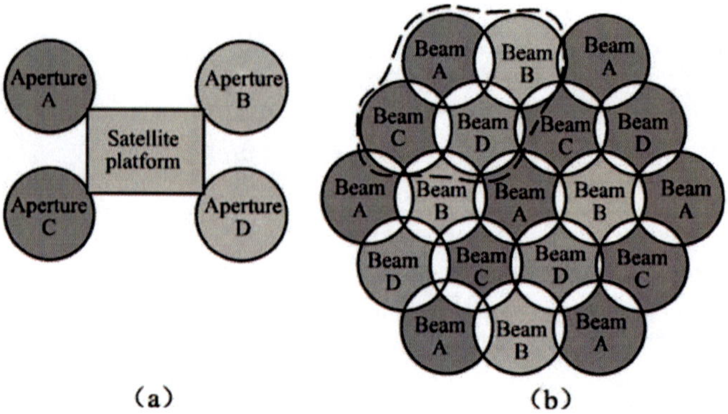

Fig. 6.26 Schematic diagram of four-aperture multi-beam antenna system. **a** Aperture; **b** spot beam formed by each aperture

6.3.3 Main Technical Parameters of Multi-beam Antenna

The main technical parameters of multi-beam antenna include beam coverage gain, frequency, bandwidth, number of beams, side lobe level, frequency reuse mode and *C*/*I*, as shown in Table 6.4.

Among these parameters, *C*/*I* performance needs to be focused on, because the isolation between beams is an important indicator of multi-beam communication system, and *C*/*I* of the system is the main parameter reflecting this indicator.

C/*I* of multi-beam antenna is defined as the ratio of carrier power to interference power. Figure 6.27 depicts the analytical models of seven co-frequency beams *C*/*I*.

Suppose that the transmitting power of the user is P_c, the loss of the link from the user to the satellite is L_c and the deviation angle of the user's position relative to the beam axis is θ_c, then the useful signal power received by the satellite is.

$$C = \frac{P_c}{L_c} \cdot G(\theta_c). \tag{6.10}$$

Suppose that the transmitting power of the co-frequency jamming user is P_i, the loss of the link from jamming user to the satellite is L_c, and the axial deflection angle of the jamming user's position from the beam is θ_i, then the power of the jamming signal received by the satellite is.

$$I = \sum_i \frac{P_i}{L_c} \cdot G(\theta_i), \tag{6.11}$$

where $G(\theta)$ is the gain function of the antenna. Therefore, the *C*/*I* value received by the satellite is.

Table 6.4 Main technical indicators of satellite-borne multi-beam antennas

No	Technical parameter	Descriptions	Remark
General main technical indicators			
1	Coverage area	Beam coverage area	
2	Operating Frequency	Antenna operating frequency band	
3	Operating bandwidth	Antenna uplink and downlink bandwidth	
4	Gain	Beam overlapping gain	
5	Side lobe level	The highest side lobes of the antenna	
6	Standing wave	Voltage Standing wave ratio of antenna	
7	PIM	Passive inter-modulation products	
8	Multiplication	Secondary electron multiplication breakdown in vacuum passive components	
Unique technical indicators			
9	Frequency reuse	Antenna frequency reuse strategy	
10	C/I	Ratio of carrier power to interference power	
11	Feed array port isolation	Beam end isolation caused by beamforming network structure	Single-aperture multi-feed
12	Beamforming network insertion loss	Loss caused by a large number of complex passive components cascading the network	Single-aperture multi-feed
13	Beam pointing calibration accuracy	Calibration accuracy of each beam pointing	Multi-aperture single-feed

$$\frac{C}{I} = \frac{P_c G(\theta_c)}{\sum_i P_i \cdot G(\theta_i)}. \tag{6.12}$$

As can be seen from the above formula that the main factors affecting the C/I performance are the side lobe level of the beam, frequency reuse times and frequency reuse strategy.

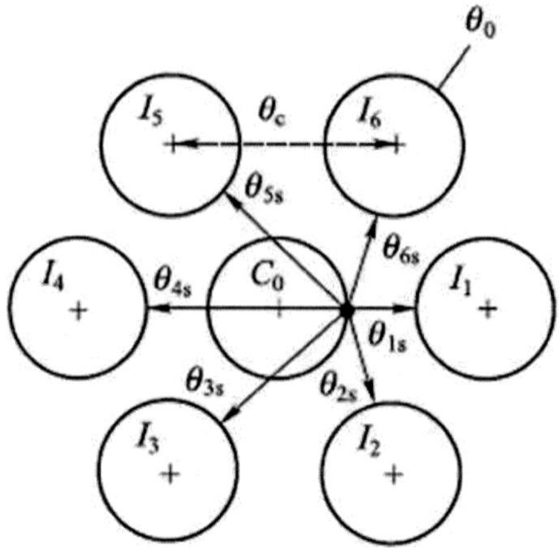

Fig. 6.27 Multi-beam antenna *C*/*I* analysis model

6.3.4 Impact and Design of Core Indicators

For three types of multi-beam antennas, the accuracy of the reflector has a great influence on the antenna gain and *C*/*I* performance.

In the design process, it is necessary to estimate the processing errors and the allowable surface-shape errors of the antenna accurately and reasonably. During the processing, corresponding design measures should be taken to ensure the accuracy of the reflector surface.

The key indicator of single-aperture multi-feed multi-beam antenna is the port characteristics of beamforming network. The realization degree of optimizing the amplitude and phase excitation coefficient directly affects the final coverage gain and *C*/*I* performance of the antenna.

In the design process, the passive components of the feed array need a lot of optimization work in the electrical design to obtain good broadband amplitude and phase characteristics. In the structural design, the multi-component of the feed array should be specially analyzed, and the unique connection and positioning methods should be adopted to ensure the accuracy of the antenna during the assembly stage.

The beam of multi-aperture single-feed multi-beam antenna is formed by the beam with interval arrangement of multiple reflectors, which requires high-calibration accuracy of antenna pointing. Therefore, the corresponding calibration beam should be considered in design stage, and the corresponding observation and calibration stations should be set up on the ground to ensure the pointing accuracy of the antenna.

6.4 Adaptive Nulling Antenna

The adaptive nulling antenna is mainly aimed at high-power blocking interference, which forms pattern null in the direction of interference and reduces the interference signal level received by the communication system, thus realizing spatial anti-interference.

Military satellite communication system plays an important strategic role in communication command. Since military communication satellites are mostly in geosynchronous orbit and their positions are fixed relative to the ground, they are susceptible to ground disturbances, especially by intentional electromagnetic interference from the ground. Man-made interference is mainly used to reconnaissance downlink signal and interfere uplink signal. The purpose of interference is to disrupt the normal operation of transmitter, transponder, sensor or ground receiving system. The ways of man-made interference are ground, ship-borne and satellite-borne interference. In practical application, ground and ship-borne interference are the main interference sources.

The characteristics of electromagnetic interference suffered by military satellite communications are as follows:

1. Both Earth stations and sources of interference are in the view of satellites, and the distances between satellites and Earth stations or sources of interference are not much different or even equal. Therefore, the interference sources have obvious advantages over other communication methods.
2. For transparent transponder, uplink interference saturates them with high power, which will cause serious loss of signal-to-noise ratio, thus disrupting the normal operation of the system; for processing transponder, uplink interference can disrupt the normal working state of transponder, deteriorate the satellite-borne demodulation function, or even out of operation.

Interference to communication satellites can be classified into different modes according to different classification criteria:

1. According to the form of interference signal, it can be divided into continuous wave- (CW), monotone-, multitone-, blinking-, narrow band noise-, broadband noise-, comb spectrum-, pulse- and scanning interference, etc.
2. According to the interference method, it can be divided into aiming, jamming, forwarding, scanning and tracking interference, etc.
3. Electromagnetic Pulses (EMPs): a microsecond single pulse with field strength of 50,000 to 100,000 V/m is used to destroy electronic components and elements by large-area irradiation and surge current induced along wires, especially for digital integrated circuits and semiconductor memory.

For the interference modes, the main anti-interference methods of communication satellite are frequency domain method (such as spread spectrum communication), space domain method (such as nulling antenna) and time domain method (such as burst transmission). Among them, the spatial method, mainly aimed at blocking interference, uses the directivity of satellite communication antenna to separate the signals and interference located in different spatial domains through the spatial isolation characteristics of the beam. There are two main spatial methods, passive mode and active mode. The passive mode is to "avoid" interference, such as adjusting the pointing of steering spot beam antenna and closing the beams affected by interference in multi-beam antenna, while the active mode is to use nulling antenna to form pattern nulling at the interfering position.

According to the difference of amplitude and position between signal and interference, the adaptive nulling antenna can optimize the pattern of antenna array by adaptive weighting processing on each element of antenna, form the pattern nulling in the interference position, and the "zero depth" can generally reach more than 20 dB, which effectively reduce the interference power received by the communication system, thus suppressing interference.

6.4.1 Application and Characteristics of Nulling Antenna

Since the 1970s, the interference suppression technology using spaceborne multi-beam antenna has been studied internationally. Since the 1980s, some effective anti-interference measures have been put forward, such as spread spectrum, adaptive antenna nulling, spaceborne processing and reinforcement technology.

In the early 1980s, the U.S. military successfully applied the adaptive nulling antenna technology to the Defense Satellite Communications System (DSCS III) [7]. The adaptive nulling antenna is a waveguide lens multi-beam antenna with 61 beams. The beamforming network is used to form a zero point in the direction of the interference source, which has certain anti-interference ability.

In the mid-1990s, the United States applied the research results of anti-interference technology of satellite communication system to military strategy and tactical nuclear relay satellite (MILSTAR). MILSTAR I was equipped with three adaptive nulling antennas. Later, the United States equipped two adaptive nulling antennas [8] on its improved MILSTAR II and AEHF satellites [9], respectively.

The application of adaptive nulling antenna has the advantages of fast response and strong suppression when it is susceptible to blocking interference. It can still maintain the normal functions when blocking interference occurs, and it is an effective tactical communication means. However, the nulling antenna also has the disadvantage of nulling indiscriminately, that is, within the nulling area, it will be suppressed both jammers and signals. The pattern of a nulling antenna is shown in Fig. 6.28, in which the color represents the system margin. The darker the color, the larger the margin is.

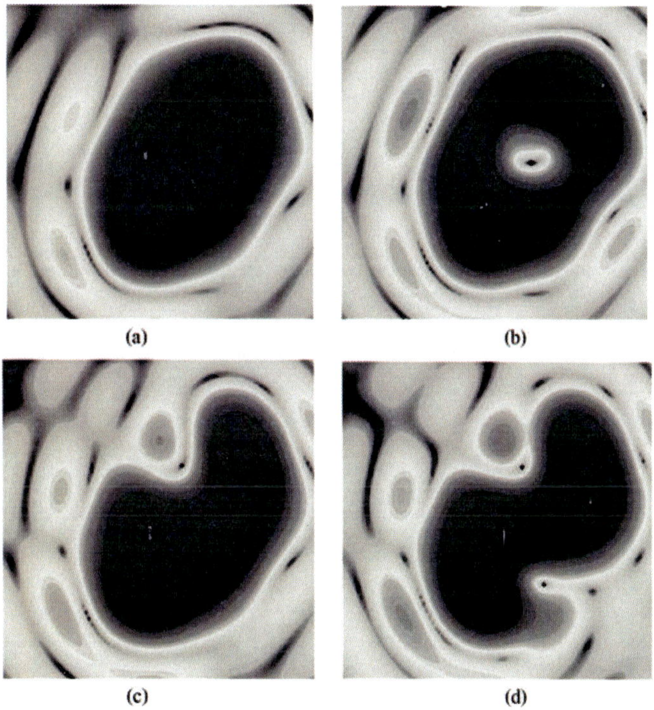

Fig. 6.28 Beam coverage of a nulling antenna before and after nulling. **a** Before nulling; **b** single jammer at center position after nulling, **c** single jammer at edge position after nulling; **d** two jammers after nulling

6.4.2 Basic Principle of Nulling Antenna

Nulling antenna, in the form of array antenna or multi-beam antenna, uses beam-forming network to change the amplitude and phase of each signal so that the desired zero point can be formed in the synthetic pattern. The binary array antenna shown in Fig. 6.29 forms zeros in the desired direction by changing the amplitude and phase excitation coefficients of each element.

Assuming that the array element pattern is A (ignoring array element delay), the array element spacing is $d = \lambda/2$, and the incident angle of the signal is θ, the received signals of the two array elements are

$$x_1 = Ae^{j\frac{\pi d \sin\theta}{\lambda}}, \quad x_2 = Ae^{-j\frac{\pi d \sin\theta}{\lambda}}, \tag{6.13}$$

The output signal y of binary array antenna (hereinafter referred to as binary array) is equal to the weighted sum of received signals x_1 and x_2:

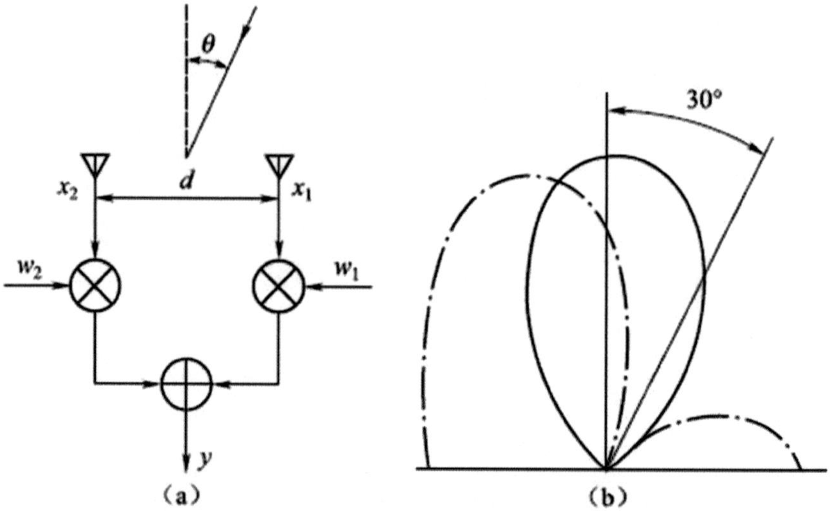

Fig. 6.29 Binary array antenna and its synthetic pattern. **a** Binary array antenna; **b** synthetic pattern

$$y = A\left(w_1 e^{j\frac{\pi d \sin\theta}{\lambda}} + w_2 e^{-j\frac{\pi d \sin\theta}{\lambda}}\right) = A\boldsymbol{w}\boldsymbol{a}(\theta), \tag{6.14}$$

where $\boldsymbol{w} = [w_1, w_2]$ is the weight vector; for lossless systems, $w_1 + w_2 = 1$; $a(\theta) = [a_1(\theta), a_2(\theta)]^{\mathrm{T}}$ is the antenna array vector, also known as array steering vector or array manifold vector.

If $d = \lambda/2$, the binary array output signal is

$$y = A\left(w_1 e^{j\frac{\pi \sin\theta}{2}} + w_2 e^{-j\frac{\pi \sin\theta}{2}}\right). \tag{6.15}$$

The output signal of the uniform binary array is.

$$y = A\cos\left(\frac{\pi}{2}\sin\theta\right). \tag{6.16}$$

At this time, the binary array pattern is shown by the solid line in Fig. 6.36b. If it is desired to form a zero point in the 30° direction, then

$$\begin{cases} y = A\left(w_1 e^{j\frac{\pi \sin 30^\circ}{2}} + w_2 e^{-j\frac{\pi \sin 30^\circ}{2}}\right) = 0 \\ w_1 + w_2 = 1 \end{cases}. \tag{6.17}$$

So,

$$w_1 = (1+j)/2,\ w_2 = (1-j)/2. \tag{6.18}$$

Thus, the output of the binary array is:

$$y = A\cos\left(\frac{\pi}{2}\sin\theta + \frac{\pi}{4}\right). \quad (6.19)$$

The pattern of the binary array is shown in the dotted line of Fig. 6.36b. As can be seen from the figure, the binary array produces a zero point in the direction of 30°. Some interference in this direction will be completely suppressed. When the interference direction is changed, the weight vector can be calculated according to the new interference direction to achieve the goal of nulling.

In practice, the interference position is not fixed, but changing. Therefore, the zero point of antenna pattern needs to change with the interference position. So, an adaptive nulling antenna is used to calculate the weight vector adaptively according to the changing interference direction to realize the adaptive nulling.

6.4.3 Principle of Space-Borne Nulling Antenna

Space-borne adaptive nulling antenna can be divided into the following two categories according to the feedback requirement in the process of nulling: closed-loop nulling antenna (see the feedback loop in Fig. 6.29) and open-loop nulling antenna. Closed-loop nulling antenna needs feedback loop, while open-loop nulling antenna does not. In Fig. 6.30, the signals received by the antenna feeder subsystem include useful signals, interference signals and noise, which are sent to the transponder after weighted synthesis of the radio frequency channel. Nulling antenna determines whether to start nulling according to the received signal, generates nulling weights through some nulling algorithm and puts them into beamforming network to form zero point in the direction of interference signal, so as to suppress interference effectively.

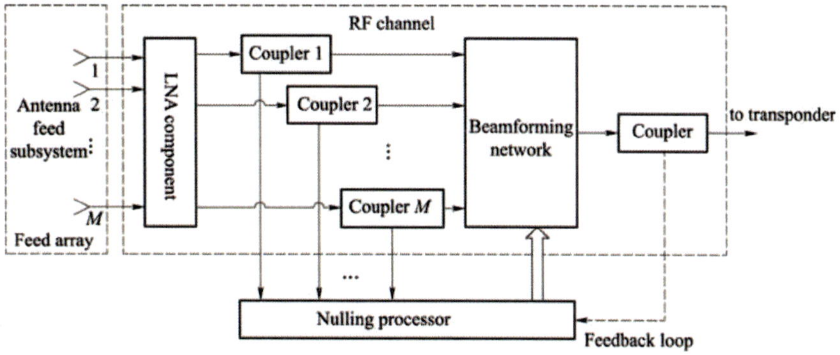

Fig. 6.30 Block diagram of nulling antenna

6.4.3.1 Closed-Loop Nulling Antenna

The closed-loop nulling antenna uses an iterative algorithm to update the weights according to the information of synthetic power feedback, previous weights and pattern changes, and finally forms a null in the direction of interference. Initially, closed-loop nulling antenna used an adaptive side lobe cancelation technique proposed by Howells in 1965. And then Applebaum proposed an adaptive interference cancelation algorithm for this antenna, named Howells-Applebaum algorithm, which is based on the steepest descent/gradient search algorithm. In 1974, Reed et al., proposed a direct Sample Matrix Inversion (SMI) algorithm.

Because gradient method can be easily implemented in hardware and software, this algorithm and its variants are mostly used in nulling antenna. In the following, a conformal gradient algorithm is described.

Firstly, the gradient method is given:

$$\boldsymbol{w}(n+1) = \boldsymbol{w}(n) - \frac{1}{2}\mu\nabla_{\boldsymbol{w}}\varepsilon^2, \tag{6.20}$$

where $\nabla_{\boldsymbol{w}}\varepsilon^2$ is the gradient of the cost function, ε^2 is related to the weight vector $\boldsymbol{w}$ (n) and the step factor μ can be a scalar or a matrix. The objective is to minimize the gradient of the cost function, which is as follows:

$$\varepsilon^2 = \boldsymbol{w}^H\boldsymbol{\Phi}_n\boldsymbol{w} + k\iint\limits_{\Omega}\left(|y_A(\theta,\phi,\boldsymbol{w})|^P - |y_Q(\theta,\phi)|^P\right)^2 d\Omega, \tag{6.21}$$

where $\boldsymbol{w}$ is an adaptive weight, $y_A(\theta, \varphi, \boldsymbol{w})$ is an adaptive pattern; $y_Q(\theta, \varphi)$ is a quiescent pattern, $\boldsymbol{\Phi}_{\mathrm{n}}$ is a covariance matrix (interference + noise), k is a nulling parameter (when the value of k is larger, the pattern is preferably conformal and the nulling is not very deep), P is a nulling parameter (different P values correspond to different application scenarios) and Ω is the covering area.

The first part of the cost function is to reduce the interference using gradient approach. The second part is to preserve the quiescent pattern by making the two patterns as close as possible over the coverage area using the difference in pattern magnitude approach.

6.4.3.2 Open-Loop Nulling Antenna

Open-loop nulling antenna combines spatial spectrum/DOA technology with nulling algorithm. Firstly, the interference direction is determined, and then the zero point is set in that direction. Therefore, this kind of nulling antenna does not need iteration.

Through spatial spectrum estimation techniques, also referred to as a super-resolution algorithm, spatial resolution can be much smaller than the antenna beam width. In 1969, Capon used the power maximum likelihood estimation method to solve the minimum variance undistorted response of the antenna array. DOA and the nulling weight were obtained by searching the power spectrum at the same time. In 1979, Reddi proposed the Minimum Norm Method (MNM), which optimizes the weight vector by solving the optimization problem, makes the weight vector orthogonal to the subspace of the signal eigenvector, and obtains the DOA by searching the pseudo-spectrum. In 1986, Schmidt proposed the MUltiple SIgnal Classification (MUSIC) algorithm, which uses the orthogonality of noise subspace and signal subspace to obtain the DOA by searching pseudo-spectrum. In 1989, the Estimating Signal Parameters Rotation Invariance Technique (ESPRIT) is proposed by Roy and Kailath, utilizing the rotation invariance of signal subspace, in which the signal subspace is generated by two antennas with parallel translation invariance structure.

Because MUSIC algorithm does not require rigorous array manifolds, it only needs to determine the geometric structure of the array in advance and has high resolution, so it has been widely studied and become the most popular spatial spectrum estimation algorithm. An introduction to the algorithm is described as follows.

(1) MUSIC spatial spectrum estimation

MUSIC algorithm uses the geometric structure of the array to estimate the spatial spectrum. It can unbiasedly estimate the number of incident signals, the DOA and the signal intensity. The algorithm relies heavily on the array structure and the array structure must be accurately understood in advance. Therefore, in practice, array calibration is required to be done.

MUSIC algorithm assumes that the noise in each channel is irrelevant, so the correlation matrix of noise is a diagonal matrix; the incident signal may be correlated, so the correlation matrix of signal is not a diagonal matrix. When the signal is strongly correlated, the traditional MUSIC algorithm may fail, and it needs to be compensated by other algorithms.

On the premise of narrow band signal, if D signals are incident on the M-array antenna, the signal received by the M-th-array is.

$$x(t) = \sum_{i=1}^{D} a(\theta_i)s_i(t) + n(t), \tag{6.22}$$

expressed in matrix form as.

$$\boldsymbol{x}(t) = \sum_{i=1}^{D} \boldsymbol{A}(\boldsymbol{\theta})\boldsymbol{s}(t) + \boldsymbol{n}(t), \tag{6.23}$$

where $\boldsymbol{s}(t) = \left[s_1(t),\ s_2(t),\ \ldots,\ s_M(t)\right]^{\mathrm{T}}$ is the incident signal vector, $\boldsymbol{A}(\boldsymbol{\theta})$ is the array manifold, $a(\theta_i)$ is the array steering vector corresponding to the ith signal, and $\boldsymbol{n}(t) = \left[n_1(t),\ n_2(t),\ \ldots,\ n_M(t)\right]^{\mathrm{T}}$ is the noise vector.

MUSIC algorithm should meet the following prerequisites:

(a) The number of elements is larger than the number of signals $M > D$, and the steering vectors of different arrays are linearly independent.
(b) Noise is irrelevant.
c) Signal is not strongly correlated.
(d) Noise is irrelevant with signal.

Based on the above premises, the autocorrelation matrix of the received signal by array element can be obtained.

$$\boldsymbol{R}_{xx} = E\{\boldsymbol{x}\boldsymbol{x}^H\} = \boldsymbol{A}E\{\boldsymbol{s}\boldsymbol{s}^H\}\boldsymbol{A}^H + E\{\boldsymbol{n}\boldsymbol{n}^H\} = \boldsymbol{A}\boldsymbol{R}_{ss}\boldsymbol{A}^H + \boldsymbol{R}_{nn}. \tag{6.24}$$

The autocorrelation matrix $\boldsymbol{R}_{\text{xx}}$ of the received signal is decomposed into eigenvalues, so.

$$\boldsymbol{R}_{xx} = \boldsymbol{U}_s\boldsymbol{\Lambda}_s\boldsymbol{U}_s^H + \boldsymbol{U}_n\boldsymbol{\Lambda}_n\boldsymbol{U}_n^H = \boldsymbol{U}_s\boldsymbol{\Lambda}_s\boldsymbol{U}_s^H + \sigma^2\boldsymbol{U}_n\boldsymbol{I}_n\boldsymbol{U}_n^H, \tag{6.25}$$

where $\boldsymbol{U}_s$, is a $M \times D$ matrix, which is formed by eigenvectors corresponding to D largest eigenvalues; $\boldsymbol{\Lambda}_\text{s}$ is a $D \times D$ diagonal matrix and the diagonal elements are D largest eigenvalues; $\boldsymbol{U}_n$, a $M \times$ (M-D) matrix, is formed by eigenvectors corresponding to $(M - D)$ smallest eigenvalues; $\boldsymbol{\Lambda}_\text{n}$ is a $(M - D) \times$ (M-D) diagonal matrix, of which the diagonal elements are $(M - D)$ smallest eigenvalues, and equal to σ_n^2.

Theoretically, $(M - D)$ eigenvalues corresponding to noise are equal to zero, but practically, the autocorrelation matrix is generated by using a limited number of sampling samples, so the eigenvalues corresponding to noise are neither equal to zero nor equal to each other, but a group of relatively small values. As the samples increases, the noise variance decreases gradually, and these small eigenvalues will be a group of very close numbers. Therefore, by determining the number of small eigenvalues K, the number of signals D (equal to M–K) can be determined.

Eigenvalue λ_i and eigenvector $\boldsymbol{u}_i$ satisfy

$$(\boldsymbol{R}_{xx} - \lambda_i\boldsymbol{I})\boldsymbol{u}_i = 0. \tag{6.26}$$

Because $\boldsymbol{R}_{nn}$ characterizes noise variance, for small eigenvalues and small eigenvectors (M-D), we have

$$(\boldsymbol{R}_{xx} - \sigma_n^2\boldsymbol{I})\boldsymbol{u}_i = (\boldsymbol{A}\boldsymbol{R}_{ss}\boldsymbol{A}^H + \boldsymbol{R}_{nm} - \sigma_n^2\boldsymbol{I})\boldsymbol{u}_i = \boldsymbol{A}\boldsymbol{R}_{ss}\boldsymbol{A}^H\boldsymbol{u}_i = 0. \tag{6.27}$$

Since matrix $\boldsymbol{A}$ is a full rank matrix and matrix $\boldsymbol{R}_{ss}$ is nonsingular, then

$$\boldsymbol{A}^H\boldsymbol{u}_i = 0 \; i.e. \; \boldsymbol{A}^H\boldsymbol{U}_n = 0 \tag{6.28}$$

The above formula shows that the steering vector $\boldsymbol{a}_i(\theta_i)$ corresponding to each direction of the incoming signal in the array manifold $\boldsymbol{A}(\theta)$ is orthogonal to the noise subspace $\boldsymbol{U}_n$, while the steering vector corresponding to other directions is not orthogonal to the noise subspace. With this feature, DOA estimation can be realized.

According to the above analysis, using the orthogonality of the guidance vector of the incoming wave signal and the noise subspace, the whole space is traversed, and the zero point is the direction of the incoming wave. Because of the limited sample size and the influence of noise, the directional vector of the incoming signal will be not completely orthogonal to the noise subspace, and the zero point search must be replaced by the minimum-value search. Then the DOA estimation formula is obtained:

$$\theta_{\text{music}} = \min\left\{\boldsymbol{a}^H(\theta)\boldsymbol{U}_n\boldsymbol{U}_n^H\boldsymbol{a}(\theta)\right\}. \tag{6.29}$$

In the process of using, DOA can be estimated by searching the peak value of MUSIC spatial spectrum (pseudo-spectrum):

$$P_{\text{music}}(\theta)=\frac{1}{\boldsymbol{a}^H(\theta)\boldsymbol{U}_n\boldsymbol{U}_n^H\boldsymbol{a}(\theta)} \tag{6.30}$$

or

$$P_{\text{music}}(\theta) = \frac{\boldsymbol{a}^H(\theta)\boldsymbol{a}(\theta)}{\boldsymbol{a}^H(\theta)\boldsymbol{U}_n\boldsymbol{U}_n^H\boldsymbol{a}(\theta)} \tag{6.31}$$

by substituting θ with (θ, φ) in the above equation, it can be extended to an area array or a multi-beam antenna.

(2) Nulling weight calculation.

According to the spatial spectrum estimation technique, the DOA result is obtained, and the pattern zero point can be formed in the direction of arrival by using the conformal nulling algorithm.

Since the zero point of the nulling antenna is not steeply dropped, the communication area may be reduced after nulling. In order to improve the ratio of the communication area after nulling, it is necessary to make the nulling pattern as close as possible to the original static pattern on the premise of effectively suppressing interference, i.e., on the premise of conforming the main beam. The whole problem is solved by the following constraint optimization:

$$\begin{cases} \min\limits_{\boldsymbol{w}} \left\| \boldsymbol{w} - \boldsymbol{w}_q \right\|^2 \\ s.t.\boldsymbol{a}^H(\theta, \phi)\, \boldsymbol{w} = f \end{cases}, \tag{6.32}$$

where $\boldsymbol{a}^{\mathrm{H}}(\theta, \varphi)$ is the steering vector corresponding to the interference direction, and f is the constraint condition. When $f = 0$, it is a problem of solving the optimal nulling weight problem, and $\boldsymbol{w}_q$ is the beam static weight when it works properly (not nulling), $\boldsymbol{w}$ is the nulling weight. The above-mentioned constraint zeroing weight optimization problem can be solved by the Lagrangian operator, and the optimal weight can be obtained thereby:

$$\boldsymbol{w}_{\text{opt}} = \boldsymbol{w}_q - \boldsymbol{a}(\theta, \phi)\left[\boldsymbol{a}^H(\theta, \phi)\boldsymbol{a}(\theta, \phi)\right]^{-1}\boldsymbol{a}^H\boldsymbol{a}(\theta, \phi)\boldsymbol{w}_q. \quad (6.33)$$

From the above formula, the optimal weight can be regarded as the static weight minus its projection vector which is the projection of the static weight in the orthogonal complement space of its constrained subspace. Since the generation of the nulling weight is closely related to the steering vector corresponding to the interference direction, the DOA result of the interference signal directly affects the nulling effect.

6.4.3.3 Comparison of Two Types of Nulling Antennas

Both types of nulling antennas have their own advantages and disadvantages. At present, military communication systems mostly work in Ku-band and above, and the working bandwidth is at least several hundred MHz. Limited by the digital device level, especially the *A/D* sampling rate, the nulling antenna using the digital acquisition system will collect the interference information in the entire working frequency band by multiple acquisitions, so the nulling time is relatively long. The closed-loop nulling antenna adopts an iterative algorithm, and the pattern zero point is directly obtained through multiple iterations, while the open-loop nulling antenna first detects the interference direction, and then sets the pattern zero point in this direction. So the nulling effect and conformal effect of the open-loop nulling antenna is better than that of the closed-loop nulling antenna. However, interference detection of open-loop nulling antenna relies entirely on the antenna array structure, so the tolerance for errors is lower. The comparison of two types of nulling antenna is shown in Table 6.5.

Table 6.5 Performance comparison of two types of nulling antennas

No.	Item	Closed-loop nulling antenna	Open-loop nulling antenna
1	Number of interference suppression	Similar	
2	Nulling depth	Similar	
3	Nulling time	Analog system time is short; digital system takes longer	Longer
4	Nulling resolution	Low	High
5	Communication coverage	Low	High
6	Error tolerance	High	Low
7	Interference direction finding	Inaccurate	Accurate

6.5 Large Mesh Reflector Antenna

For mobile communication satellites supporting handheld terminals, the EIRP on ground is required to be much larger than that of ordinary fixed communication satellites. Therefore, the power density of the uplink and downlink is achieved by using large-diameter antennas and multi-beam feeds to meet the requirements of the handset, which is the best choice for mobile communication satellites. For example, the Inmarsat-4 with more than 100 beams uses a 12 m-diameter deployed antenna to support mobile terminal communication with medium bit rate. The large-aperture reflector antennas with deployable structure mainly include annular mesh antenna, umbrella antenna, truss antenna, and space synthetic mesh antenna (such as the Japanese ETS-8 communication satellite). This chapter mainly addresses the first three.

6.5.1 Annular Mesh Antenna

Annular mesh antenna refers to a deployable antenna formed by a flexible reflective surface with an annular truss support. When the antenna is folded, the surrounding truss is in a folded state too, which will then be unfolded after on-orbit, and drive the reflective mesh into work state. An annular mesh antenna mainly includes reflector, unfolding arms and feed assembly. Figure 6.31 shows its structural composition and its installation on the satellite. This structure is suitable for an antenna with unfolding aperture of 6.50 m due to its special unfolding structure, high reliability,

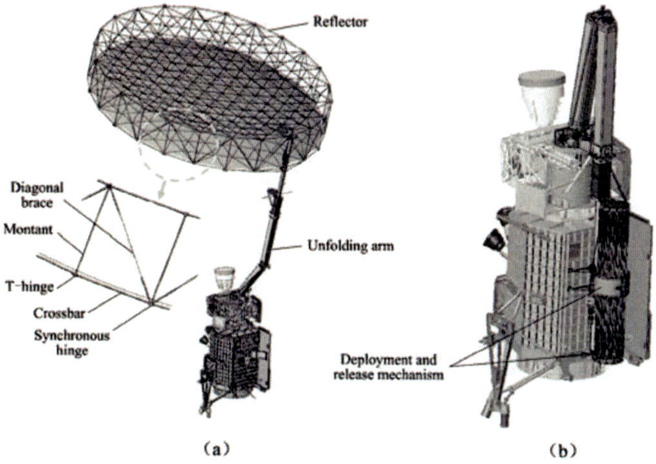

Fig. 6.31 Annular mesh antenna structure and its installation on the satellite [10]. **a** Unfolded state; **b** folded state

high storage ratio, and lightweight. Major international aerospace agencies have conducted in-depth research and developed a series of products of the antenna, which are widely used in many fields, such as military electronic reconnaissance, mobile communication, and deep space exploration.

The annular mesh antenna is mainly developed in both directions of high accuracy and large aperture. In terms of high accuracy, the deployed aperture is usually less than 10 m, and its working frequency can reach Ka-band. In terms of large aperture, it is reported that the maximum aperture can be about 100 m, but when the aperture is larger, the folded size is too huge to the requirements of the rocket envelope. Therefore, for the large aperture, it is necessary to increase the contracted rate and reduce the geometrical size in the folded state. Astromesh-2, the second generation Astromesh has been developed, whose main structure of the antenna is the same as that of Astromesh and Astromesh-1 with a ring composed of multiple quadrilateral elements. The difference is the movement mode of the quadrilateral member bars in the unfolding/folding process. In the Astromesh structure, the connecting quadrilateral is unfolded by shortening diagonal brace, and in the Astromesh-2 structure, the length of the montant of the connecting quadrilateral is variable, and a connecting point of the cross rod and the montant slides along the montant to unfold the antenna.

Astromesh-2 is compact in size in folded state and suitable for annular mesh antennas with a diameter of 18–50 m. At present, Astromesh-2 is in the pre-research stage and has completed the development of the prototype. Figure 6.32 shows typical structure of the ring truss deployable reflector. Figure 6.33 shows the physical unfolding process of the ring truss deployable reflector antenna.

After years of development, Northrop–Grumman has developed four generations of Astromesh family of annular truss unfoldable antennas, which are all suitable for 3–50 m antennas. The unfolding forms of Astromesh, Astromesh-1 and Astromesh-Lite are all much of a muchness. Astromesh-1 has a 40% smaller folded diameter than Astromesh and is suitable for antennas with a diameter of 6.25 m. Astromesh-Lite is 50% lighter than Astromesh-1 and suitable for antennas

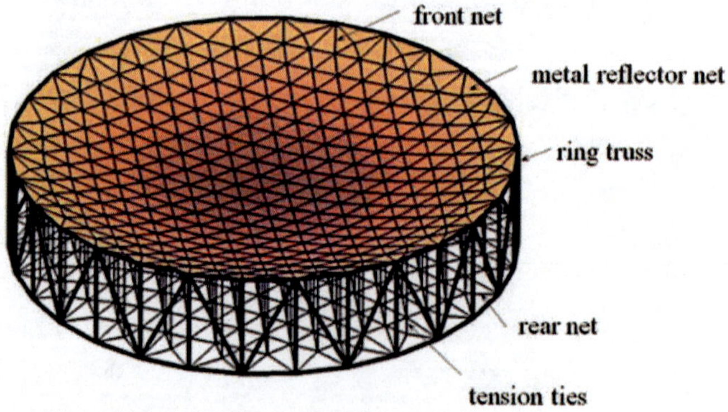

Fig. 6.32 Typical structure of the ring truss deployable reflector [11]

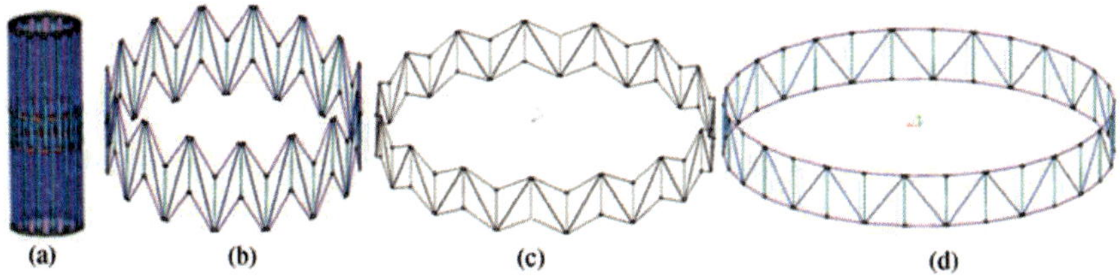

Fig. 6.33 Physical deployment process of Astromesh-2 [12]. **a** Folding state; **b** pre-unfolding; **c** semi-unfolding state; **d** fully unfolding and locking state

with 3–8 m diameter. Astromesh-2 is further optimized on deployment methods compared to the above three antennas, and its height in the folded state is 40% lower than that of Astromesh-1.

According to relevant reports, the Atemesh family of annular truss unfoldable antennas is used in the Large Deployable Antenna (LDA) with a diameter of 150 m of the American signal intelligence satellite "advanced folding chair". Compared with other antenna structures, the annular truss unfoldable antenna with a simple structure has a relatively large application space. The increase of the aperture within a certain range does not change the structural form of the antenna, and the weight does not increase proportionally. Therefore, it is an ideal structural form of a large spaceborne deployable antenna.

China also developed some annular mesh antenna, Fig. 6.34 shows the 16 m reflector prototype.

Fig. 6.34 16 m reflector prototype

6.5.2 Umbrella Antenna

The deployment principle of the umbrella antenna (also known as the radial rib antenna) is similar to that of an umbrella, and generally consists of radial ribs, radial rib deployment mechanism, a locking mechanism and a metal reflection mesh. The unfolding rib is connected to the cylindrical structure at the center of the bottom of the antenna through a hinge, and the metal reflection mesh is laid on the unfolding rib. When folded, the rib is rotated around the hinge to form a column, and the antenna is similar to an open umbrella after the deployment. The umbrella antenna has small folded volume, lightweight and high precision, and can be used in communication satellites, data relay satellites, deep space exploration as well as remote sensing application. The aperture size of the umbrella antenna is usually 4 to 6 m, which meets the requirements of the L ~ Ka band. The deployment of the radial ribs uses a spring deployment mechanism or an electric deployment mechanism.

The umbrella unfoldable antenna was first developed by Harris Corporation of the United States. An antenna of this type used in the Galileo Jupiter detector has a diameter of 5 m, a diameter of 0.9 m after folding, a height of 2.7 m, and an aperture contracted efficiency of 18% [13]. Since 2011, Russia has launched a series of Tracking and Data Relay Satellite (TDRS), as shown in Fig. 6.35, each carrying two umbrella antennas. The research of umbrella antenna technology has also been carried out in China and put into on-orbit application. Figure 6.36 shows a 4.2 m umbrella antenna of China.

Early radial rib antennas were operated in the lower frequency band, and the accuracy of the reflector was not high. The reflection mesh was laid directly on

Fig. 6.35 Layout schematic diagram of umbrella antenna on Russian TDRS satellite series [14]

Fig. 6.36. 4.2 m umbrella antenna of China

the radial ribs. With the gradual improvement of the accuracy requirements for the shape-surface, more and more umbrella antennas are arranged with a cable mesh structure between the two ribs (Figs. 6.37 and 6.38) for improving the accuracy of the shape-surface and further reducing the anti-pillow effect of the metal mesh.

As the diameter of the umbrella antenna increases, the rib becomes thicker and the mass of the entire antenna increases sharply in order to ensure that the entire reflector has sufficient rigidity. In addition, as the aperture of the antenna increases, the cable nodes will also increase greatly to ensure sufficient shape accuracy, that will cause the cable mesh structure to become extremely complicated, and the power required for antenna deployment will further increase. Therefore, the maximum diameter of the umbrella antenna is generally less than 10 m.

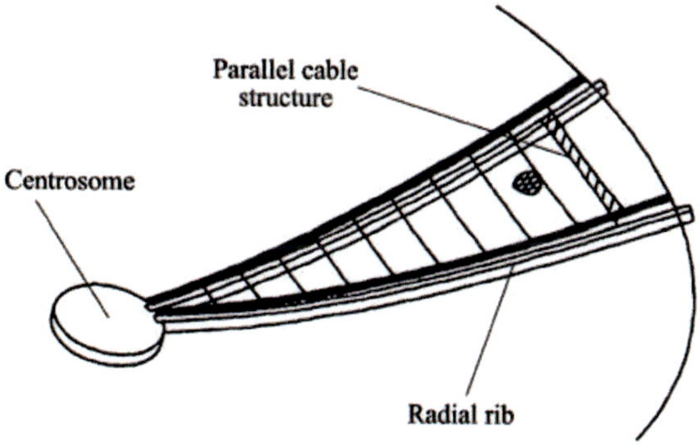

Fig. 6.37 Structure distribution of parallel cable nets

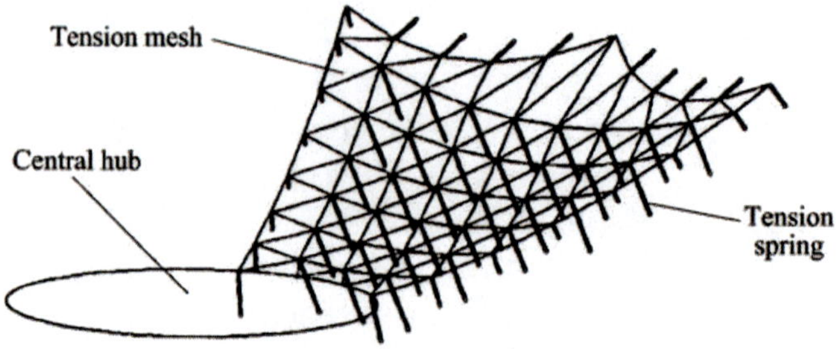

Fig. 6.38 Structure distribution of three-directional mesh net

6.5.3 Truss Antenna

The full name of the truss antenna is Tetrahedral Truss Reflector Antenna (TTRA), which is composed of a metal mesh and an unfoldable truss. The metal mesh is laid on the truss by means of multi-point connection, which is used to reflect the electromagnetic wave; the unfolding truss is composed of tetrahedral units staggered the beginning and the end to share the vertices, and constitutes a support structure in combination with the bevels. Each tetrahedral unit includes four disc chucks (vertices), three web members (bevels) and three synchronizing rods (folding and unfolding bottom edge), as shown in Fig. 6.39. In order to realize the folding and unfolding function, a spring-driven synchronous hinge is arranged in the middle of the synchronizing rod. The spring element of the synchronizing rod stores energy when the tetrahedron is folded, and the stored energy is converted to driving force to unfold tetrahedral unit when unfolding.

The truss antenna is an important spaceborne large deployable antenna structure, and is widely used on orbit with a diameter of 4–8 m. It can be used to implement

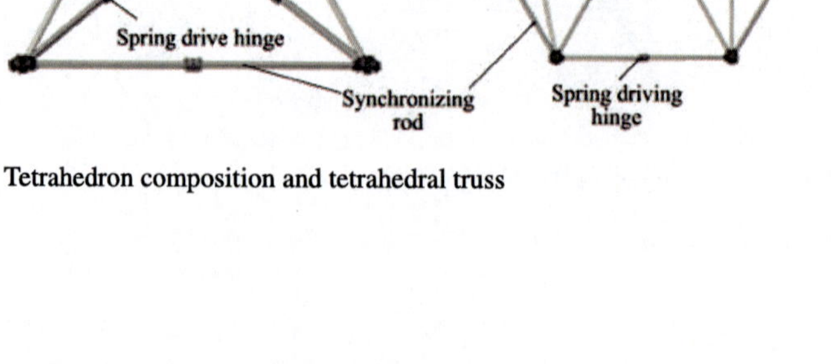

Fig. 6.39 Tetrahedron composition and tetrahedral truss

Fig. 6.40 Truss antenna of Russian resources satellite series [15]

forward-feeding or offset-feeding installation. Its reflector is compact in structure and small in volume in folded state.

The truss antenna was first developed by the Russian Aerospace Agency. Since 1985, it has been successfully applied to spacecraft such as the "Nature", "Union" spacecraft and the "Peace" space station. The Synthetic Aperture Radar (SAR) antenna of the "Peace" space station uses a 6 m × 2.8 m truss antenna, which operates at 1.28/3.28 GHz with a folded size of 0.6 m × 0.4 m × 0.58 m, a total weight of 35 kg, and an area density of 2.1 kg/m^2.

SAR antenna on Russia's resource series Resurs-O also uses a truss antenna (Fig. 6.40) with an aperture of 6 m × 7 m, and the feed can be rotated as a whole to achieve linear polarization transformation.

There is no report on the on-orbit application of truss antennas in the United States, but relevant research institutions have carried out research on ground.

A number of research institutions and universities in China have carried out related research on truss deployable antennas, mainly including CAST Xi'an, Zhejiang University, and Electronics Institute of Chinese Academy of Sciences. The truss antenna developed by CAST Xi'an has the experience of flight test and is the most representative. A 6 m-aperture truss antenna for the satellite HJ-1C was also developed by CAST Xi'an, which was launched into orbit on November 19, 2012 and successfully unfolded. The unfolded state of the truss reflector on the zero gravity rack is shown in Fig. 6.41.

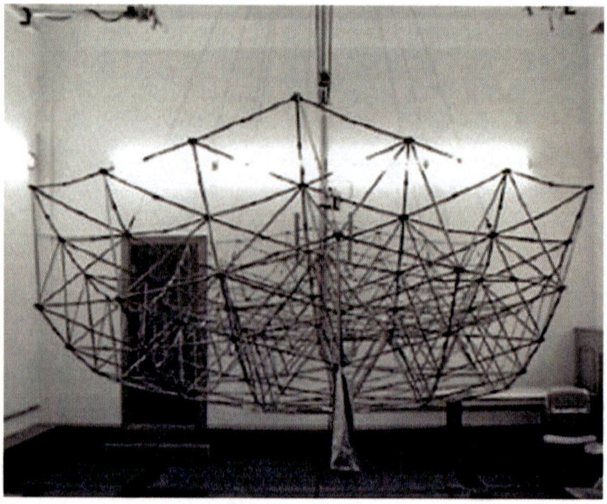

Fig. 6.41 Satellite HJ-1C SAR antenna in unfolded state

References

1. Rao SK (2015) Advanced antenna technologies for satellite communications payloads. IEEE Trans. Antennas Propag. **63**(4):1205–1217
2. Wu, Y. (2003) Precision beacon-assisted attitude control for Spaceway. In: 2003 AIAA Guidance, Navigation, and Control Conference and Exhibit, Austin, 2003, pp 1–7
3. Schneider M, Ratkorn N, Hartwanger C, Wolf H (2013) Design and space qualification of the Medusa feed. Antennas Propag. Wireless Commun. 1077–1080
4. Fenech H (2013) High throughput satellites: an analytical approach. In: 31st AIAA International Communications Satellite Systems Conference, pp 1–19
5. KA-SAT series. Jane's Space Systems and Industry (2018), 1–6
6. ViaSat series. Jane's Space Systems and Industry, Apr 2018
7. DSCS III (Defense Communications Satellite System), Jane's Space Systems and Industry, Aug 2015
8. Hinshilwood DJ, Dybdal RB (2001) Adaptive nulling antennas for military communication. AEROSPACE, 20010000
9. Advanced EHF (AEHF) system, C4ISR & Mission Systems: Joint & Common Equipment, JANES_ZB, 26 Aug 2016
10. Mehran M, Steve K, Chris S, Eric S (2012) Design and performance of Astromesh reflector onboard soil moisture active passive spacecraft. IELCONF, IEEE-2012, pp 1–10
11. Liu Wa (2013) Research on nonlinear dynamic modeling and surface error characteristics of spaceborne large cable-network antenna structures. National University of Defense Technology
12. Yang Q (2015) The deployment process analysis of loop truss antenna. TianJin University
13. Johnson Michael R (1994) The Galileo high gain antenna deployment anomaly. NASA 94N33319
14. www.russianspaceweb.com/luch5a.html.
15. INTERFAX.Kondor—E satellite launched from Baikonur reaches orbit. Russia&Cis Military Newswire (2014)

Chapter 7
Microwave Remote Sensing Satellite Payload Antenna

Abstratct Since microwave can penetrate clouds, rain and snow, microwave remote sensing can provide with all-weather all-time detection capability. This chapter provides design of passive remote sensing antennas including traditional antenna of radiometer and synthetic aperture radiometer antenna in detail. Moreover, active remote sensing antenna such as synthetic aperture radar (SAR) and scatterometer are also involved.

Remote sensing is the acquisition of information about an object or phenomenon without making physical contact with the object and thus in contrast to on-site observation. Satellite-to-Earth remote sensing obtains various pieces of information by using artificial Earth satellites on orbit through remote sensors, which are the payloads of the to-Earth remote sensing satellites. Currently, space has become a "dominant strategic position" for national security and interests. Space remote sensing satellite has become an important measure to control space, acquire space control and information control power in modern information warfare.

From different perspectives, the payloads used in Earth remote sensing spacecraft can be divided into the following six parts:

1. According to the principle of remote sensor used in payload, it can be divided into film camera, CCD camera, multi-spectral scanner, imaging spectrometer, scanning radiometer, microwave radiometer, synthetic aperture radar, microwave scatterometer, microwave altimeter, laser altimeter and laser scatterometer.
2. According to the received electromagnetic radiation spectrum of the target by the remote sensor used in payload, it can be divided into visible light, infrared, ultraviolet and microwave remote sensors.
3. According to the imaging ability of the remote sensors used in payload, it can be divided into imaging and non-imaging remote sensors. Non-imaging remote sensor refers to payloads such as missile early warning, electronic reconnaissance (including reconnaissance of ground radar and communication signal) and radar altimeter.
4. According to the applications of remote sensor used in payload, it can be divided into military remote sensor, including military imaging reconnaissance (low-resolution sensing and high-resolution sensing), electronic reconnaissance,

R. You et al., *Technologies for Spacecraft Antenna Engineering Design*, Space Science and Technologies, https://doi.org/10.1007/978-981-15-5833-7_7

missile early warning, ocean surveillance (surveillance warships, aircraft carriers and submarines), military mapping and military meteorology, etc.; and civil remote sensor, including Earth resources exploration, meteorological observation, ocean observation, environmental monitoring, natural disaster monitoring, geographic mapping, etc.

5. According to the radiation status of remote sensor used in payload, it can fall into active mode (such as microwave altimeter, microwave radar, synthetic aperture radar) and passive mode (such as camera, scanner, imaging spectrometer, microwave radiometer, electronic reconnaissance payload).
6. According to the satellite orbit, it can be divided into geosynchronous orbit (such as the U.S. missile early warning satellite DSP, the U.S. electronic reconnaissance satellite Magnum, China's meteorological satellite "FY-2" and "FY-4"), solar synchronous orbit (such as most transmission remote sensing satellites in 500–800 km orbital altitude), low-orbit constellation (such as the payload of the US missile early warning satellite SBIRS) and very low orbit (such as the film cameras which are the payloads of the early US, Soviet Union and China's returning satellites).

Active mode remote sensor detects the reflection response of the object to artificial radiation (such as radar electromagnetic wave); passive mode remote sensor detects the electromagnetic wave spontaneously emitted or scattered (reflected) from the surface of the object.

In the remote sensing technology system, visible light and infrared remote sensing are the earliest and most widely used electromagnetic remote sensing. Compared with optical remote sensing, microwave remote sensing can penetrate clouds, rain and snow, and has all-weather detection capability; microwave remote sensing can work all the time with its microwave signal independent of sunlight; since microwaves can certainly penetrate the ground objects, microwave remote sensing can detect ground objects that cannot be detected by visible light and infrared rays. Microwave remote sensing has two forms: active and passive. It can not only receive and record the amplitude and phase information of the reflected wave of the measured object, but also measures its height and change by interferometry. This is beyond the reach of optical remote sensing. Microwave remote sensing is an essential complement to optical remote sensing, but it cannot be completely replaced by optical remote sensing.

7.1 Technical Characteristics of Microwave Remote Sensing Satellite Payload Antenna

Microwave remote sensing payloads can be classified into two groups: passive and active. Passive remote sensor, such as radiometer, mainly measures the microwave temperature of targets; active remote sensor, such as Synthetic Aperture Radar (SAR), can obtain two-dimensional or three-dimensional (interferometric SAR) images of targets. Payload antenna of microwave remote sensing satellite is very different from platform antenna. Compared with satellite-borne communication antenna, it has the following characteristics that the indicator of the antenna is closely related to the microwave remote sensing payload indicator even user's indicator:

1. Diversity of detection targets. In addition to acquiring remote sensing information such as soil and vegetation, microwave remote sensing payloads also detects meteorological, oceanic and atmospheric components, oxygen and hydrogen molecules in the universe, and cosmic microwave background.
2. The frequency of microwave remote sensing payloads almost covers the whole wireless spectrum (up to several terahertz), such as the satellites Planck and Hershell with the highest frequency of 850 GHz launched by the European Space Agency (ESA) in 2009, so higher requirements are put forward for antenna accuracy and structural stability.
3. The equivalent temperature of antenna is very crucial for passive microwave remote sensing satellite. Therefore, such antennas should avoid the radiation source which causes the temperature change of antenna, e.g., sunshine and moon, and some need special shielding layer to ensure that the antenna is in low-temperature state.
4. This antenna features a large electrical size. Compared with TT&C and communication antennas, the remote sensing antenna is basically an electrically large aperture antenna. Generally, D/λ is greater than 100, and the absolute size D of the aperture is also very large. For example, as a SAR antenna, planar phased array antenna is longer than 10 m, or the aperture of parabolic antenna is greater than 6 m. In order to cut down the complexity of antenna in some applications, interferometric synthetic aperture microwave remote sensing antenna has been developed. The effect of a large real aperture antenna is obtained using pairwise correlation convolution calculations of the antenna elements and requiring a high-speed processor, but at the cost of measuring the amplitude and phase for each element.
5. In addition to large size, most antennas are required to have beam scanning capability. Whether electronically controlled beam or mechanically scanned beam, the design complexity is increased on the premise of ensuring reliability. In 2011, China launched the microwave remote sensing satellite HY-2 to measure the ocean wind field, ocean surface temperature and ocean wave height. The working principle and design method of the satellite-borne radiometer and scatterometer antenna of HY-2 are taken as an example below.

7.2 Ocean Microwave Radiometer Antenna

The driving force of the rapid development of ocean remote sensing is that scientists expect more convenient and fast access to obtain ocean information such as sea surface wind field, sea surface height, sea salinity and tide, so as to apply them to the related research of ocean science and even global environmental change. Before the 1980s, the above-mentioned factors were mainly acquired by means of surveying ships, sea buoys and aircraft, which were inefficient, costly and often difficult to meet the needs of large-scale marine scientific research due to the scattered data obtained. Moreover, in some areas, such as the North and South Poles, and storm surge centers, these measurement methods are difficult to carry out. Satellite-borne ocean remote sensing with high spatial resolution, high temporal resolution and the ability of continuous observation overcomes the weaknesses of traditional data acquisition methods. In this context, spaceborne microwave radiometer and scatterometer came into being.

7.2.1 Ocean Observation Microwave Radiometer

As an important payload of China's ocean dynamic environment satellite, ocean observation microwave radiometer has two major functions: one is to obtain the parameters of ocean surface wind speed and direction, sea surface temperature, atmospheric cloud water content and water vapor content; the other is to provide calibration parameters for other microwave remote sensing such as radar altimeter and microwave scatterometer.

As shown in Fig. 7.1, the main observation beam of the microwave radiometer has a fixed angle α (called the incident angle or zenith angle, which is related to the sea surface microwave radiation emissivity, generally assumed as 40° to 50°) with the sub-satellite point. The observation beam performs both a conical scanning motion (rate υ) around the sub-satellite point and a linear scanning motion along the motion

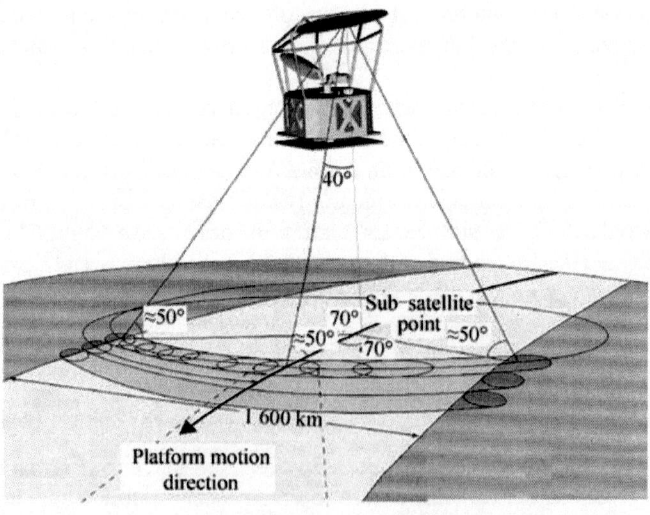

Fig. 7.1 Observation principle of ocean microwave radiometer

direction of the satellite platform. Suppose that the area with a certain angle along the direction of motion is the sea surface observation area ($\pm\beta$), the observation swath is defined as the strip region of the beam along the direction of motion in the range of $\pm\beta$. The seamless observation of the sea surface within the swath can be realized by choosing an appropriate υ and beam width.

Ocean microwave radiometer receives microwave radiation from the Earth's surface, and gets the brightness temperature of the Earth's surface through data processing. Reliable brightness temperature of ocean surface can be measured by accurate calibration of the microwave radiometer.

At present, the common calibration method of microwave radiometer is two-point method, that is, the cold calibration source and the hot calibration source are observed once in each observation period, and the calibration equation curve is formed. The accurate brightness temperature data of the ocean can be obtained by the corresponding calculation with the observed data of main observation antenna and the equation curve. The cold calibration source is selected from the cosmic background, whose radiation is very stable. The temperature observed by NASA satellite COBE is about 2.72 K. The hot calibration sources use radiators with known temperatures.

Ocean exploration microwave radiometer is classified as passive microwave remote sensing detector with typical multi-band common antenna. The classical ocean exploration microwave radiometer works in five frequency bands: 6.925, 10.7, 18.7, 23.8 and 37 GHz. In recent years, the detection frequency has gradually been developed toward millimeter wave and sub-millimeter wave. In general, ocean microwave radiometer antenna is composed of a main observation antenna, a cold calibration antenna, a rotating mechanism and a high-frequency box carrying the above components. Main observation antenna and cold calibration antenna share a common feed, as shown in Fig. 7.2.

The ground microwave radiation is received by the Earth observation antenna, the cosmic background microwave radiation by the cold space calibration antenna, and the microwave radiation of the hot radiation source directly by the feed. The antenna rotates continuously driven by the rotating mechanism, and every 360° is an observation period. In each observation period, the Earth observation antenna performs Earth observation once, and hot and cold sources are calibrated, respectively.

Microwave radiometer is required to rotate 360° continuously at a uniform speed (such as the payload of HY-2). The microwave signals received by microwave radiometer are transmitted to the payload module for data processing. Since the number of the microwave signal frequency and the occupied channel are relatively large, the multi-channel microwave signal transmission cannot be realized only by the rotary joint. Therefore, the microwave signal received by the microwave radiometer is converted into the medium or low-frequency signal through the signal processor and the transducer, and the data are exchanged with satellite payload module through the shunt ring and packaged for downlink transmission. Therefore, the microwave radiometer not only includes the antenna assembly, but also integrates the receiver and the medium–low-frequency processing equipment. It is structurally designed as a whole to realize on-orbit continuous Earth observation driven by the driving mechanism.

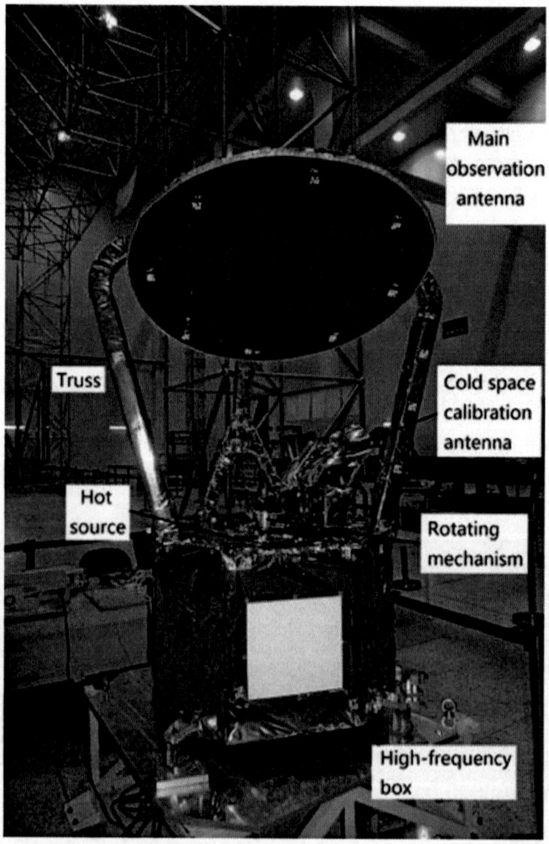

Fig. 7.2 China ocean exploration microwave radiometer

7.2.2 Design of Earth Observation Antenna

Microwave radiometer is a passive microwave remote sensing detector, which receives the weak microwave radiation energy from the ground to retrieve the brightness temperature of the ground. Because the ground microwave radiation energy is very small, it is sensitive to interference from other sources, which requires the stray energy entering through the antenna side lobe to be as small as possible. Therefore, the main beam efficiency is the most important parameter of the microwave radiometer antenna, which represents the proportion of the main beam energy of the antenna to all the energy received by the antenna. Generally, the main beam energy of the antenna is required to be more than 90%.

Microwave radiometer uses the received microwave radiation energy in specific directions of the ocean surface and the difference between vertical polarization and horizontal polarization energy to retrieve the brightness temperature and roughness of the ocean surface. Therefore, the cross-polarization performance of the Earth observation antenna must be considered.

The 3 dB beam width of the antenna is related to the detection sensitivity. The narrower the beam width, the higher the detection sensitivity is. However, greater rotational speed of the antenna is required to ensure seamless observation within the detection swath.

The formula (spherical coordinate system) for calculating the efficiency of the main beam of the Earth observation antenna is

$$\eta = \frac{p_{\text{co}}(\theta_1)}{p_{\text{co}}(\pi) + p_{\text{xp}}(\pi)} \times 100\% \tag{7.1}$$

$$p_{\text{co}}(\theta) = \int_0^{\theta} \int_0^{2\pi} |G_{\text{co}}(\theta, \varphi)|^2 \sin\theta \, \mathrm{d}\theta \, \mathrm{d}\varphi \tag{7.2}$$

$$p_{\text{xp}}(\theta) = \int_0^{\theta} \int_0^{2\pi} \left|G_{\text{xp}}(\theta, \varphi)\right|^2 \sin\theta \, \mathrm{d}\theta \, \mathrm{d}\varphi, \tag{7.3}$$

in which p_{co} is the main polarization energy of the beam, p_{xp} is the cross-polarization energy of the beam, p_{co} (θ_1) is the main beam energy, and angle θ_1 is the main beam width, it is defined as the angle interval between first nulls. In order to avoid the null ambiguity caused by the absence of sharp nulls, a better approximation is to use 2.5 times the beam width of 3 dB as the main beam width for the beam of reflector antenna.

$$\theta_1 = 2.5\theta_{3\text{dB}}. \tag{7.4}$$

As shown in Fig. 7.3, the Earth observation antenna is a single-offset standard parabolic reflector antenna. The focal diameter ratio of the antenna should be large enough to improve the cross-polarization performance of antenna under the consideration with the limitation of installation of satellite platform and the carrier fairing cover. Generally, the reflector projection aperture is about 1 m and the focal length is about 0.85 m. The narrower beam width is about 0.7° (37 GHz) and the wider beam width is about 3.1° (6.9 GHz), which can meet the requirements of detection sensitivity of microwave radiometer. The cross-polarization of the worst section under this condition is less than −20 dB, which meets the application requirements.

Main observation antenna reflector and high-frequency box are fixed by truss. When microwave radiometer antenna works, the main reflector rotates together with high-frequency box and feed, which drives the main observation beam to rotate. The rotation axis of high-frequency box points to the Earth center. The angle between the electric axis of main observation antenna and the rotation axis of high-frequency box is the observed incidence angle α.

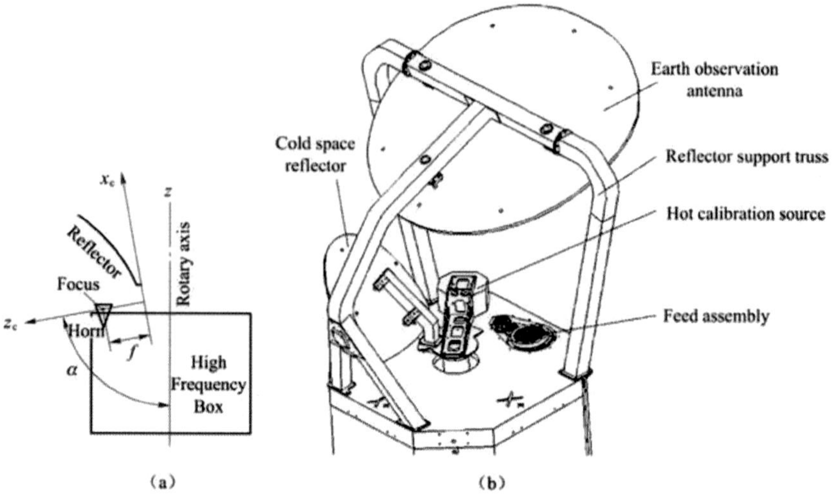

Fig. 7.3 Schematic diagram of Earth observation antenna. **a** Principle; **b** structure

It is necessary to consider the blockage of cold calibration antenna reflector on the main reflector in antenna simulation.

Reflector unit includes reflector with carbon fiber aluminum honeycomb structure and support truss with carbon fiber layered tubular structure, the later is supported on high-frequency box, to ensure relatively correct position relationship between reflector and feed. This design helps to reduce the weight of antenna reflector, and maintain sufficient support stiffness of reflector.

7.2.3 Design of Cold Calibration Antenna

The position of cold calibration antenna reflector is required to be fixed because its function is fixed point and calibration. When the reflector and feed of the Earth observation antenna are turned directly above the cold space reflector by the driving mechanism, the cold space reflector obscures the Earth observation antenna reflector. The cold space reflector and the feed constitute a cold calibration antenna. The antenna beam points to the cold space for low-temperature calibration, as shown in Fig. 7.4.

Because the polarization characteristics of cosmic microwave background radiation in all directions are identical, the cross-polarization and beam shape are not considered in the design of cold calibration antenna as long as it can effectively receive the energy of cold-space radiation. The cold-space antenna should take into account the main beam efficiency, which is more than 85%. The width of the cold-space beam is less than 12°, and the incidence angle of the beam for the Earth

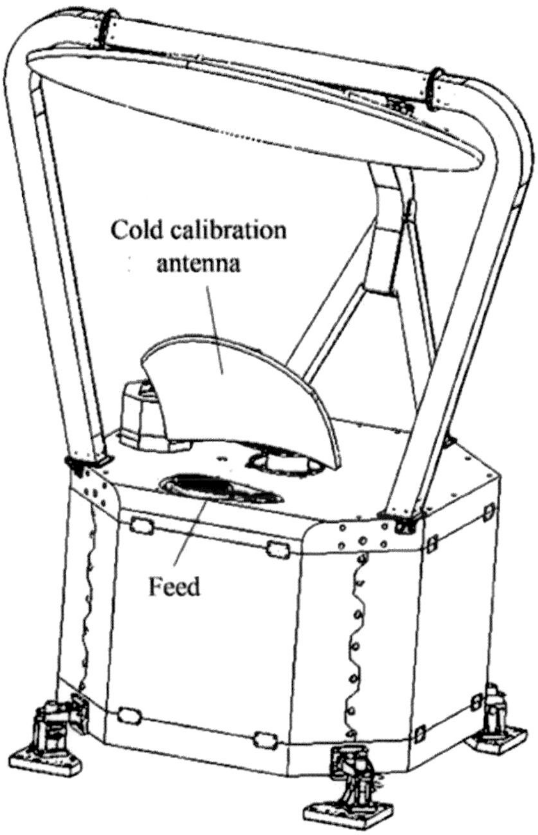

Fig. 7.4 Cold calibration antenna of ocean microwave radiometer

observation is more than 80°, so as to ensure that the main beam of the cold calibration antenna can effectively point to the cold space and avoid the influence of the Earth.

Ideally, the cold calibration antenna receives only the energy from the cold space and the microwave radiation from other celestial bodies shall be suppressed. The sun and the moon are very far away from the satellite. From the perspective of the antenna, it can be considered that the radiation brightness temperature is uniform and undifferentiated. Its influence can be easily corrected according to the trajectory of the moon and the sun. The Earth is only a few hundred kilometers away from the satellite. The field angle of the satellite to the Earth is more than 100°, which is much larger than the beam width of the cold calibration antenna. If the side lobe of the cold calibration antenna is too high in the Earth direction, significant impact will be made on the cold calibration results. Therefore, restraining the side lobes of the cold calibration antenna pattern in the Earth direction is the key to the design of the cold calibration antenna. Generally, the normalized toward-Earth side lobe of cold space antenna should be less than −20 dB.

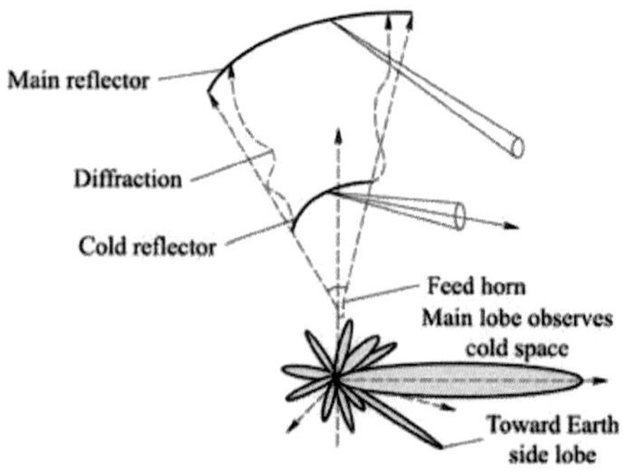

Fig. 7.5 Schematic diagram of toward-Earth side lobe formation of cold calibration antenna

7.2.3.1 Cause Analysis of Toward-Earth Side Lobe of Cold Calibration Antenna

The detector head of microwave radiometer mainly includes main observation antenna reflector (main reflector), cold calibration antenna reflector (cold reflector), high-frequency box and feed horn. Both cold calibration antenna and main observation antenna are single-offset parabolic. The cold reflector and the main reflector share a common focus. The cold reflector is fixed on the rotating axis of the high-frequency box. When the high-frequency box, the main reflector and the feed rotate, the cold reflector remains motionless, and the electric axis always points to the cold space.

When the microwave radiometer works, the main reflector rotates around the rotating axis together with the feed and the high-frequency box, and the cold reflector is fixed on the rotating axis. The feed will enter the focus of the cold reflector in turn, and the radiometer receives the cold space signal.

The main reflector is located on the back of the cold reflector when the cold calibration antenna is working. Because of the directly leaked radiation and edge diffraction, the energy radiated by the feed is not completely intercepted by the cold reflector, and some of radiated energy hits the main reflector and forms radiation. The radiation pattern on the main reflector is superimposed on the radiation pattern of the cold reflector, forming parasitic radiation. Because the electric axis of the main reflector points to the Earth, most of the energy intercepted by the main reflector must radiate to the Earth, forming the toward-Earth side lobe of the cold calibration antenna, as shown in Fig. 7.5. Due to the large aperture and high gain of the main reflector, even less energy can cause stronger radiation to Earth.

7.2.3.2 Measures to Suppress the Toward-Earth Side Lobe

It can be seen from the forming reasons of toward-Earth side lobe of cold calibration antenna that the toward-Earth side lobe is contributed by the main reflector. The directly leaking wave from the edge of the cold reflector on the main reflector is a

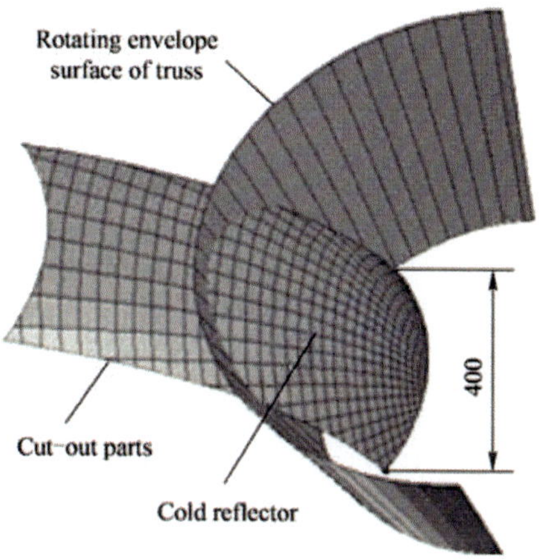

Fig. 7.6 Formation of cold reflector

contributing source to the toward-Earth side lobe. After the feed is designed, the best way to suppress this part of energy is to increase the size of the cold reflector.

Because the cross-polarization characteristics do not need to be considered in designing the cold calibration antenna, so it doesn't matter whether the shape of cold reflector is regular, if it does not interfere with other structures. However, from the perspective of main observation antenna, it is expected that the cold reflector is small to avoid the negative influence on the radiation pattern of the main observation antenna when the main observation antenna is working. Therefore, the choice of the aperture and surface of the cold reflector is based on the compromise of the above considerations.

Enough safe clearance must be provided to ensure that high point of the cold reflector does not interfere with the support truss of the main reflector. The detailed method is to translate the support truss nearest to the cold reflector to the rotation axis by about 10 mm, rotate around the rotation axis to form a surface, and then the cold reflector (parabolic surface) is cut by this surface, as shown in Fig. 7.6. The cold reflector formed in this way maintains a safe clearance of at least 10 mm from the support truss everywhere during rotating, and can maximize the aperture of the cold reflector.

7.2.4 Feed

Generally, ocean microwave radiometers are multi-band, and the classical bands include C, X, Ku and Ka. At present, sub-millimeter band ocean microwave radiometer is being developed. The polarized Orthogonal Mode Transducer (OMT) + horn (corrugated or slick-wall shaped horn) is used as the feed. The arrangement of the feed horns is the most important in the feed design. Most of the electrical interfaces are waveguide or radio frequency coaxial plugs. The appearance of the feed is shown in Fig. 7.7.

The basic concepts of horn arrangement design are: (1) to ensure that multiple horns do not occlude each other, (2) and the maximum radiation direction of the horn basically points to the center of the main reflector.

To ensure that multiple horns do not occlude each other, it is necessary to ensure that all surfaces of horn opertures are flush, this is a necessary condition for the arrangement of horns. In order to obtain a better beam shape, the lateral offset-focus of the horn must be as little as possible, preferably with only the necessary structural clearance between the horns. Because of the close distance between the horns, if one horn exceeds the other two, it is bound to block the other two. Because the size of low-frequency horn is larger than that of high-frequency horn and medium-frequency horn, if the low-frequency horn exceeds the other two, it will cause serious occlusion. In order to ensure that the three horns do not occlude each other, in general, higher frequency horns should be as close to the focus of the reflector as possible. The aluminum alloy feed assembly is easy to be processed by the mature processing technology, such as WEDM and EDM.

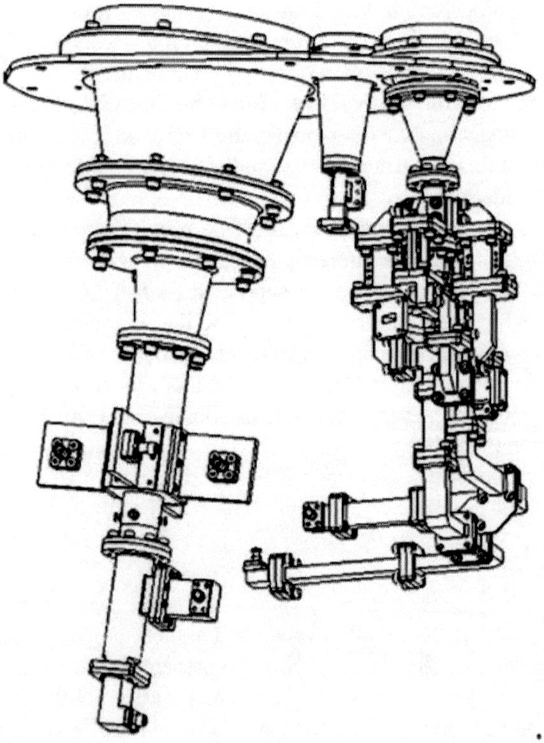

Fig. 7.7 Feed appearance

7.2.5 Scanning Mechanism

Scanning mechanism, an important component of antenna, is used as a driver of microwave radiometer antenna to receive microwave radiation from Earth.

The on-orbit rotation of remote sensing antennas features:

1. Starting acceleration time as long as possible, starting acceleration as small as possible, an adjustable acceleration time and a relatively small starting torque;
2. Stable and uniform on-orbit working state, no acceleration and deceleration, only overcoming the friction moment of the mechanism itself in normal operation.

There is a wide range of mechanism used in antenna. At present, motor and reducer assembly is widely used in the aerospace, which can obtain larger driving moment. Reducers can be further divided into gear reducer and harmonic reducer. Because the motor directly drive the reducer, the motor must run at a very high-rotation speed for rapidly scanning antenna used in remote sensing satellite, and the motor life is required up to tens of billions rotations. So, it is very difficult to develop the motor.

Based on the requirement of motor life and the working characteristics of remote sensing antenna itself, the advantages of the scheme that the scanning mechanism directly drives the antenna without the reducer are significant. There are only bearings and no other contact parts between the rotor and the stator of the mechanism itself to support and rotate. The life of the mechanism depends on the bearings. Therefore, the lifetime of the scanning mechanism is effectively improved to satisfy the requirement of on-orbit lifetime of remote sensing satellites.

The scanning mechanism acts both as a driving component to drive the antenna to rotate, and as an installation support of the cold reflector and the hot calibration source due to the intermediate shafting fixed relative to the satellite, thus achieving the positioning on-orbit relative to the Earth observation antenna. The installation positions of the cold reflector and the hot calibration source on the scanning mechanism are shown in Fig. 7.8.

The scanning mechanism consists of motor, servo resolver, conductive slip ring, bearing and fixed parts, as shown in Fig. 7.9. The motor and resolver of the scanner mechanism with the separate-mounted structure have no individual bearings, they both share bearings with the mechanism, which reduces the usage of bearings, and facilitates the assembly of the mechanism, to improve the rotational reliability of the mechanism. Torque motor with the outer rotor structure belongs to a brushless motor. The outer rotor drives the antenna to rotate uniformly. The middle stator is fixed relative to the satellite. Since the power source of the scanning mechanism motor is provided by the satellite, the motor stator is designed as a motor winding. Slip ring channel is saved because the input of motor power does not participate in rotation. Similarly, the resolver used to measure rotation angle information also uses inner stator windings and outgoing wires, which not only acts as the original side of the transformer to accept the excitation voltage, but also acts as the secondary side of the transformer to obtain the induced voltage through electronic coupling; the rotor part only has windings, which can induce the excitation voltage, at the same

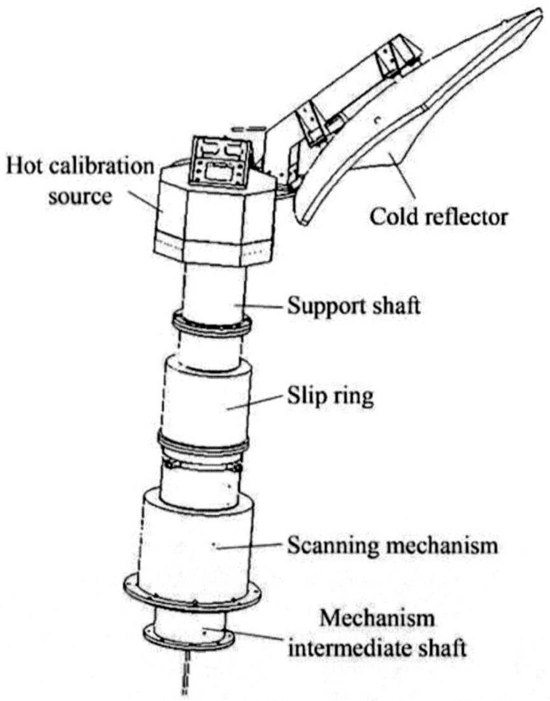

Fig. 7.8 Installation position of scanning mechanism

time couples the induction voltage with the stator and outputs through the stator lead wire. The middle spindle is not only the supporting structure of antenna and high-frequency box, but also the bearing part of the whole rotating mechanism. The end face of the spindle is installed on the satellite deck. The top part is a conductive slip ring, the middle part is a resolver, and the bottom part is an electric motor and a bearing which supports and bears the load.

The slip ring (also known as conductive sliding ring) requires a very long rotational life (more than 50 million rotations). In China, there is no flight-proven development experience and products of such slip rings, thus it is mainly relying on imports. With the increasing demand for long-life slip ring, some Chinese manufacturers or research institutes of slip ring have started the R&D work, and have achieved some success and experience. At present, the slip-rings used in microwave radiometer antennas and other remote sensing antennas are mainly dependent on imports.

The calculation parameters are as follows:

The friction moment of single bearing: 0.05 N m;

The total friction moment: about 0.2 N m (considering the rotational resistance of resolver);

The friction moment of the slip ring: 0.2 N m;

The required moment for generating acceleration: 0.03 N m;

The calculated total resistance moment: 0.43 N m;

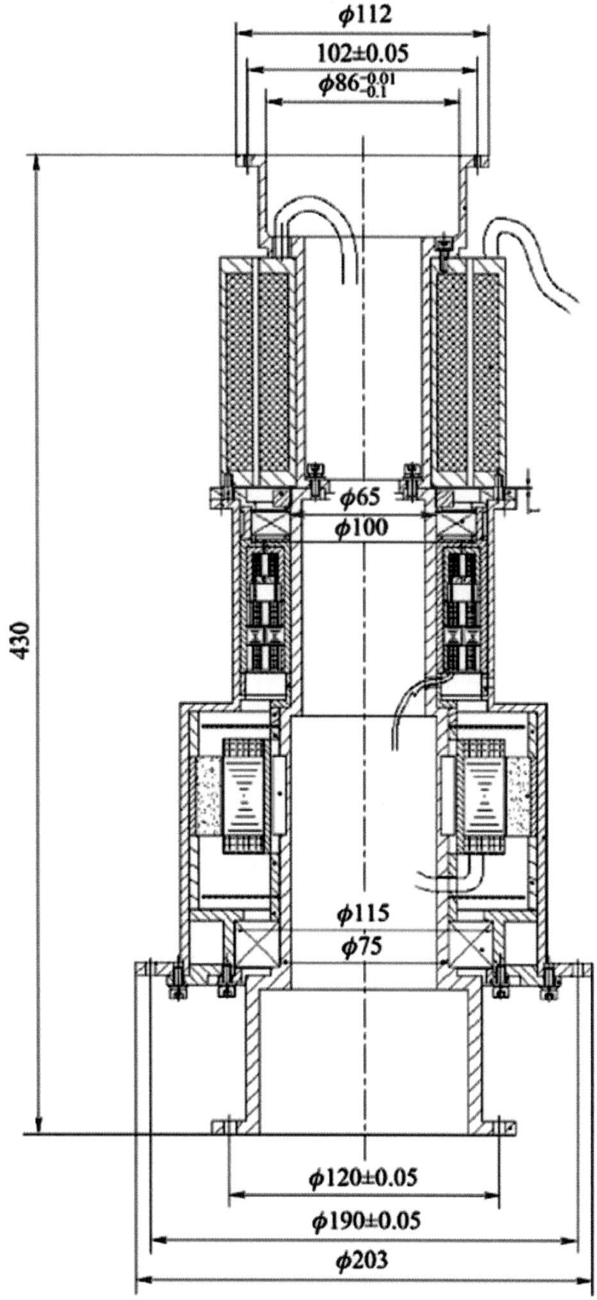

Fig. 7.9 Structure of scanning mechanism

The output torque of the motor: 2 N m;

The driving margin design requirement of the scanning mechanism is greater than 3.

The calculation method of driving margin of scanning mechanism is as follows:

Safety margin of mechanism design can be identified by dynamic ratio coefficient *R*, i.e.,

$R =$ total power moment/total resistance moment $= (2\ \text{N m})/(0.43\ \text{N m}) = 4.65$, which meets the requirements.

7.2.6 Lock-Release Device

In order to realize the antenna locked in the launch state and the free axial rotation of antenna after release on orbit, the lock-release device should be properly arranged. Because of the continuous rotation of the antenna, after release, the lock-release device should not interfere with the rotating part of the antenna, so that the traditional device cannot achieve this function. Therefore, the flipping plate, a new type of structure, is used to separate two parts of the lock-release device through flipping a certain angle under the drive of the torsion spring after release, thus ensuring the rotation of the antenna.

The lock-release device compresses the antenna on the module of the satellite in the launch stage. The antenna receives the remote control signal of pyrotechnical device after on-orbit, and detonates the cutting device of pyrotechnical device to unlock the antenna. The locking and releasing states are shown in Fig. 7.10.

The other components, such as high-frequency box, are customized. Different microwave radiometers have different component configurations because of different antenna operating frequencies, geometric sizes and functions, which will not be discussed here in detail.

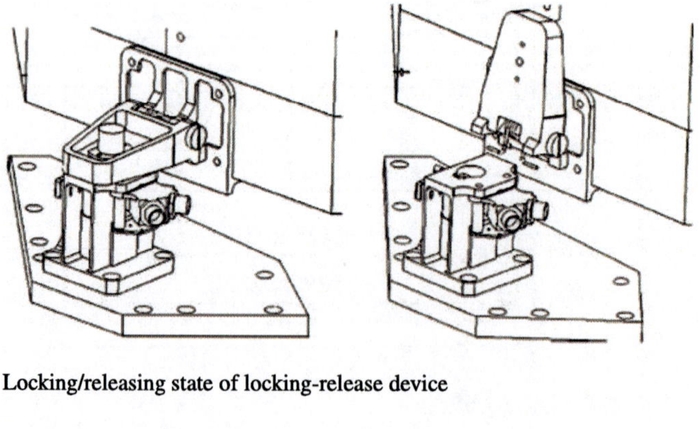

Fig. 7.10 Locking/releasing state of locking-release device

7.3 Ocean Microwave Scatterometer Antenna

Ocean satellite microwave scatterometer is an active radar system, which uses different responses of sea surface roughness to radar backscattering coefficient under different wind speeds and indirectly retrieved sea surface wind field information from multi-angle observations. In addition, the scatterometer can also be used for soil water retrieval, vegetation cover and sea ice change monitoring. The distribution of wind speed determines the distribution of wave height and the propagation direction of ocean surges, and predicts the impact of surges on ships, offshore buildings and coastal zones. Therefore, the monitoring of sea surface wind field is very important for understanding the interaction between ocean and atmosphere and carrying out related research in the field of ocean and atmosphere. Spaceborne microwave scatterometer is the only satellite sensor system that has been proved to be able to detect wind speed and direction at the same time. Since the all-weather, all-day and high-coverage observation capabilities of the scatterometer, it plays an important role in obtaining information on the sea surface wind field. The acquired wind field information is widely used in weather forecasting, storm surge monitoring, etc. Continuous scatterometer observation data over the past 20 years are used as an auxiliary to other high-resolution sensors (such as SAR) data, and can be used to monitor changes in vegetation, soil water, sea ice and other fields over a long time scale.

7.3.1 Principle of Ocean Microwave Scatterometer

Ocean microwave scatterometer is an active and non-imaging radar system, which detects the information about the target by transmitting microwave pulse signals to the sea surface and land surface and receives the energy of the backscattered echo signal. The energy of the echo signal depends on the roughness of the target surface and the dielectric properties of the material itself. Both of them can be used to describe the characteristics of sea ice and vegetation cover.

The principle block diagram of ocean microwave scatterometer is shown in Fig. 7.11. The basic working process of ocean microwave scatterometer is as follows: the microwave transmitter is connected to the antenna through a duplexer (not depicted in the figure), and transmits RF pulse to the sea surface. The pulse signal is scattered by the target, and the scattered echo signal is received by the antenna, then sent to the receiver by the duplexer, and recovered to video signal by receiving and processing. Hence, the video signal is sent to signal processor. The internal calibration device couples a part of the transmitter power to the receiver to form a closed loop, so as to achieve internal calibration and eliminate the measurement error caused by any change in the transceiver system. The signal processor processes the echo signal, the internal calibration signal and the pure noise signal without echo to obtain the normalized radar backscattering coefficient σ^0 of the sea surface scattering source. The ground processing system calculates the sea surface wind speed and direction

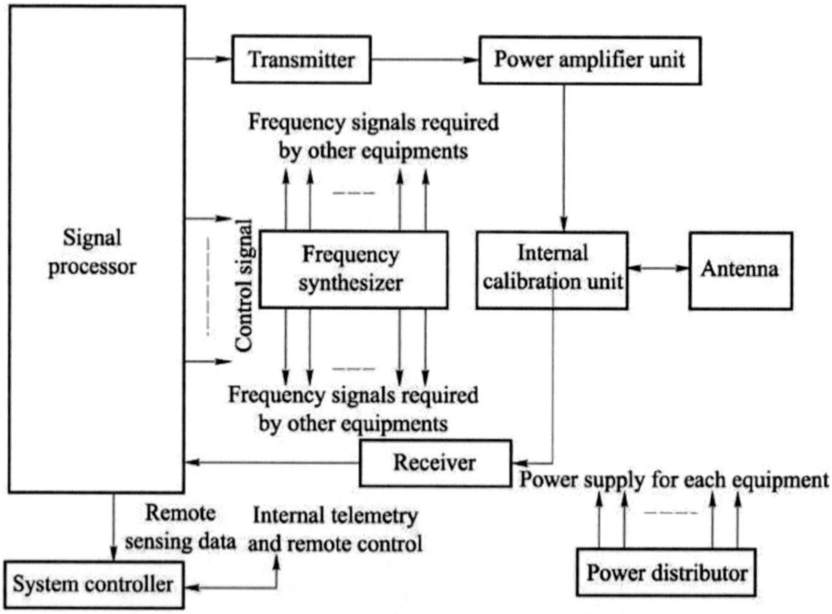

Fig. 7.11 Principle block diagram of ocean microwave scatterometer

through a certain mathematical model based on σ^0 measured at different incident angles, different azimuths and different antenna beams.

The basic principle of ocean microwave scatterometer for measuring sea surface wind speed and direction is that the sea surface capillary wave generated by sea surface wind field can produce Bragg scattering by centimeter-wavelength microwave, to which the corresponding small-scale orientation and slope distribution are related to the direction of wind. The amplitude of wind-induced capillary wave increases with the increase of wind speed, so does the intensity of Bragg scattering. Because wind-induced capillary waves can cause small-scale slope of sea surface, that is, the direction of scattering elements corresponding to Bragg scattering is related to the direction of wind, so that the intensity of backscattering is also related to the direction of wind. According to the relationship between σ^0 and sea surface wind speed and direction (e.g., Moore model and Wentz model describing this relationship), the sea surface wind vector can be obtained by measuring σ^0 of the same resolution unit from multiple azimuths and calculating according to model operation. Among them, wind speed measurement will be retrieved according to the value of radar scattering coefficient, while wind direction measurement needs scattering information of multiple incident azimuths. The principle and ability of the spaceborne scatterometer to measure the sea surface wind field have been fully demonstrated in the satellites of Seasat-1, ERS-1/2, ADEOS-1/2, QuickSCAT, Shenzhou-4 and HY-2.

7.3.2 Antenna System of Ocean Microwave Scatterometer

According to the principle of ocean microwave scatterometer for measuring sea surface wind field, the main requirement of the scatterometer is to obtain scattering coefficients of multiple incident azimuths based on certain measurement accuracy, so as to retrieve wind speed and direction. The ocean microwave scatterometer can be divided into two systems according to the antenna structure: fan beam system and pencil beam system. The realization of multi incidence measurement includes the following three systems:

1. Fixed fan beam system
   Fixed fan beams are generated by multiple rod antennas, as shown in Fig. 7.12.
   Advantages: No beam scanning, no servo and rotation mechanism, no disturbance to satellite attitude.
   Disadvantage: Complex antenna layout and a blind area in the sub-satellite point.

2. Pencil beam conical scanning system
   Conical scanning is performed using a point beam antenna, as shown in Fig. 7.13.
   Advantages: high-antenna gain, fixed incidence angle, compact equipment, small size, lightweight.
   Disadvantage: The antenna scans conically, which disturbs the satellite platform.

3. Fan beam conical scanning system
   A rod antenna is used to generate a narrow fan beam. During the operation, the beam is controlled to conically scan around the nadir axis, as shown in Fig. 7.14.
   Advantages: Wide swath, high overlap, no coverage blind area for measurement area.
   Disadvantage: The antenna scans conically, which disturbs the satellite platform.

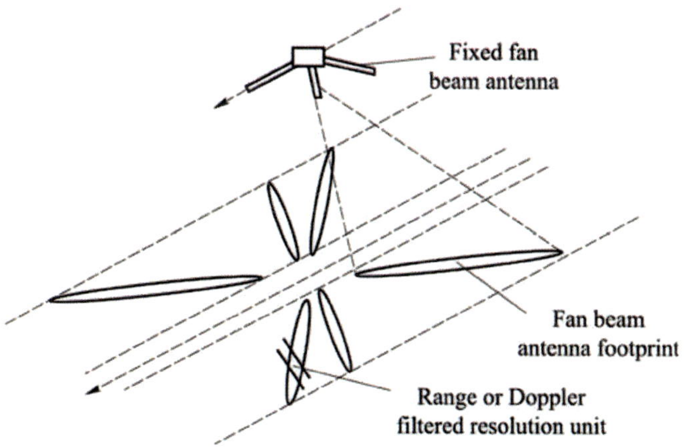

Fig. 7.12 Schematic diagram of fixed fan beam systems

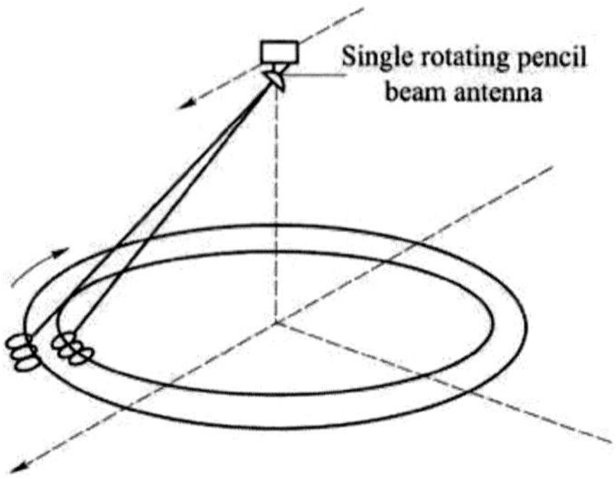

Fig. 7.13 Schematic diagram of pencil beam conical scanning system

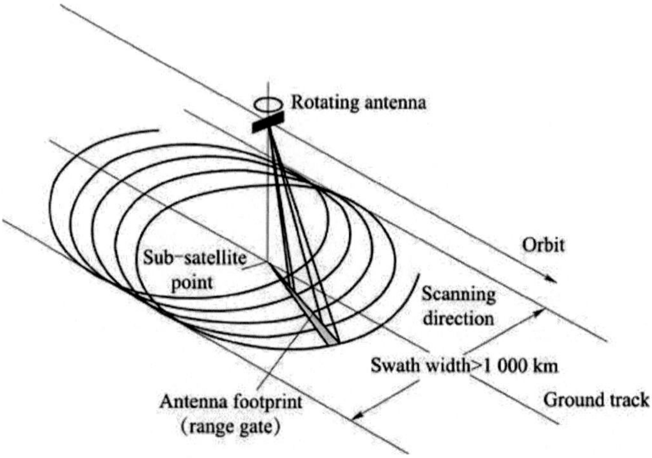

Fig. 7.14 Schematic diagram of fan beam conical scanning system

Fixed fan beam system and pencil beam conical scanning system have been applied in satellites, for example, the fixed fan beam system has been used in Seasat-1, ERS-1 and ERS-2, and NSCAT of ADEOS-1, the pencil beam antenna system has been used in the Seawinds scatterometers in U.S. QuickSCAT and Japan ADEOS-2, and the multi-mode microwave remote sensor in China spacecraft Shenzhou-4 and the microwave scatterometer in satellite HY-2.

The fan beam conical scanning scatterometer is a new type of microwave scatterometer, which combines the advantages of fixed fan beam system and pencil beam conical scanning system. The fan beam conical scanning scatterometer uses a rod antenna to produce a narrow fan beam, and controls beam to scan conically around the apex axis in the course of operation. With the motion of the satellite, the fan beam conical scanning scatterometer can measure the area with wide swath, high overlap and no coverage blind area on the sea surface. The size of scanning swath is determined by the maximum incident angle of the antenna beam. The coverage of antenna ground footprint (6 complete scanning cycles) during continuous scanning is shown in Fig. 7.15.

The main advantage of the fan beam conical scanner is that the number of observations of the measurement unit and that of independent samples are relatively large, so it has superior accuracy of wind retrieval and ability of wind direction ambiguity resolution. The observation area has a high overlap. During the period of the satellite flying over the target, the measurement units in some positions can be observed as many as 10–11 times, and the spatial resolution of the measurement units can be further improved using resolution enhancement processing technology. For example, the developing scatterometers of satellites CFOSAT and FY-3(05) adopt fan beam conical scanning system.

The main parameters of the three systems of the scatterometer are compared as shown in Table 7.1. The patterns of antenna are mainly pencil beam and fan beam. Pencil beam scanning antenna is actually a point beam scanning antenna. Therefore, parabolic reflector antenna is preferred, which has many advantages, such as high gain, mature technology, easy to achieve dual polarization, internal and external beams. Because the antenna pattern requires narrow at azimuth and wide at range, the waveguide slot array antenna is preferred for fan beam, which has also many advantages, such as easy control of the aperture field distribution, high-aperture efficiency, stable performance, simple and compact structure, low profile and high intensity, and easily achieving narrow beam, shaped beam, low side lobe and even ultra-low side lobe, low cross-polarization. So this form of antenna has become the preferred form of fan beam. The design theory of waveguide slot array antenna is relatively complete.

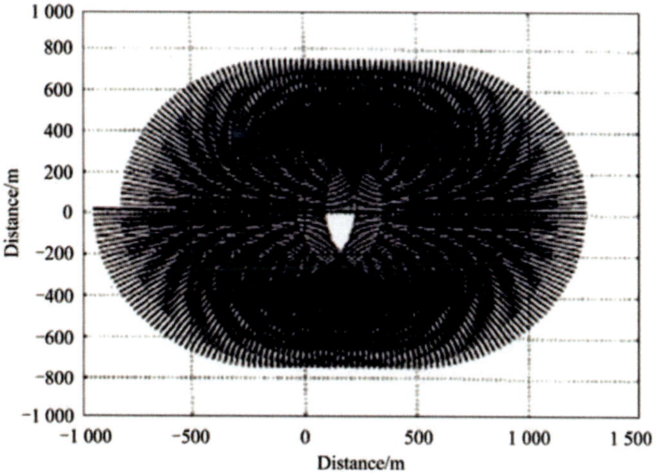

Fig. 7.15 Ground coverage of fan beam conical scanning scatterometer with multiple scanning periods

Table 7.1 Comparison of main parameters of microwave scatterometers with different systems

Parameter	Fan beam	Pencil beam	Fan beam scanning
Typical representative	NSCAT ASCAT	Seawinds Seawinds-1B	Ku scatterometer
Beam	Fan	Pencil	Fan
Swath	Narrower	Wide	Wide
Antenna form	6-rod antenna	Parabolic antenna	1-rod antenna
Antenna layout on the satellite	Complex	Simple	Simple
Power	Big	Small	Big
Scanning method	Do not scan	Conical scanning	Conical scanning
Observation azimuth	4	4	6 times on average
Mapping swath	There is a blind area on sub-satellite points	No blind zone	No blind zone

7.3.3 Examples of Ocean Microwave Scatterometer Antennas

Scatterometers, which work mainly in C-band (5.3 GHz) and Ku-band (13.5 GHz), can be basically divided into fan beam scatterometer and pencil beam scatterometer according to scanning modes. For example, the scatterometer emitted by NASA uses Ku-band, while the scatterometer emitted by ESA uses C-band. C-band has a long wavelength and is less affected by cloud and rain, but it is not suitable for small satellite platforms. Ku-band has a high frequency, which is more sensitive to the change of target characteristics and is conducive to the detection of low-speed wind fields.

SASS scatterometer on the US Ocean Satellite Seasat-1 proves that the spaceborne microwave scatterometer can efficiently measure wind field from space to provide global, all-weather, high-precision, high-resolution and short-period sea surface wind data. SASS scatterometer has four dual-polarization antennas transmitting fan beams. Four active beams are directed at satellite trajectories along azimuth directions (±45°) and (±135°), and form cross-shapes to cover the swath with each side of 475 km and a gap of 400 km. Each beam with a width of about 0.5° and a length of about 25° illuminates the Earth surface at an incidence angle of 25° to 55°. The repetitive observation period is 3 days. Because SASS scatterometer uses two polarized antennas for a single swath, it can measure the same area on the sea twice in different azimuth with the sea surface resolution of 50 km, the wind speed measurement range of 4–26 m/s, and the accuracy of wind speed measurement of 2 m/s or 10% of the wind speed.

The microwave scatterometer of China's satellite HY-2 adopts a pencil beam conical scanning system based on reflector antenna, as shown in Fig. 7.16. It has the characteristics of wide observation range and no sub-satellite blind area. It can quickly observe the global sea surface and cover 90% of the global sea area every day.

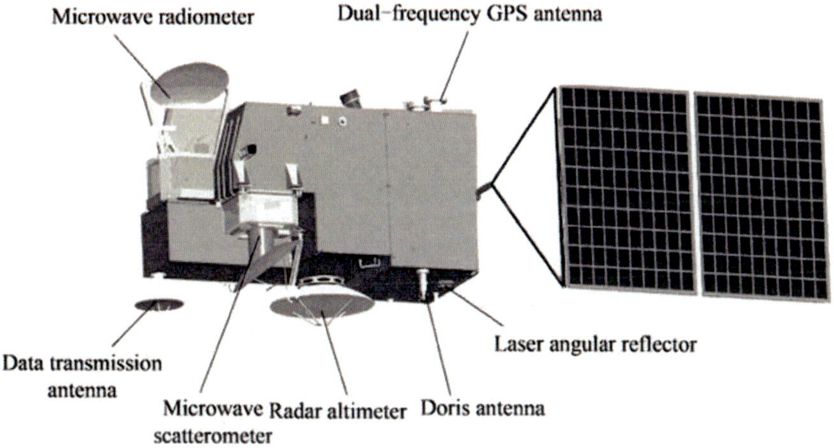

Fig. 7.16 Layout of satellite HY-2

7.4 Synthetic Aperture Microwave Radiometer Antenna

7.4.1 Principle of Synthetic Aperture Microwave Radiometer

The angular resolution of the traditional microwave radiometer depends entirely on the physical aperture of the antenna. Taking the satellite HY-2 launched by China in 2011 as an example, the antenna aperture of the microwave radiometer is 1 m, the beam width of −3 dB is 0.7° at 37 GHz, and the corresponding ground resolution is about 12 km. Therefore, to increase the resolution, the antenna aperture must be increased first. However, large aperture space antenna will lead to more complex structure. Inspired by the sparse array antenna of ground radio astronomical observation, Christopher et al. proposed a scheme of interferometric synthetic aperture microwave radiometer. Its basic principle is to use the sparse antenna array (Fig. 7.17) to obtain the equivalent resolution corresponding to the real aperture antenna size, in order to reduce the difficulty of design with large antenna. After pairwise correlation operation (visibility function) of different baseline elements, Fourier transform is performed on the visibility function of different baselines and the temperature distribution of the target is obtained.

The elements in the synthetic aperture array antenna can be arranged in cross shape, T shape, Y shape, or X shape, etc. The baseline is the distance between a pair of elements, so there are many baselines and correlators. The antenna system composition is as shown in Fig. 7.18.

The basic concept of synthetic aperture microwave radiometer is to transform the difficulty of large antenna into the complexity of circuit and data processing. The high resolution of the antenna depends on the longest baseline between the antenna elements, which is equivalent to the resolution of the real aperture antenna with the

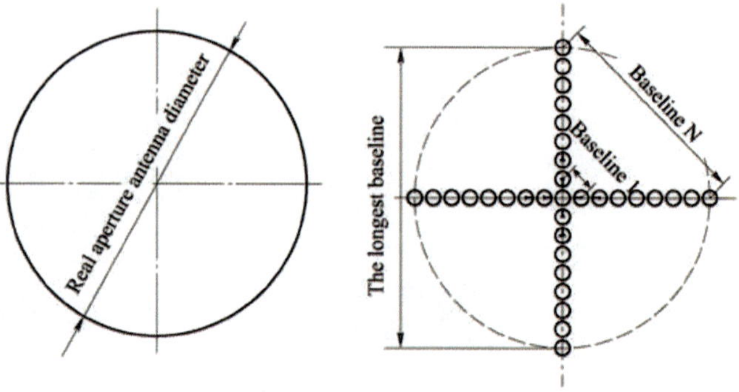

Fig. 7.17 Comparison of real aperture antenna (left) and synthetic aperture array (right)

Fig. 7.18 Block diagram of synthetic aperture array antenna composition

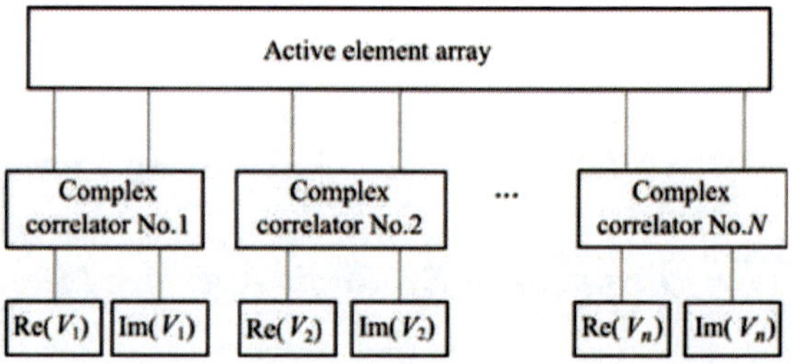

longest baseline as its diameter.

$$\theta_{-3\mathrm{dB}} = \frac{51^{\circ}}{D_{\max}}, \tag{7.5}$$

where $D_{\max}$ is the longest distance between two elements. The received voltage between two antenna elements is processed as shown in Fig. 7.18 to obtain the complex visibility function V. The relationship between the received voltage and the brightness temperature T of the target is as follows:

$$V(u, v) = \int_0^{2\pi}\int_0^{\pi} T_B(\theta,\ \varphi)\mathrm{e}^{\mathrm{j}2\pi(u\sin\theta\cos\phi + v\sin\theta\sin\phi)} \cdot \sin\theta\ \mathrm{d}\theta\ \mathrm{d}\varphi, \tag{7.6}$$

where $u = D_x/\lambda$, $v = D_y/\lambda$, and D_x and D_y are projections of baselines of two antenna elements on the X- and Y-axis in the plane coordinate system of the antenna, respectively.

The brightness distribution of the target can be obtained by inverse Fourier transform of Eq. (7.6).

7.4.2 Synthetic Aperture Radiometer Antenna

CAST Xi'An has developed a synthetic aperture radiometer, which will be installed on subsequent ocean satellites for ocean salinity remote sensing. The envisaged satellite is shown in Fig. 7.19. The developed synthetic aperture antenna is shown in Fig. 7.20. The antenna operates in L-band with dual-line polarization. Its main parameters are as the follows:

The longest baseline: 5.4 m;
The deployment accuracy of the antenna: $\leq \pm 0.5°$;
The element spacing: 0.82 λ;
The beam width of the element antenna: >63°;
The phase accuracy of the antenna pattern: $\leq \pm 3°$.

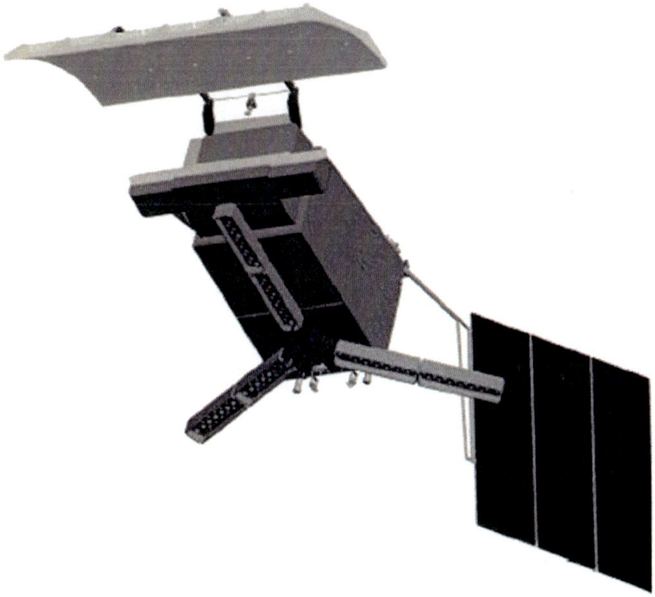

Fig. 7.19 Envisaged diagram of China subsequent ocean satellite with synthetic aperture radiometer

Fig. 7.20 Synthetic aperture radiometer antenna (the right picture shows the test status)

Theoretically, antenna arrays can adopt various configurations, such as Y-shaped, U-shaped, T-shaped, O-shaped and random-shaped, if baselines with different lengths can be formed. However, Y-shaped antenna array can cover the most sampling points in the spatial frequency domain with the least number of antenna elements, and is easy to install on the satellite, so the system uses Y-shaped antenna array.

The antenna consists of three foldable arms and a center array. Each antenna arm includes 16 antenna elements, and the center array includes 7 elements. So, there are 55 antenna elements.

The block diagram of the system is shown in Fig. 7.21. Antenna elements receive both horizontal and vertical polarization signals simultaneously. Receive Front Ends amplify and modulate the microwave signals to light signals. The light signals are transmitted into the center receivers in the payload carbin. Then the signal is divided into I- and Q-path by orthogonal demodulator. After band-pass filtering, the data is collected and output. The data acquisition bit is 1-bit, and the data are transmitted to the processor and operated correlatively.

Before imaging, noise signals are generated by standard noise sources, and injected into the system, then the stability of the amplitude and phase of the system is calibrated by the response of the complex correlator.

The data procession and retrieval calculation software is developed. The result is shown in Fig. 7.22. The imaging effect is satisfactory.

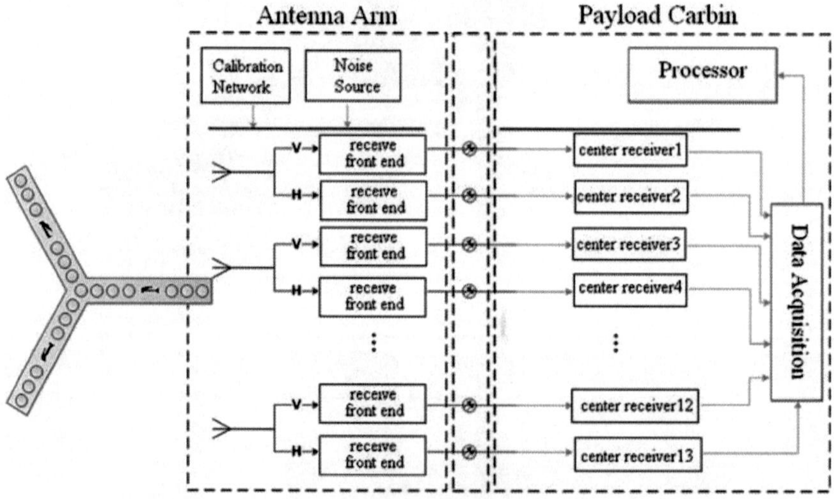

Fig. 7.21 Block diagram of synthetic aperture radiometer

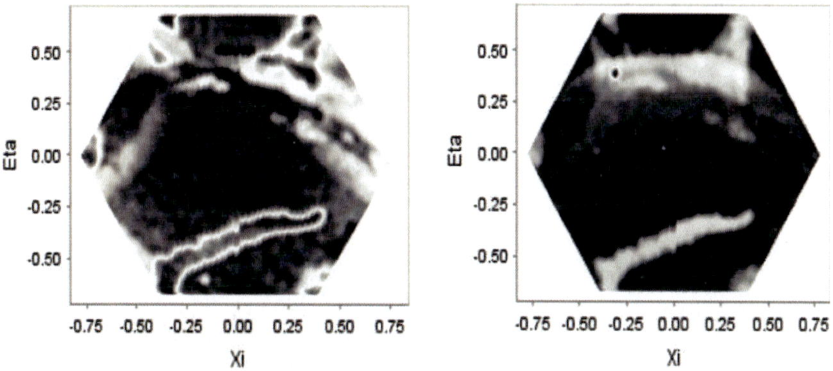

(a) (b)

Fig. 7.22 Retrieval simulation results of brightness temperature map. **a** V polarization; **b** H polarization

7.4.3 U.S. GeoSTAR

Geostationary Synthetic Thinned Aperture Radiometer (GeoSTAR) is developed by Jet Propulsion Laboratory (JPL) of the United States (Fig. 7.23). It takes advantage of the high-altitude advantages of geostationary orbit to continuously observe the Earth, which is significant for weather and meteorological observation and prediction, especially for effectively enhancing the ability to predict transient atmospheric phenomena such as hurricanes.

In order to cover the entire Earth's disk, the GeoSTAR with Y-shaped antenna arrays should meet the following requirements:

1. No aliasing field of view: 17.5°;
2. Minimum antenna spacing: 3.5 times wavelength (20 mm at 50 GHz, 6 mm at 183 GHz);
3. System geometric resolution: 50 km;
4. Maximum baseline length: 4.7 m at 50 GHz, and 1.1 m at 183 GHz;
5. The number of antenna elements on each antenna arm: about 100;
6. The total number of antenna elements: about 300, which can form about 30,000 different baselines, corresponding to 60,000 sampling points in spatial frequency domain (u, v).

Obviously, the requirements of antenna aperture, antenna spacing and antenna layout are different for different frequency bands. According to NASA's plan, observations at 50 and 183 GHz on one satellite are required by GeoSTAR. Therefore, it is necessary to design a dual-band composite antenna array.

The dual-band composite antenna array is improved based on the original Y-shaped array.

Fig. 7.23 50 GHz prototype [1]

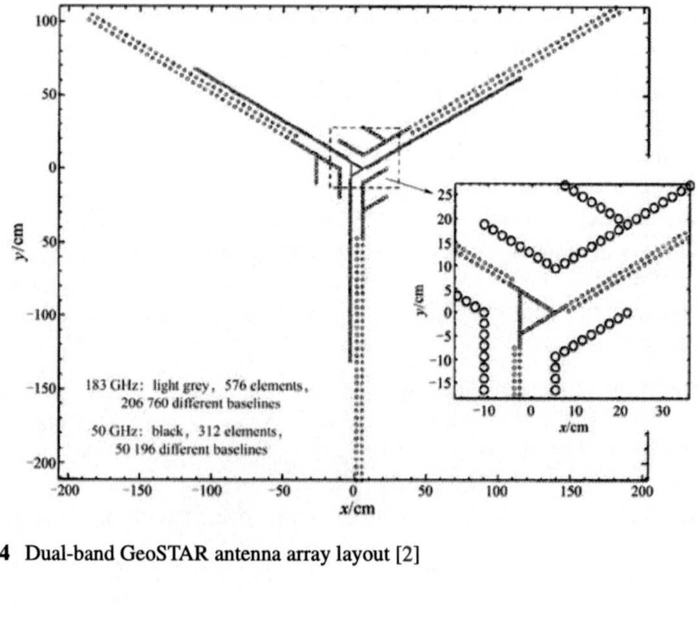

Fig. 7.24 Dual-band GeoSTAR antenna array layout [2]

1. Eight elements are moved out of the 50 GHz antenna array, leaving space for 183 GHz array.
2. The movement will result in the absence of sampling points in the spatial frequency domain (u, v), so eight antenna elements should be added to the 50 GHz array to fill in the absence of (u, v) sampling points, and, finally an F-shaped antenna array is formed.
3. 183 GHz arrays are embedded in the center and keep Y-shape in dislocation arrangement.
4. The whole antenna array consists of 206,760 different baselines with 576 antenna elements of 183 GHz, and 50,196 different baselines with 312 antenna elements of 50 GHz.

The dual-band GeoSTAR antenna array layout is shown in Fig. 7.23.

7.4.4 European SMOS

The European Space Agency (ESA) Soil Moisture and Ocean Salinity (SMOS) explorer was successfully launched in November 2009. It is the first satellite in the world to observe global soil moisture and ocean salinity.

The payload of SMOS is L-band Microwave Imaging Radiometer using Aperture Synthesis (MIRAS). Because the L-band wavelength is very long and the antenna aperture is very large, SMOS uses the synthesis aperture system to form a large-aperture antenna array with 69 antenna elements and to retrieve image of a hexagon with a diameter of 1000 km.

MIRAS payload consists of several subsystems, such as:

1. Lightweight Cost-Effective Front End (LICEF) integrated with 66 antennas and receivers.
2. 3 Noise Injection Radiometers (NIR) for highly stable radiation calibration sources.
3. CAlibration System (CAS), on-orbit calibration subsystem, including internal calibration and external calibration.
4. MIRAS Optical HArness (MOHA), the optical transmission subsystem, reduces the electromagnetic interference introduced by traditional cable transmission, has high phase stability and lightweight for bending and temperature variation; it consists of 74 solid-state lasers, 168 optical receivers, 800 m optical fibers, several optical splitters and photoelectric converters.
5. Correlator and Control Unit (CCU) processes the signals received by the front end and monitors the thermal control effect of 12 thermal control loops.
6. Control and Monitoring Node (CMN), CCU monitoring node.

SMOS consists of three foldable antenna arms and a hub. The spacing between the arms is equal (120°). Each arm is divided into three sections, the arm length is 3.3 m (3 × 1.1 m), and each section includes 6 antenna elements (LICEF). The central hub

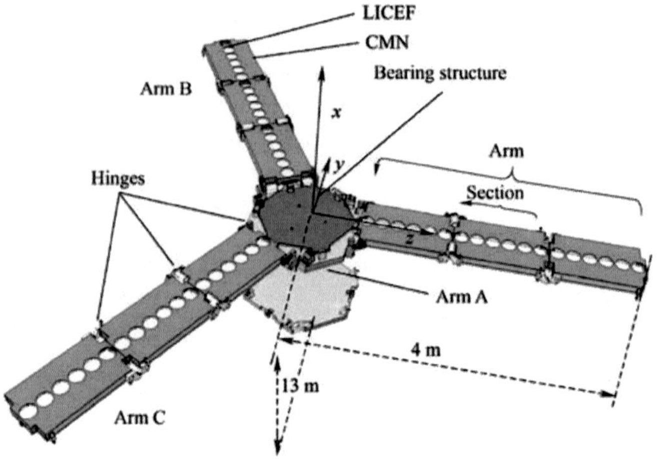

Fig. 7.25 Schematic diagram of the SMOS antenna array [3]

is 1.3 m in diameter, with 12 antenna elements and 3 highly stable Noise Injection Radiometers (NIRs), each of which is connected to two receivers. Therefore, SMOS consists of 69 antenna elements (66 LICEFs and 3 NIRs) and 72 receivers (66 LICEFs and 6 NIRs).

SMOS antenna array layout is shown in Fig. 7.25.

7.4.5 Swedish GAS

Geostationary Atmospheric Sounder (GAS) is a geostationary orbit synthetic aperture radiometer funded by ESA and developed by Sweden. Its main objective is to retrieve the temperature and humidity of the Earth disk. The main parameters of the system are shown in Table 7.2.

The basic principle, system composition and engineering implementation of this system are similar to those of United States GeoSTAR. The main differences are as follows.

Because GAS observes radiations in four frequency bands and different horn antennas are required for different frequency bands, the antenna array of GAS consists

Table 7.2 GAS satellite system parameters

Indicator	Parameter
Frequency band/GHz	53, 118, 183, 380
Geometric resolution/km	30
Payload weight/kg	300
Payload power consumption/W	300

Fig. 7.26 Picture of GAS prototype [5]

of four horns with different sizes. The total number of antenna elements is 140–150. The configuration of the antenna array is very complex [4].

The deployment size of each antenna arm is more than 4 m. In order to facilitate the launch, the antenna arm of GAS is designed as a foldable structure.

Limited by payload and antenna arm size, the number of antenna horns in four bands is limited, resulting in limited independent baselines and limited sampling points in spatial frequency domain, and ultimately affecting the accuracy of retrieved images. In order to solve this problem, GAS adopts the concept of antenna array rotation to obtain enough baselines and improve the resolution of the retrieved image.

A momentum wheel is equipped in the satellite body, which rotates opposite to the rotation direction of the antenna array. The simulation results show that the image accuracy can be greatly improved by rotating the antenna when the number of antennas is limited [4, 5].

In 2010, Sweden completed the development of GAS prototype with a frequency of 53 GHz and 21 antenna elements. The diameter of antenna array is about 0.75 m, as shown in Fig. 7.26.

7.5 Synthetic Aperture Radar Antenna

Radar was first invented for locating and ranging moving targets (aircraft). However, microwave radar cannot image targets on radar display (PPI). Because the radar resolution is not high enough, it can only see the position, height and speed information of the measured target, but the "shape" of the target. Radar antenna aperture (D/λ) is much smaller than optical lens. Taking the most basic 50 mm/F1.4 lens as an example, its D/λ can reach 6×10^4, the lens resolution angle is 0.045°. If an antenna reaches to the same resolution angle in X-band, the diameter of antenna must be up to 1785 m, which is almost impossible to achieve in hardware.

The sharp pattern can be obtained by the virtual antenna array formed after multiple measurements to same ground target on the radar flight path. Follow this idea, Carl Wiley of Goodyear Company first proposed the method of improving radar azimuth resolution by Doppler analysis in 1951. Sherwin et al.. of Illinois University obtained the first non-focusing radar image by experimental method in 1953. In the academic seminar of Michigan University in 1953, many scholars put forward the concept of Synthetic Aperture Radar (SAR), which can use the motion of the aircraft to form a small antenna array into a large antenna array. In August 1957, Cutrona and Leith et al. of Michigan University cooperated with the military in the first SAR system and obtained the first SAR image with large area and full focus. From then on, SAR technology developed from theory to practice and entered the development period.

Synthetic Aperture Radar (SAR) is an all-weather and all-day high-resolution microwave remote sensing imaging radar. Since its inception in the 1950s, it has achieved rapid development. In June 1978, the United States launched the world's first SAR satellite, Seasat-1, working in L-band. In September 1983, the Soviet Union launched an ocean Side-Looking Radar (SLR) satellite COSMOS 1500, which operates in X-band. Satellite-borne SAR has been widely used and achieved outstanding results. SAR plays an increasingly important role in military reconnaissance, Earth observation (topographic mapping, vegetation analysis, ocean and hydrological observation, environment and disaster monitoring), resource survey and crustal micro-deformation monitoring.

At present, China has successfully launched four SAR satellites. In 2006, L-band SAR satellite (Remote Sensing Satellite No. 1) was launched with microstrip patch phased array antenna; in 2010, X-band SAR satellite was launched with microstrip patch phased array antenna; and in November 2012, S-band SAR satellite (HJ-1C) was launched with horn array fed reflector antenna, as shown in Fig. 7.27. In August 2016, the C-band full polarization SAR satellite (GF-3) was launched with the dual polarization waveguide slot element phased array antenna.

7.5.1 Requirements for Synthetic Aperture Radar Antennas

The imaging principle of SAR, which can be considered as the principle of antenna array, is that the pulse signal is transmitted at a certain time interval (PRF) through the linear motion of the moving carrier (satellite), the parameters such as amplitude and phase, position and direction of the echo of each pulse are recorded, and then array beams are synthesized as long as the flight length of a satellite upon the real beam covered area, forming a much narrower beam width than that of the real aperture antenna, thus high-azimuth resolution detection is achieved. In the direction of motion (azimuth), the antenna can obtain a very large equivalent length L_S, i.e., synthetic aperture, as shown in Fig. 7.28.

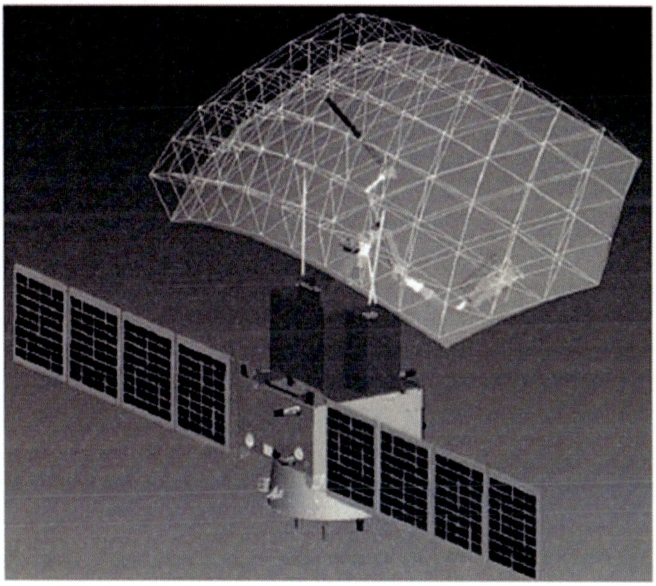

Fig. 7.27 China's SAR satellite HJ-1C (S-band, single polarization VV)

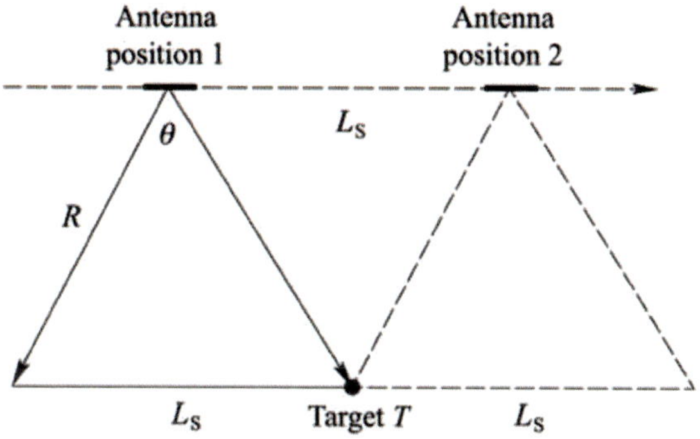

Fig. 7.28 Schematic diagram of synthetic aperture radar imaging

In the range direction, high resolution cannot be obtained with ordinary pulse due to the limitation of pulse width, so it is necessary to compress the pulse. Narrow pulse means broadband, so the broadband requirement of antenna is improved. The purpose of SAR is to obtain the two-dimensional image of the target, and ground range resolution is the projection of slant range resolution. In the range direction, the antenna beams of SAR satellites and the sub-satellite point should form an angle, which is called the side looking angle, as shown in Fig. 7.29. Early SAR radars were also called side-looking imaging radars.

There are many specifications of SAR system, among which the antenna-related specifications include incidence angle and scope (relative to the normal direction

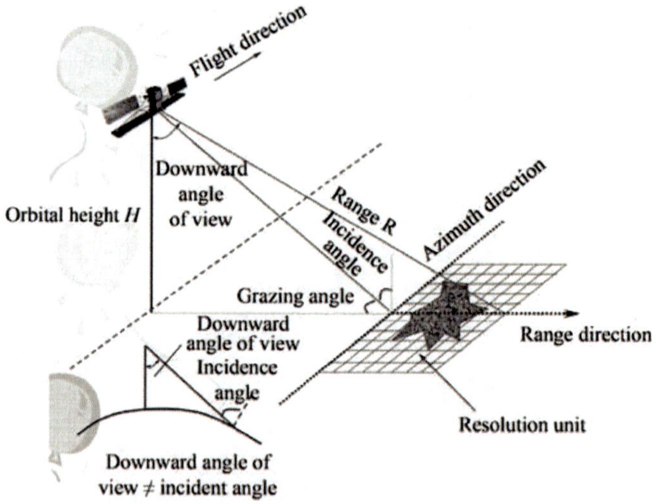

Fig. 7.29 Geometrical relationship of SAR beam

of antenna array), range resolution (relative to the antenna bandwidth), azimuth resolution and signal-to-noise ratio (relative to the antenna aperture, antenna area and antenna beam scanning range along flight direction), etc. In synthetic aperture radar system, antenna is the core component, whose index is closely related to the index of SAR satellite system, and directly affects the quality of imaging.

Figure 7.30 illustrates the relative relationship between spaceborne SAR and the Earth during on-orbit flight, where θ_{i} is the incident angle of radar wave, assume that the normal direction to the ground is 0°, and θ_{L} is the downward angle of view of satellite, assume that the direction of the sub-satellite point is 0°, and θ_{E} is the angle of the center of the Earth to the radar wave and to the Earth's reflection point, R_{E} is the radius of the Earth, H_{S} is the altitude of satellite orbit, and ρ_{gr} is the ground range resolution of the SAR. The relationship between these parameters is shown in Eqs. (7.7)–(7.9).

$$\rho_{\mathrm{gr}} = \frac{C\tau}{2\sin\theta_{\mathrm{i}}}, \tag{7.7}$$

where C is the speed of light and τ is the pulse width.

As can be seen from Eq. (7.7), in order to obtain high range resolution ρ_{gr}, the antenna beam of SAR must be slant incidence (side looking) to form a certain incidence angle (generally greater than 10°) in addition to reducing the pulse width (actually by transmitting linear frequency modulation pulse and receiving narrower pulse with matched filter). The downward angle of view θ_{L} is generally between 30° and 50°.

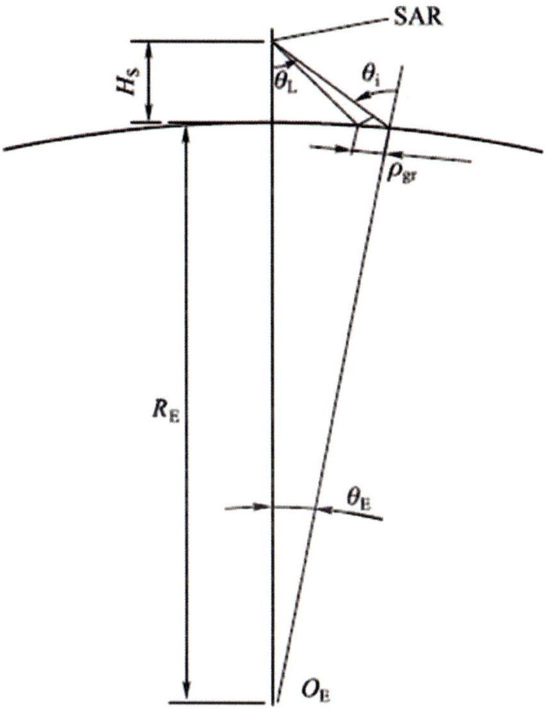

Fig. 7.30 Angular relation between SAR satellite in orbital flight and Earth (The flight direction is vertical to the paper)

$$\frac{\sin\theta_L}{\sin\theta_i} = \frac{R_E}{R_E + H_S} \tag{7.8}$$

$$\theta_i = \theta_L + \theta_E. \tag{7.9}$$

The measured width S_T on the ground is.

$$S_T = R_E(\Delta\theta_E), \tag{7.10}$$

where $\Delta\theta_E$ is the field angle of the difference between the reflector points of the maximum and minimum incident angles to the center of the Earth. If the minimum incident angles θ_{imin} and S_T are given, the range of downward angle of view can be obtained from Eqs. (7.8)–(7.10), so that the range aperture of the antenna can be estimated. For example, assume that the orbital height is 750 km, the frequency is 5.4 GHz, the minimum incident angle is 20°, the imaging swath is 100 km, and the radius of the Earth is 6371 km, then the size D_{AT} in the range direction of the antenna is calculated as 1.23 m. This is a preliminary estimate. In practice, factors such as PRF are also considered, see Eq. (7.11).

$$D_{AT} \geq \frac{2\,\mathrm{PRF}\cdot\rho_m\lambda\,\tan(\theta_{im})}{C}, \tag{7.11}$$

where ρ_m and $\theta_{im,}$ respectively, denote the distance and incidence angle of the farthest end of the swath to the satellite platform.

In the most basic SAR stripmap mode of a low-orbit satellite, if the resolution of a given azimuth is ρ_{Ar}, then the antenna azimuth size is

$$D_{AV} \leq 2\rho_{Ar}. \tag{7.12}$$

It can be seen that the azimuth resolution is related to the antenna azimuth size, and is independent of frequency and distance.

The pulse power received by the antenna is.

$$\frac{S}{N} = \frac{P_r}{KBT_e} = \left(\frac{\lambda}{4\pi\rho}\right)^3 \frac{C\tau}{2\sin\theta_i} \frac{P_t G_A^2 \sigma^0}{D_{AV} KBT_e}, \tag{7.13}$$

where P_r is the radar received power, K is the Boltzmann constant, B is the signal bandwidth, T_e is the equivalent noise temperature; P_t is the radar transmit power, G_A is the antenna gain (assuming the same gain in transmitting and receiving) and σ^0 is the cross-sectional area of the radar scattering, ρ is the beam slant range. If the received signal-to-noise ratio S/N and σ^0 are given, the azimuth dimension D_{AV} must satisfy the above several constraints, which is a multiple iteration process. For example, a 10 m antenna can be used to achieve an azimuth resolution of 5 m, and a higher resolution can be achieved by different beam scanning modes.

Figure 7.31 shows several common operation modes of SAR. In addition to stripmap mode, the spotlight, mosaic, and sliding spotlight modes require the antenna beam to scan in the azimuth direction, thereby increasing the target's synthetic aperture time and the resolution in the azimuth direction, but at the cost of reducing the area of swath on the ground. To increase the swath, the wide-range mapping can be achieved by the beam scanning of the antenna in the range direction, which is called ScanSAR mode, but at the expense of the azimuth resolution. The principles of these modes can be specifically referred to the literature on the design of SAR radar systems.

Therefore, if the aperture size of the spaceborne SAR antenna is determined, the core work is to complete the design and verification of various operation modes.

7.5.2 Several Main Forms of Synthetic Aperture Radar Antennas

Since the introduction of the synthetic aperture radar system in 1951, antenna has become a very important subsystem of SAR. The performance of the antenna mainly affects the SAR sensitivity, range and azimuth resolution, image ambiguity and signal-to-noise ratio, and the width of the swath. In SAR research such as multi-mode, multi-polarization, multi-band, multi-channel, high-resolution, and

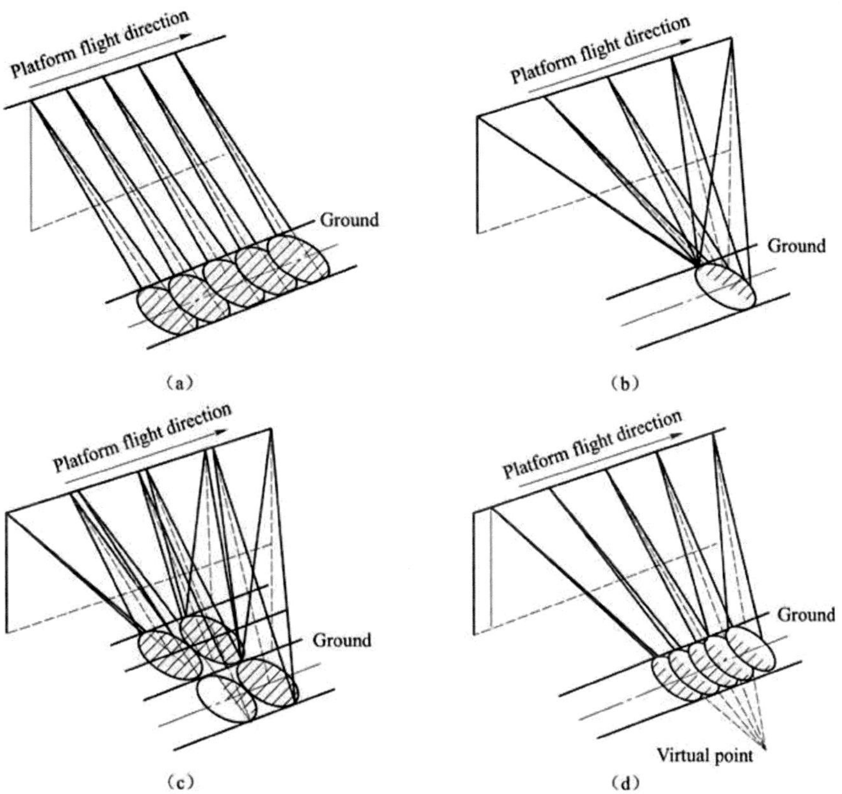

Fig. 7.31 Four most commonly operating modes used by SAR. **a** Stripmap mode; **b** spotlight mode; **c** mosaic mode; **d** sliding spotlight mode

wide swath, antenna research and design plays an important role. Almost every new mode of operation is accompanied by the application of a new beam form of the antenna.

SAR antenna technology is rapidly evolving with the development of SAR technology. Based on calculation of the azimuth and range aperture sizes of the antenna in the previous section, the antenna system can be designed. The waveguide slot array, the microstrip antenna array and the parabolic antenna are the most commonly used form according to the spaceborne SAR antennas that have been launched.

Waveguide slot array antennas, used in early SAR passive array antennas, mostly operate in the S, C and X bands with higher frequencies. This antenna has a large power capacity, good cross-polarization and low-side lobe level, and the technology is relatively mature. However, this kind of antenna has a complicated structure and high processing precision, and small size limited by the space application environment. Recently, in some SAR systems, a waveguide slot sub-array is used as a radiating element in phased array antennas, which is a preferred solution for improving radiation efficiency. Figure 7.32 shows the composition of a SAR antenna in China, which consists of four deployable arrays. Each array is composed of six sub-arrays. The waveguide slot is used as the radiating element, and the entire SAR antenna is composed of more than 700 radiation waveguides and more than 300 four-channel T/R components, the total length of the array is about 15 m after it is deployed on orbit.

Microstrip antenna arrays are mostly L-band SAR antenna arrays with lower frequencies. The microstrip antenna has the advantages of small size, lightweight, thin profile, simple manufacturing process, low cost, easy integration with microwave active devices, suitable for large planar phased arrays, and can easily realize dual polarization and dual-frequency bands functions. With the development of microwave integration technology, solid-state active phased array microstrip antennas have become the mainstream of spaceborne SAR antennas, such as Canada's Radarsat-2 and ASAR, and China's remote sensing satellite No. 1.

Reflector antennas with lightweight, such as Russia's Almaz-2, are suitable for small satellites, but its beams are relatively fixed and are not easy to achieve flexible beam scanning. Also, the reflector surface needs high processing technology, reducing the microwave energy penetrating the reflector mesh, and the surface geometric tolerances should not be too large, so the general operating frequency is low. In addition, high-power microwave devices are required, which bring technical problems such as multipactor discharge and heat concentration in space applications. However, with the development of satellite platform technology, the antenna beam scanning can be realized by the whole satellite swing, which makes the reflector antenna in the small satellite-borne SAR more active.

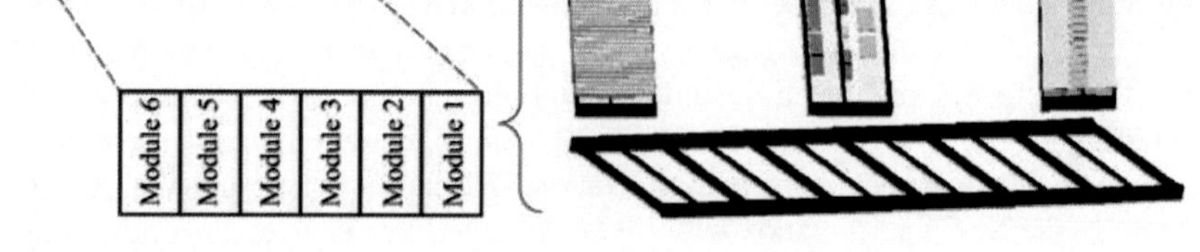

Fig. 7.32 Composition of a Chinese SAR antenna

7.5.3 Synthetic Aperture Radar Antenna Application Example

7.5.3.1 German SAR-Lupe System

The SAR-Lupe satellite is the first satellite to provide space-based military reconnaissance for Germany. The space segment of the SAR-Lupe system consists of five satellites running on three orbital planes, two of which are deployed on two orbital planes [6]. The SAR-Lupe constellation responds very quickly and has high system redundancy. According to the analysis, only four satellites on orbit can meet all system requirements, and the fifth satellite is mainly used as an orbital backup to reduce risks. The satellite began launching in 2005 and was fully launched in 2006.

The SAR antenna uses a fixed parabolic antenna with a size of 3.3 m × 2.7 m. The satellite uses a single solar wing with an average power consumption of 250 W and a design life of 10 years.

In general, the satellite has two imaging modes: stripmap mode and spotlight mode. When the satellite is working in the stripmap mode, the antenna is pointed at a fixed angle, and an imaging strip is formed as the satellite flies, and the imaging width is 8 km × 60 km. Since the satellite uses a fixed antenna, the entire satellite swings in the spotlight mode, so that the antenna is aimed at the target area, and the integration time is increased. In this mode, the imaging width is 6 km × 6 km, and the resolution is better than 1 m (There were also reports with a resolution of 0.5 m.).

In addition, the satellite SAR-Lupe provides products such as ground elevation model, 3D image of the target, target change detection information and radiation resolution enhancement using interferometric measurements by multiple flying over the target.

7.5.3.2 German Satellite TerraSAR-X

TerraSAR-X is the first German satellite implemented jointly by the German Aerospace Center and EADS Astrium. It was launched from Baikonur on June 15, 2007 via a Russian Dnepr-1 rocket. The orbit of the satellite is at 500 km × 517 km with angle of tilt of 97.45°. TerraSAR-X radar data can be used in many fields, such as surveys, large-scale topographic maps, disaster management, forestry and agriculture, and the satellite's highest resolution is better than 2 m, which has extremely high military application potential, especially its Digital Elevation Model (DEM) that can complement the capabilities of the SAR-Lupe satellite system. The TerraSAR-X is equipped with an X-band active phased array antenna that collects high-quality X-band radar data.

The satellite support structure adopts carbon fiber composite material, and SAR antenna, solar battery, and heat radiator are, respectively, installed on six sides of the prism of the support structure. The SAR antenna is an active phased array antenna with the size of 4.784 m × 0.704 m × 0.15 m, and mounted on one of the sides, which offsets 33.8° from the nadir point. S-band TT&C antenna, an X-band data transmission antenna, and a laser transmitter for precision orbit determination are mounted on another toward-Earth side. The data transmission antenna is mounted on a 3.3-m-long stretchable truss to avoid SAR interference and transmits data synchronously during SAR imaging. A Laser Communication Terminal (LCT) and a heat radiator are mounted on the backward-Earth side. The entire satellite mass is about 1230 kg. The satellite control system uses a set of high-torque reaction momentum wheels as fine-pointing control components, which can rapidly maneuver attitude to image high-priority targets, with a rolling maneuver of 67.6° in less than 180 s.

The SAR payload of TerraSAR-X has dual-channel receiving capability. The phased array antenna can operate in two sub-arrays. The radiating element is a dual-polarized waveguide slot array, which can operate in dual-polarized mode. Therefore, Active Target Indication (ATI) can be used to measure ocean current motion, and full polarization reception can be tested.

7.5.3.3 Canadian Satellite Radarsat

In cooperation with United States NASA, Canada launched the satellite Radarsat-1 in 1995 with a resolution of 8 to 100 m. Radarsat-2 is a spaceborne synthetic aperture radar system developed jointly by Canadian Space Agency (CSA) and MacDonald, Dettwiler and Associates Ltd (MDA). Radarsat-2 was launched into space on December 14, 2007 via a Soyuz rocket from the Kazakhstan launch site. The satellite launch quality was 2200 kg and the design life was 7 years. In order to maintain the continuity of the data, Radarsat-2 inherits all the operating modes of Radarsat-1, and adds multi-polarization imaging, 3 m resolution imaging, dual-channel imaging and Moving Object Detection Experiment (MODEX). Radarsat-2 has the same orbit as Radarsat-1, but lags 30 min behind Radarsat-1 in order to obtain two-satellite interference data. Radarsat-2 is designed to provide commercial users with high-resolution, full polarization spaceborne SAR images that are highly practical in terms of topographic mapping, environmental monitoring, ocean and glacial observations.

The satellite Radarsat-2 itself has a bilateral side looking function that further increases the swath width of the SAR system and shortens the revisit cycle. Radarsat-2 has up to 10 operating modes, and the corresponding beam illumination range is very flexible [7]. The image has a maximum resolution of 3 m and each operating mode has an optional polarization function, seven of which are available in dual or full polarization mode.

The function of the SAR radar satellite dictates the requirements of the antenna. The SAR system works with multiple imaging modes, such as standard stripmaps, fine stripmaps, wide-angle imaging and spotlight. In order to improve the recognition rate of the target, it needs to operate in different polarization modes, such as HH, VV, HV and VH. Each operation mode requires the antenna to have the appropriate function.

7.5.3.4 Israel Satellite TecSAR

TecSAR was the first radar satellite mission issued by the Israeli Ministry of Defence. TecSAR was launched in the fall of 2006 at the Satish Dhawan Space Center (SDSC) base in Sriharikota, India.

The satellite operates on an orbit of 550 km with an inclination of 143.3° and an orbital period of approximately 90 min. The whole satellite has a mass of only 300 kg, of which the SAR payload is only 100 kg. The satellite has a design life of 4 years, but the life expectancy can reach 8 years, depending on solar activity and orbital height.

The SAR antenna is a lightweight umbrella-shaped deployable antenna [8], which is fed by an X-band waveguide horn array, and 8 horns are fixed in staggered form along the range direction for forming 8 fast scanning beams for Scan-SAR mode. In addition, a horizontally polarized waveguide horn is provided with HH, VV, HV, VH polarization imaging capabilities.

7.5.3.5 China SAR Satellite GF-3

The SAR satellite GF-3 is China's first multi-mode fully polarized SAR satellite (Fig. 7.33), and has 12 operating mode, which exceeds that of Canada's satellite Radasat-2. The satellite can monitor and supervise global ocean and land environmental information all-weather and all-day, and expand the range of Earth observation and improve the rapid response capability through left and right attitude maneuvers. The acquired C-band multi-polarized microwave remote sensing information can be used for the marine, disaster reduction, water conservancy and meteorological fields to serve many correlative industries and business sectors. The obtained information is important for the implementation of marine development, terrestrial environmental

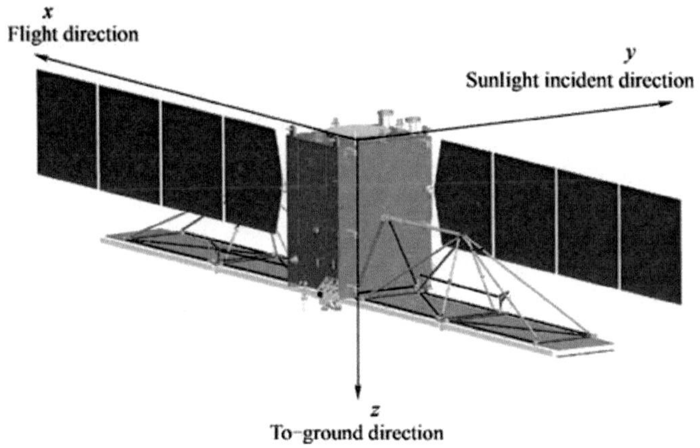

Fig. 7.33 China SAR satellite GF-3

resource monitoring and disaster prevention and mitigation in China. Launched in August 2016, the satellite operates on an orbit of 750 km with an inclination of 98.5°. After the launch of this satellite, it obtained a large amount of image information and made an important contribution to the national economy.

The SAR antenna of the GF-3 with multi-polarization and multi-operation mode adopts an deployable planar two-dimensional scanning solid-state active phased array antenna system to realize various SAR imaging modes such as spotlight, stripmap, scansar, and to quantitatively detect ocean and land information in high precision.

The antenna operates in the C-band. In the transmit mode, its transmit chain amplifies the power of the input LFM signal, radiates horizontal or vertical polarized electromagnetic energy to the designated space. In the receive mode, the antenna array receives horizontal or vertical polarized echo signals, or dual-polarized echo signals simultaneously and send them to the SAR central electronic device via a low-noise amplification chain. In order to achieve antenna performance detection, fault detection and isolation, the antenna has an independent calibration network, and the calibration of transmit and receive chain can be completed under the control of the central electronic device. The antenna has multi-polarization, high-polarization isolation, and a large amount of shaped beam. It has a high-peak output power of 15,360 W, and the lightweight design technology represented by T/R components and thin-walled waveguide and high-strength mechanical design techniques for complex structures such as unfolding mechanisms are adopted.

The antenna deployment mechanism can satisfy the functions of folding, unfolding, and supporting of the antenna array. The thermal control of the array realizes the temperature control capability of the SAR antenna array, satisfies the requirements of the use environment of each electrical single machine, and ensures the service life of the electrical equipment.

References

1. Tanner B, Gaier TC, Lambrigtsen BH (2008) GeoSTAR performance demonstration. In: IEEE Aerospace Conference AC paper # 1421, Version 3, Updated 2008, pp 1–3
2. Gaier TC, Lambrigsten BH, Tanner AB (2007) A dual-gain antenna option for GeoSTAR. In: IEEE International Geoscience and Remote Sensing Symposium, IGARSS 2007, pp 227–230
3. McMullan KD, Brown MA, Martín-Neira M et al (2008) SMOS: the payload. IEEE Trans Geosci Remote Sens 46(3):594–605
4. Carlström A, Christensen J et al (2009) Geostationary Atmospheric Sounder (GAS) Demonstrator Development, 3rd European Conference on Antennas and Propagation, 2009, EuCAP 2009, pp 2036–2040
5. A R Laganà, Iero DAM, Isernia T et al (2010) On the design and optimization of the array elements in the GEO atmospheric sounder instrument: a new design procedure. In: 2010 IEEE International Geoscience and Remote Sensing Symposium, IGARSS2010, pp 578–581
6. Braun HM, Merkle F (2009) The German way to SAR-lupe, a very high resolution SAR. In: 60th International Astronautical Congress
7. www.asc-csa.gc.ca/eng/satellites/technical-feature/radarsat-comparison.asp
8. Sharay Y, Naftaly U (2006) TECSAR: design considerations and programme status. IEE Proceedings of Radar Sonar and Navigation, Apr 2006, vol 153(2), pp 117–121

Chapter 8
Low-Frequency Antenna and Small Satellite Antenna

Abstract VHF and UHF bands are widely used in special communication satellite, small satellite, etc. This chapter provides UHF deployable helix antennas for communication satellite, and introduces some typical antennas used in small satellite.

Antennas with a frequency below 1 GHz are usually referred to as low-frequency antennas. According to ITU, there are 400 MHz for communication, 403/467 MHz for Data Collection System (DCS), 243 MHz for domestic rescue, 406 MHz for international rescue, 600 MHz for telemetry and remote control signals on the spacecraft. In addition, inter-satellite links of the United States GPS II R/II R-M use signals of 250–290 MHz frequencies.

Satellite systems operating in VHF and UHF bands are characterized by simple ground equipment, low loss of radio wave propagation and small direct shielding effect of ground objects on radio waves. Since DFH-1, low-frequency antennas of China have been used in many communication satellites, manned spaceflight, amateur radio, etc. This chapter will briefly discuss UHF communication antennas. For 243 MHz for domestic rescue antennas and 406 MHz for international rescue antennas, please refer to Chap. 11, Sect. 11.6.

Small satellite is one of the research hotspots in the field of aerospace nowadays. More and more attention has been paid on small satellites by all countries due to its many advantages, such as small size, short development cycle and low cost, and widely used in communication, Earth remote sensing, interplanetary exploration, scientific research and technical tests. Antennas on small satellites must be small in size and light in weight, but the miniaturization design of low-frequency antennas is a hard nut. Therefore, the antennas used in small satellites are briefly introduced in this chapter.

8.1 UHF Communication Antenna

UHF communication antenna is mainly composed of antenna support cone, UHF receiving antenna, UHF transmitting antenna, and antenna deployment drive control

R. You et al., *Technologies for Spacecraft Antenna Engineering Design*, Space Science and Technologies, https://doi.org/10.1007/978-981-15-5833-7_8

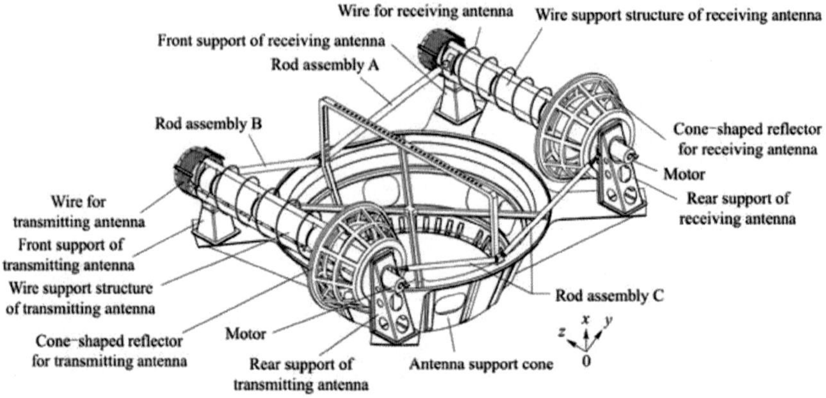

Fig. 8.1 Configuration of UHF antenna in folded state

box. The antenna is installed on the eastern board of the satellite platform of "DFH-3" and points to the Earth.

When the satellite is launched, the antenna is in a folded state to be accommodated in the envelope of the fairing, as shown in Fig. 8.1. After the satellite is on orbit, the antenna is deployed by the drive of the deployment mechanism to meet the antenna gain requirements, as shown in Fig. 8.2.

The antenna is connected with the transponder in the satellite communication module by high-frequency cable. The receiving antenna receives the RHCP signal of the ground terminal, which is amplified by the satellite transponder, and sent back to the ground through the transmitting antenna in the RHCP signal to implement a communication link.

The antenna support cone, including an isolation net bracket, front and rear rod, is a support structure which connects the UHF transmitting antenna, UHF receiving

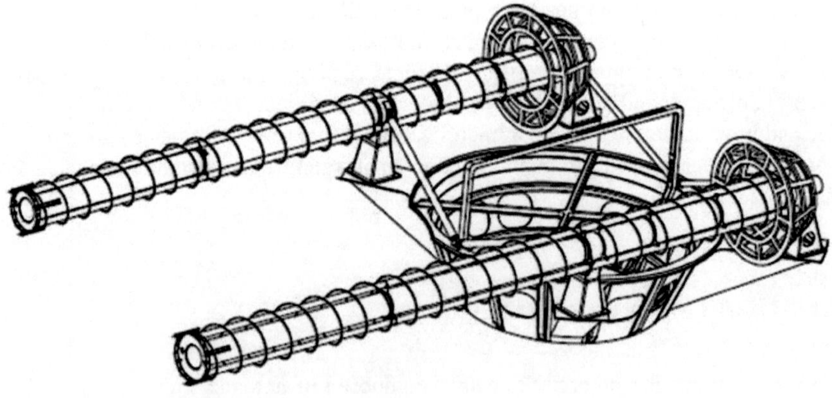

Fig. 8.2 Configuration of deployed UHF antenna

antenna and the satellite. The two UHF antennas are installed on the upper part of the support cone structure, which is connected with the satellite bearing cylinder.

UHF receiving antenna and UHF transmitting antenna are mainly composed of antenna wire, support structure, conical reflector, front and rear supports, and deployment mechanism, etc. The two antennas are identical in structure form, but different in length and size.

8.1.1 Electrical Performance Design

A three-stage long helical antenna with the maximum length after deployment of nearly 4.4 m is used to realize the high gain in the operating state and small envelope in the launch state of UHF communication antenna. The maximum antenna gain can be achieved within the corresponding operating bandwidth by optimizing the parameters of helical antenna.

Transmitting and receiving are implemented separately to avoid the influence of passive inter-modulation. The distance between receiving antenna and transmitting antenna is so designed to ensure that the isolation between these two antennas is about 50 dB. However, it is impossible for the antennas to be far apart from each other due to the envelope size limitation of the fairing. Therefore, a metal isolation net is added between the antennas, so as to ensure the isolation between the antennas.

In order to relieve the influence of the satellite on the antenna radiation performance, the antenna is set up higher than the satellite platform by using the wave-transparent support cone, but this height is limited by the anti-vibration performance requirement of the antenna system. When the helical antenna is deployed, more than half of its helical length extends beyond the satellite to further ensure that the pointing direction of the antenna beam is not affected by the satellite.

The helical parameters of UHF communication antenna, the distance between transmitting and receiving antennas and the mounted height of the antenna on the satellite are determined by theoretical calculation and scaling test.

1. Scaling model analysis

For antennas with low frequency, scaling test is a very effective method to optimize the parameters. In combination with the frustum-shaped reflectors with different sizes, the dual-winding, parasitic and single-winding antennas with different elevation angles are compared and tested on a 1:6 scaling model of the whole satellite. Using the same scaling model, the shape of the metal net that improves the isolation degree is optimized.

In the scaling test, it is found that the single-winding helical antenna with small elevation angle can be reduced to the length of folded state when the electrical performance is similar, which has obvious advantages in structure. Therefore, the scheme of the single-winding helical antenna with small elevation angle is selected.

Tests show that the side lobe of the antenna on the inner side of the satellites is decreased and the side lobe on the outer side is enlarged due to the influence of the satellites, but the shape of the main lobe remains basically unchanged. Because the operating area is within the main lobe, its influence of the satellite on the radiation performance of the antenna can be neglected. The test results also show that the gain and radiation characteristic of UHF antenna are slightly affected by the satellite body and other bands of antenna, but all of them are within the allowable range.

2. Numerical calculation analysis

In order to obtain more accurate design parameters, the moment method analysis program is used for theoretical calculation, analysis of the electrically large-size antenna pattern and the influence of the satellite body, which greatly reduces the blindness of the design and accelerates the development progress.

The gain values under different parameters can be obtained by calculating the long helical antennas with different elevation angles of rise and turns, as shown in Fig. 8.3.

It can be seen from Fig. 8.3 that the antenna gain can be increased with the reduced elevation angle under the same spiral axial length, but the bandwidth with the high gain becomes narrow. Selecting an elevation angle of 11.5° can not only increase the gain, but also meet the operating bandwidth requirements.

Moreover, the influence of the satellite body on the antenna is further analyzed, and the pattern of the helical antenna near the satellite model is calculated, as shown in Fig. 8.4. The coordinate system of the model is consistent with the satellite platform coordinate system. The left box is a satellite body model, and the right side is two helical antenna models with a frustum-shaped reflector. The helical antenna is in a deployed state, and a metal isolation net is placed between the antennas.

Figure 8.5 shows the pattern calculation results of the UHF communication antenna affected by the satellite. It can be seen that the antenna is setup away from the satellite body and extends more than one-half of the length of the UHF communication antenna. Therefore, the influence of the satellite on the main lobe of the pattern is small, and the gain index of the coverage area is not affected at all. However, the first side lobe is greatly affected, and the side lobe near the side of the satellite is depressed.

Fig. 8.3 Gain-frequency diagram calculated by the helical antenna theory (comparison of different elevation angles)

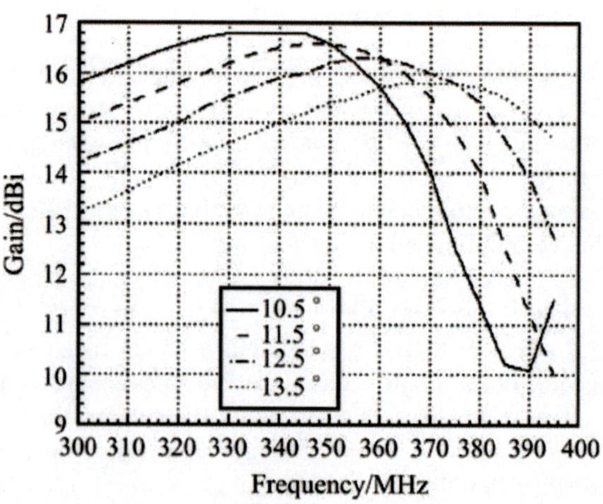

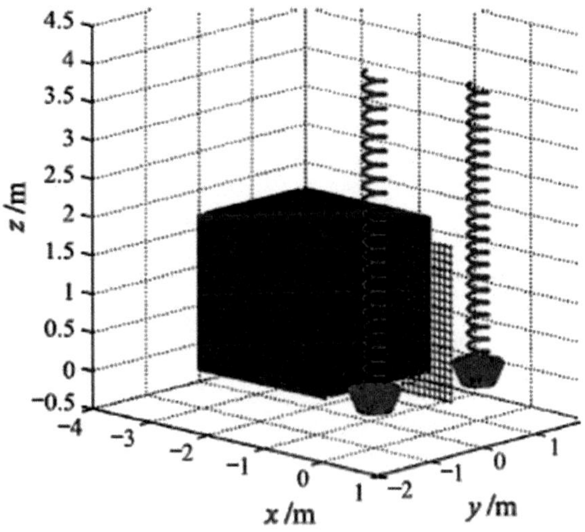

Fig. 8.4 Calculation model of UHF antenna and satellite

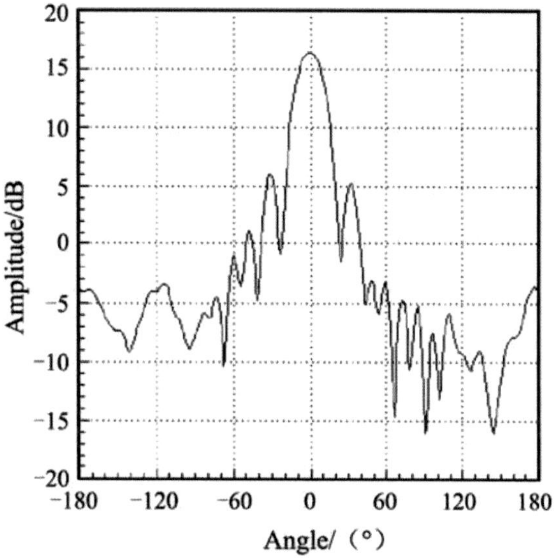

Fig. 8.5 Calculation pattern of UHF antenna with satellite

The calculation shows that the metal isolation net optimized by the scaling model has better isolation performance. The theoretical isolation value of the downlink-frequency band is greater than 52 dB, while the theoretical isolation value of the uplink-frequency band is greater than 48 dB.

The conclusions of the theoretical calculations verify the test results of the scaled model, which have also been verified in subsequent physical tests. The field test photos are shown in Fig. 8.6.

Fig. 8.6 Field test photos of UHF communication antenna

8.1.2 Deployment Mechanism Design

In order to achieve a high gain in the UHF communication antenna operating state and a small envelope in the launch, a long helical antenna is deployed by a pendulum rod or a three-stage sleeve extension.

In the launch, the antenna is folded and placed in the fairing. Once on orbit, the antenna is first pushed away from the satellite body by the swing mechanism to reduce the influence of the satellite on the antenna pattern; then the antenna is deployed successively by the three-stage sleeve under push of the extension mechanism, by which the antenna wire is stretched into position. The maximum spiral length reaches about 4 m, thus achieving high gain.

The compositions of the deployed UHF communication antenna are shown in Figs. 8.7 and 8.8. The ground deployment test is shown in Fig. 8.9.

8.2 Small Satellite Antenna

Small satellites feature high-functional density and technical performance, and their volume and weight are less than one-tenth or even one percent of traditional large and medium-scale satellites. Especially in the cubic satellites studied in recent years, the size can be as small as 10 cm × 10 cm × 10 cm, the weight is less than 1 kg. Following subsystems are generally included in small satellites: TT&C subsystem, control subsystem, propulsion subsystem, power subsystem, and payload, etc. For satellites with different functions, the payload is different, but the communication

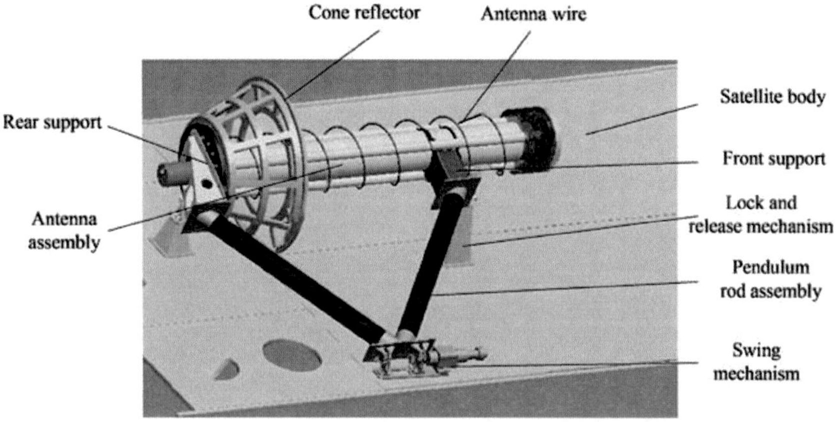

Fig. 8.7 An antenna in folded state on a satellite

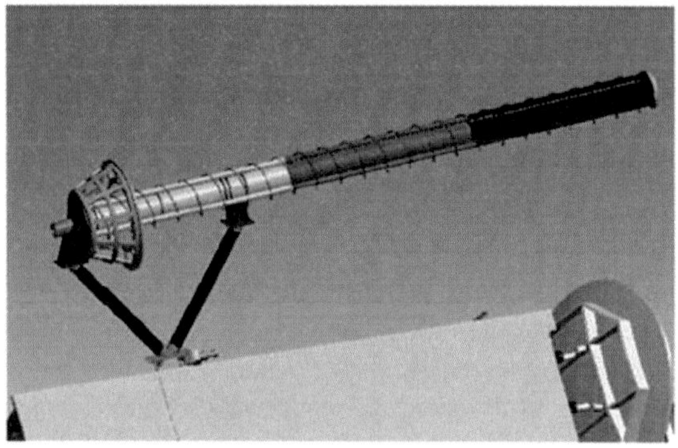

Fig. 8.8 An antenna in deployed state on a satellite

module is the only indispensable one, which is mainly composed of an antenna and a signal processor.

The antenna subsystem is mainly responsible for the mutual communication between the satellite and the ground, and the receiving and transmitting of the ground-to-satellite remote control and telemetry commands. Therefore, the stability and reliability of the antenna subsystem determine whether the satellite can successfully perform the predetermined task. The design of the satellite antenna has also become a crucial process. The key of small satellites is the miniaturization of various equipments and devices. Therefore, the miniaturization of communication antennas has become an important research subject of small satellite antennas.

For the antenna design of small satellites, the following points shall be considered:

1. The antenna must have high reliability because it is impossible to repair or replace the small satellite antenna on orbit.

(a) (b)

Fig. 8.9 Photos of deployment test of UHF communication antenna on the ground. **a** the folded state; **b** the deployed state

2. The antenna must be small in size, light in weight, high in efficiency, and low in cost. Especially, such a small size as the pico-satellite cannot provide more space on the surface to install a large antenna.
3. The mechanical strength of the antenna must be able to withstand the vibrations and shocks encountered without being damaged by launching the satellite.
4. The space environment adaptability shall be considered in the antenna design process. For example, the temperature changes greatly with the range from −150 to +150 °C. The antenna shall have certain anti-irradiation capability to withstand various ionizing radiations in space, including high-energy particles from the sun; the antenna of a low-orbiting satellite may be affected by atomic oxygen, and it shall be equipped with a protective cover or directly treated on the surface of the antenna for anti-impact of atomic oxygen.
5. Select the appropriate antenna material for vacuum and microgravity environments in space.
6. Effective data transmission requires the antenna with a certain gain. High gain often means large size, complex structure and high-pointing control accuracy. However, it is difficult to achieve ideal high gain due to room constraints on modern small satellites. Therefore, there is a trade-off between the need for gain and the limitations of the structure in antenna design.

7. Special attention should be paid to the interaction between the antennas on the satellite and between the antenna and the satellite in the antenna design process. As mentioned above, due to the small size of the satellite, when antennas are installed in a small space, the mutual coupling and electromagnetic compatibility of the antennas must be considered. Because of the electromagnetic scattering of the satellite structure itself, which will affect the radiation pattern of the antenna, the gain will be reduced and the communication between the satellite and the ground will be deteriorated.

8.2.1 Introduction to Several Common Antennas

Like large satellites, small satellites mainly use antennas for telemetry, tracking, remote control, data transmission, GNSS navigation signal reception and inter-satellite links. The implementation of these functions requires several different antennas. According to the frequency range, beam coverage requirements and other applications, the commonly used antennas mainly include TT&C antennas and data transmission antennas.

8.2.1.1 TT&C Antennas

At the beginning on orbit, the attitude of the satellite has not been stabilized since it was just separated from the rocket. Therefore, an omnidirectional beam TT&C antenna should be selected in order to effectively capture the satellite from the ground station. Generally, plenty of TT&C antennas are installed on different surfaces of the satellite to better achieve the omnidirectional coverage of the pattern and implement the command transmission when the satellite attitude is unstable.

Dipole antennas, Printed Inverted F Antennas (PIFAs), helical antennas and microstrip patch antennas can be used for TT&C. They operate in various frequency bands of satellite communication and feature low cost, simple structure and easy manufacture. They work under wide beam and have low requirements on satellite attitude.

Figure 8.10 shows a microstrip patch antenna operating in the S-band. The antenna, consisting mainly of ground plate, dielectric and radiating elements, is fed at its bottom with a 50Ω probe and operates at 2.0 GHz. The chamfer and feed bias are used to achieve circular polarization. It has a maximum gain of 6.8dBi, a size of 82 mm × 82 mm × 20 mm and a weight of less than 80 g.

8.2.1.2 Data Transmission Antennas

When the satellite operates properly on orbit, a large amount of payload data shall be transmitted back to the ground at high speed, which requires the data transmission antenna to have a certain gain. But the gain of the antenna cannot be too high, because

Fig. 8.10 S-band microstrip patch antenna [1]

it may not have high-precision attitude control system on the micro satellite, and the satellite will not transmit effectively when the attitude is deviated.

As shown in Fig. 8.11, a four-arm helical antenna operating in the S-band consists of four spiral arms that feed in phases of 0°, 90°, 180° and 270°, respectively. The antenna operates at an uplink frequency of 2.025 to 2.11 GHz and a downlink frequency of 2.2 to 2.29 GHz, and has good circular polarization characteristics and a wide beam heart pattern. It has a size of 100 mm × 100 mm × 500 mm and a weight less than 500 g and is widely used in low-orbit satellite communications.

8.2.2 Dipole Antennas

In order not to occupy the to-Earth desk, the VHF/UHF antenna for general TT&C uses a multi-pole to synthesize a circular polarization pattern. The monopole antenna is folded during launch to maintain the square outer contour of the satellite; after the injection, the nickel–chromium alloy wire of the lock and release mechanism is energized and broken, then the monopoles are, respectively, laterally expanded to form four monopole arrays with an angular difference of 90°, as shown in Fig. 8.12. The phase relationship of each monopole is adjusted so that the phase difference is 90°, i.e., the circular polarization pattern is formed.

UHF transmitting antenna and VHF receiving antenna are equipped on XW-1 popular science satellites in China for Morse code, voice and data relay communication, and data storage and forwarding communication between satellite and ground amateur radio stations.

In order to reduce the interference between antennas, the antennas layout on the satellite should make full use of the surface space of the satellite, and should be scattered as much as possible. As shown in Fig. 8.13, the UHF antenna is mounted on the +z plane of the satellite. The four UHF antennas are spaced 260 mm apart and squarely distributed. The VHF receiving antenna is mounted on the −z plane of the satellite. The four VHF antennas are spaced 410 mm apart and squarely distributed.

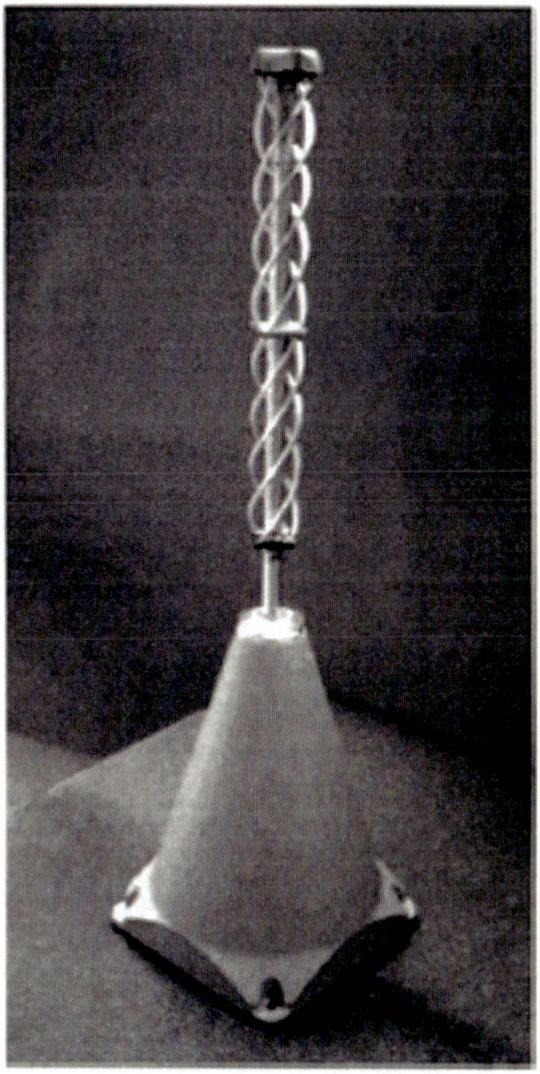

Fig. 8.11 S-band four-arm helical antenna [1]

Both the UHF transmitting antenna and the VHF receiving antenna are in the form of a monopole antenna. The monopole length is generally slightly smaller than the quarter wavelength, which can achieve good resonant radiation, as shown in Fig. 8.14. Each antenna consists of four monopole elements that form a circular polarization through their respective antenna network (power split phase shifting network). In order to adapt to the satellite envelope limitation, the VHF receiving antenna is spirally loaded at the root of the pole, and the antenna length is reduced from 500 to 370 mm, which satisfies the satellite envelope restriction requirements.

Fig. 8.12 Cube satellite PolySat (R&D by California Polytechnic) [1]

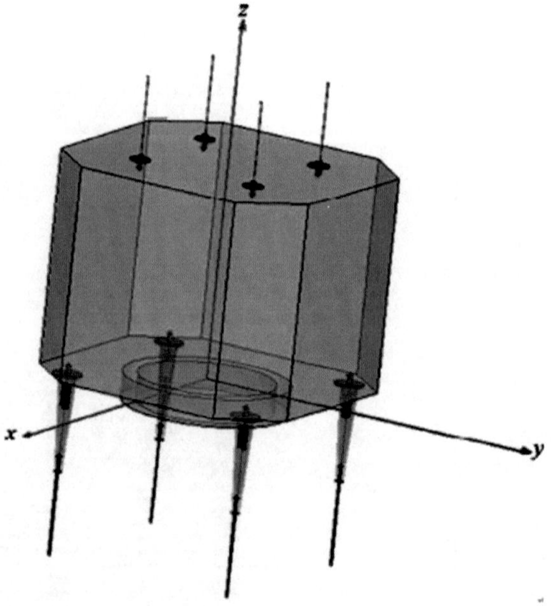

Fig. 8.13 Layout of XW-1 satellite antennas

8.2.3 *Deployable Helical Antennas*

For applications that require high gain, such as receiving Automatic Identification System (AIS) signals, a deployable helical antenna can be used, typically in the form of a single-winding or four-arm spiral.

The helical antenna is compressed into a smaller volume during the launch phase, and is deployed to its desired length by its own elasticity or deployment mechanism after the injection, so as to achieve high gain. An example of a deployable helical antenna in the L-band is shown in Fig. 8.15.

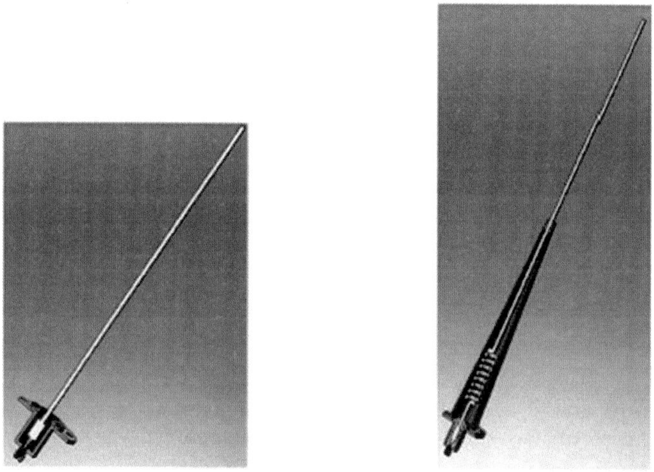

(a) (b)

Fig. 8.14 Dipole antenna of satellite XW-1. **a** UHF antenna; **b** VHF antenna

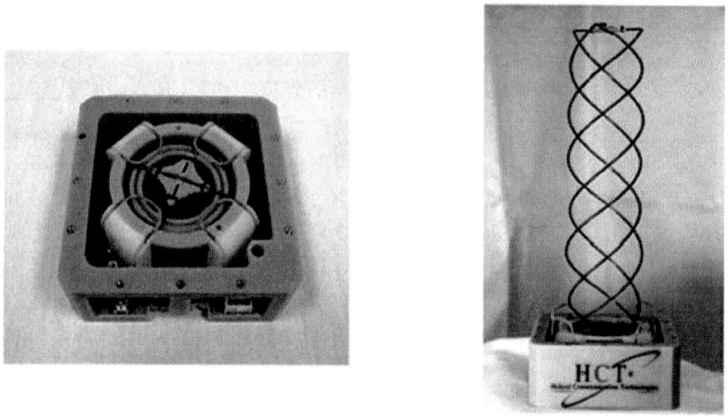

(a) (b)

Fig. 8.15 L-band deployable helical antenna [2]. **a** The folded state; **b** the deployed state

An other example is the VHF antenna on the satellite AISat [3]. The satellite main payload is the device receiving AIS signal. The navigation, location, and identification information can be exchanged between ships by AIS signals to ensure that global maritime traffic is safe. The high-gain helical antenna with the diameter of 0.57 m and 7–8 turns helix can obtain good directivity, improve resolution and accurately receive rescue signals from ships in distress.

In contrast to the antenna length (4 m) and the satellite volume, the antenna must have a very high-compression ratio during the launch phase, and the folded length of the antenna is about 10 cm. After the satellite injection, the antenna is fully deployed by its own elasticity, and the thin rope on the spiral is stretched to the required length to ensure the correct pitch.

8.2.4 Inter-satellite Link Antennas

Inter-satellite links are required in multiple small satellite formation flights or constellation operations. However, the satellites are too small, especially cubic satellites, to carry mature attitude control systems for adjusting the inter-satellite link antenna pointing. Therefore, a dynamic communication link shall be established and maintained between satellites by an electronically controlled beam scanning antenna. Unfortunately, it is unrealistic to apply traditional phased array antennas or smart antenna technologies because of the limitations of small satellite size, energy, cost, etc.

Obviously, inter-satellite signal transmission can be achieved by using the omnidirectional antennas, but such constellation is easily eavesdropped by unauthorized satellites or ground stations. And the omnidirectional antennas are not efficient, because the electromagnetic energy cannot be concentrated to the receiver, but be inefficiently spreaded into the whole space.

A retro-directive antenna can align the beam to the direction of the incoming wave, thus ensuring the security of the inter-satellite link and the efficiency of signal transmission. The retro-directive antenna is an array antenna that detects the direction of the incoming wave of the signal. The beam can be automatically aligned to the direction of the signal source by conjugate processing the received signal of each element. This technology does not require expensive phase shifters (used in phased arrays) or complex digital signal processing modules (used in smart antennas), thus significantly reducing cost, complexity and energy consumption. Generally, in order to obtain phase conjugation, a corner reflector, a Van Atta array, or an electrical heterodyne array is used.

The traditional method of electrical heterodyne array mixes the incoming wave signal (RF) of each element with the Local Oscillator (LO) signal. The LO signal frequency is twice that of the RF signal, so the inter-modulation signal (LO-RF) frequency is the same as that of the RF signal, but its phase is the conjugate of the RF signal.

The University of Hawaii conducted a study on the application of retro-directive antennas in small satellite communications, as shown in Fig. 8.16. The retro-directive antenna with eight circularly polarized patch elements, which is mounted on one surface of the cubic satellite, can implement a two-dimensional phase conjugate array with 10 GHz operating frequency. Based on the quartered harmonic mixer (Lo), the LO frequency (Lo) is half of the RF frequency, so the LO frequency is reduced. Finally, the electrical heterodyne array performance was achieved by using a 5-order inter-modulation product ($4 \times$ Lo-RF).

Frequency relationship:

$$
\begin{aligned}
f_{\mathrm{LO}} &= f_{\mathrm{RF}}/2 \\
f_{\mathrm{IF}} &= 4f_{\mathrm{LO}} - f_{\mathrm{RF}} = f_{\mathrm{RF}}.
\end{aligned} \tag{8.1}
$$

Amplitude and phase relationship [4]:

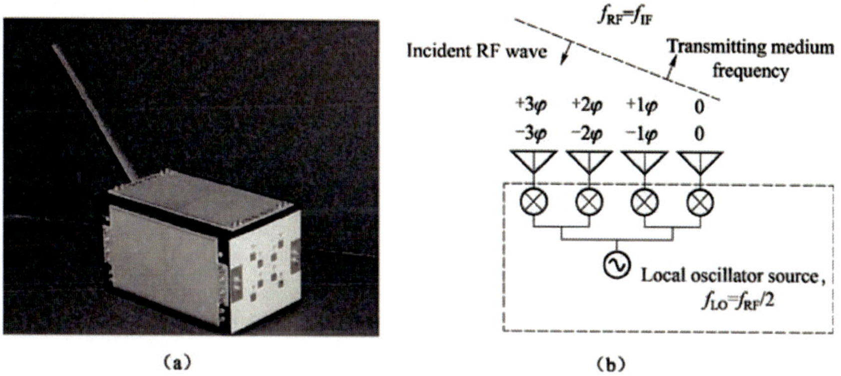

Fig. 8.16 Example of retro-directive antenna and system block diagram [4, 5]. **a** retro-directive antenna; **b** system block diagram

$$
\begin{aligned}
& V_{IF} \propto [V_{LO}\cos(\omega_{LO}t) + V_{RF}\cos(\omega_{RF}t + \theta)]^5 + \ldots \\
& \Rightarrow V_{IF} \propto \frac{5}{16} V_{LO}^4 V_{RF}\cos(4\omega_{LO}t - \omega_{RF}t - \theta) + \ldots
\end{aligned}
\tag{8.2}
$$

with $2\omega_{LO} = \omega_{RF}$, we have

$$
\begin{aligned}
V_{IF} & \propto \frac{5}{16} V_{LO}^4 V_{RF}\cos(2\omega_{RF}t - \omega_{RF}t - \theta) + \cdots \\
& = \frac{5}{16} V_{LO}^4 V_{RF}\cos(\omega_{RF}t - \theta) + \cdots,
\end{aligned}
\tag{8.3}
$$

that is, the IF and RF phase conjugate.

The test scheme of a retro-directive antenna auto-tracking pattern is shown in Fig. 8.17. The test result is shown in Fig. 8.18, which achieves the expected goals.

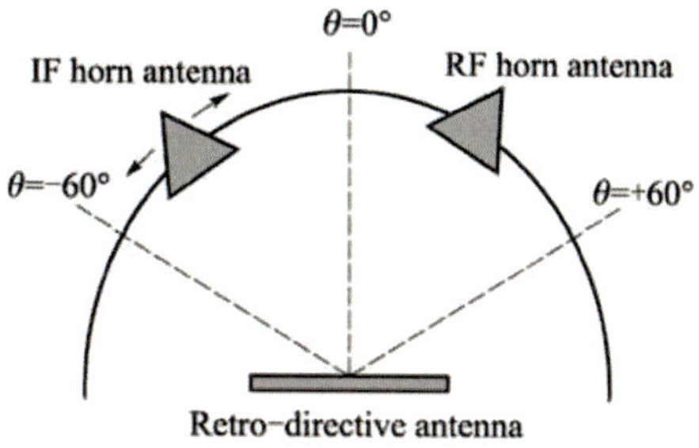

Fig. 8.17 Test scheme of a retro-directive antenna auto-tracking pattern [4]

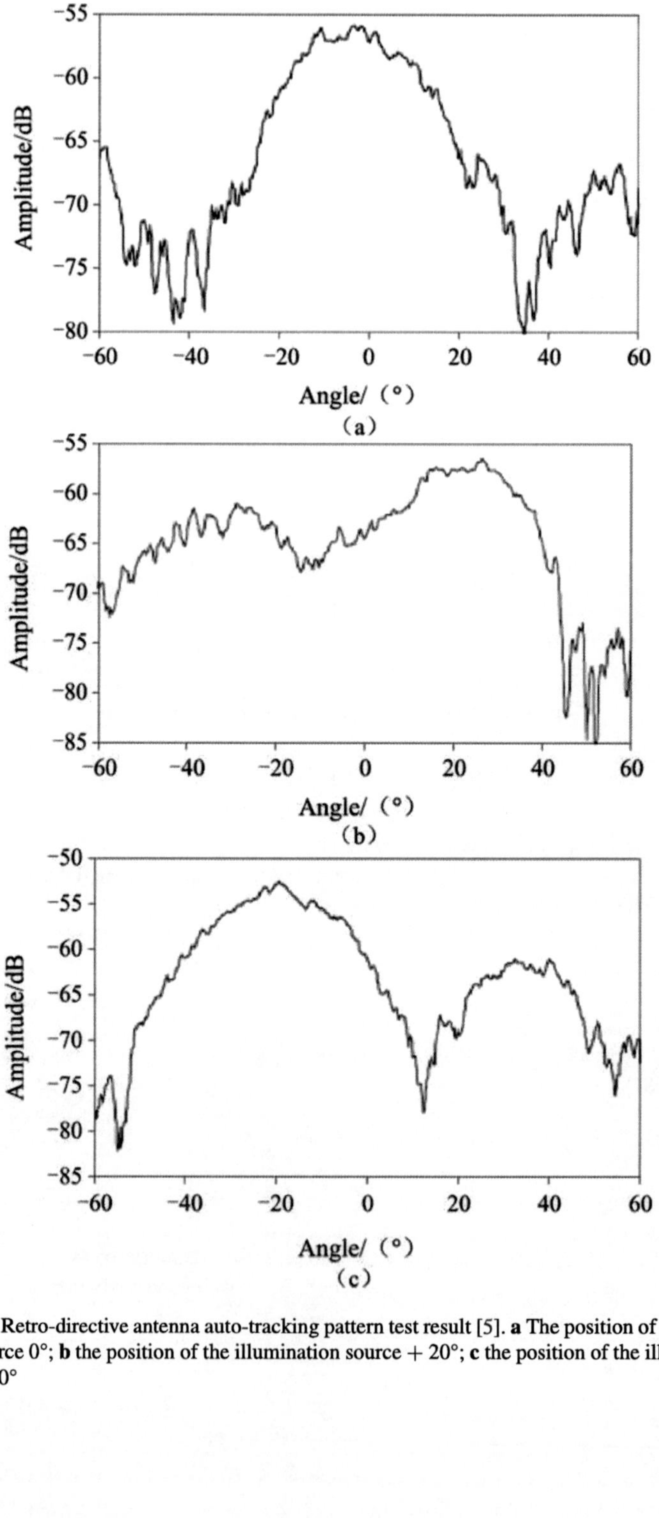

Fig. 8.18 Retro-directive antenna auto-tracking pattern test result [5]. **a** The position of the illumination source 0°; **b** the position of the illumination source + 20°; **c** the position of the illumination source −20°

References

1. Gao S, Clark K et al (2009) Antennas for modern small satellites. IEEE Antenna Propag. Mag. **51**(4): 40–56 (2009)
2. O'Neill G (2016) Deployable small satellite antenna quadrifilar helical antenna. In: IELCONF IEEE, pp 116–118 (2016)
3. Block J, Bager A, Behrens J et al (2013) A self-deploying and self-stabilizing helical antenna for small satellite. Acta Astronautica 86:88–94
4. Murakami BT, Roque JD, Sung SS (2004) A quadruple subharmonic phase-conjugating array for secure picosatellite crosslinks. In: Microwave Symposium Digest, 2004 IEEE MTT-S International, vol 3, 1687–1690
5. Mizuno TJ, Roque JD et al (2005) Antennas for distributed nanosatellite networks. In: IEEE/ACES International Conference on Wireless Communications and Applied Computational Electromagnetics, pp 604–609

Chapter 9
Reliability Design of Spacecraft Antenna

Abstract The reliability design of spacecraft antenna runs through the whole process of antenna development. Reliability design includes reliability prediction, resistant-to-mechanical-environment design, thermal design, EMC design, space environment protection design, redundancy and derating design. Structure, mechanical analysis and design, thermal deformation analysis and thermal design, venting hole design, multipaction protection design, passive inter-modulation protection design for spaceborne antennas are provided in this chapter.

For general spacecraft antenna product, reliability design mainly includes anti-mechanical environment, thermal and space environment protection design. For spacecraft antennas working in special environment, a special design should be carried out. For example, because the Shenzhou spacecraft reentry capsule passes through the atmospheric ablation environment during reentry, so anti-heat and isolation design should be carried out to ensure that the antenna can still work normally at high temperatures above 1000 °C.

In design of spacecraft antennas with deployment and rotation mechanisms, the redundancy of important and easily invalid components and adequate reliability verification tests should be provided according to their failure modes and effect analysis, so as to ensure that the antenna has sufficient reliability.

At first, taking the antenna of spacecraft platform as an example, this chapter mainly introduces the structural design, mechanical analysis, thermal design and thermal deformation analysis. Then, it introduces the protection design of multipaction for vacuum environment, and passive inter-modulation protection design for multiple operating frequencies in mobile communication.

9.1 Structure Design of Spacecraft Antenna

In the design of antennas, many iterations method is used. The typical design process is shown in Fig. 9.1. The antenna structure should be so designed that the antenna

R. You et al., *Technologies for Spacecraft Antenna Engineering Design*, Space Science and Technologies, https://doi.org/10.1007/978-981-15-5833-7_9

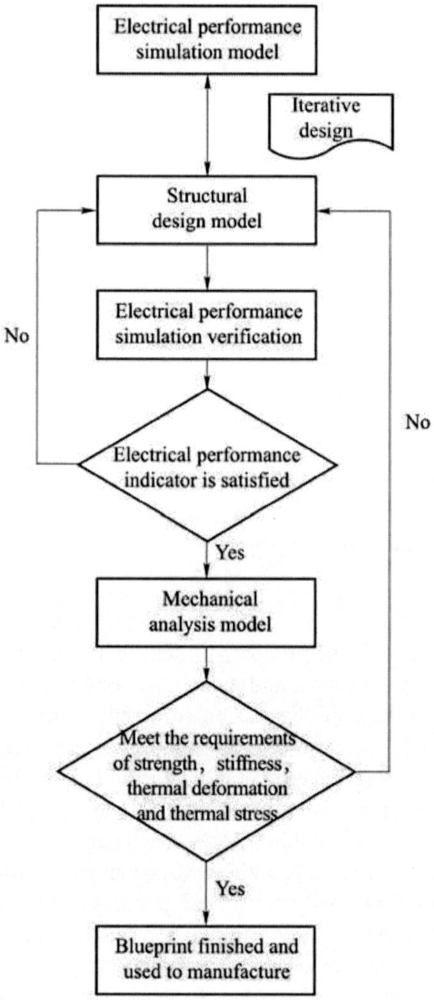

Fig. 9.1 Design flow of typical antennas

electrical performance index, structural strength and stiffness must meet requirements of satellite design and construction specifications. In addition, lightweight design should be carried out under the consideration of cost control on the premise of meeting the above requirements.

9.1.1 Material Selection for Antenna Structure

Common metal materials for antennas are shown in Table 9.1.

Beryllium bronze with high-elastic modulus, tensile strength and good weldability is widely used in the radiation part of antenna, such as helix, inner and outer conductors.

Aluminum alloy features high-specific strength and stiffness and relatively low cost is used in antenna mounting flange, choke ring and bracket. Under the higher requirement on weight reduction, the choke ring and bracket can be formed by carbon fiber composite, but its weak point is relatively high cost.

Magnesium alloy with low density can also be used in non-load-bearing parts, such as choke ring. This material can effectively reduce weight. However, due to the low-elastic modulus and tensile strength of magnesium alloy, it is not recommended to be used as bearing parts.

Titanium alloy or elastic alloy with relatively small thermal expansion coefficient can also be used when a part of an antenna requires high-thermal stability. For example, elastic alloy is often used for the feed of reflector antenna with high frequency.

Table 9.1 Common metal materials

Material name	Code/grade	Delivery status	Standard
Beryllium bronze wire	QBe2	Y2	GB 3134
Beryllium bronze bar	QBe2	Y2	YS/T 334
Beryllium bronze plate	QBe2	Y2	GB/T 5231
Brass bar	H62	Y	GB/T 4423
Brass plate	H62	Y	GB/T 2041
Aluminum wire	2 A12	T4	GB/T 3191
Aluminum bar	2 A14	T6	GB/T 3191
Aluminum plate	2 A12	T4	GB/T 3880
Titanium bar	TC4	R	GB/T 2218
Magnesium alloy	MB2	R	GB/T 5154
Elastic alloy	4J36	–	YB/T 5241
Stainless steel	1 Crl8Ni9Ti	–	GB/T 1220
Stainless steel wire	1 Crl8Ni9Ti	No requirements, Q	GB/T 4240

Table 9.2 Common dielectric materials

Material name	Code/grade	Standard
Teflon bar	SFB-1, SFB-2	QB/T 3626
Polyimide	YS20	HU Q/GHAE19
Epoxy laminated glass cloth plate	3240	GB/T 1303.1

The dielectric materials commonly used in antennas are shown in Table 9.2.

Polytetrafluoroethylene (PTFE) is usually used as the dielectric material in the antenna socket.

Polyimide is used for the dielectric screw and dielectric cylinder etc., which play an auxiliary supporting role in the antennas.

Epoxy laminated glass cloth plate is often used as the supporting material with larger size in antennas, while PTFE is used for the smaller support between inner and outer conductors.

Compared with PTFE, polyimide has higher modulus, lower density and good adhesion; compared with epoxy laminated glass cloth plate, it has better machinability, and its raw materials are in plate or bar form, which can be used in many structural forms.

Antennas are generally equipped with protective shields, which are made of the quartz fiber composite material.

9.1.2 Antenna Structure Design

9.1.2.1 Force Path Optimization Design

Space-borne antennas undergo complex mechanical environment during satellite launching, so they must have enough strength and stiffness. Since the structure of antennas often depends on electrical performance, its force path will be optimized in structural design to ensure the safety margin of the structure on the premise of ensuring the proper electrical performance. For example, a conical equiangular spiral strip antenna is composed of an outer conductor, an inner conductor, a spiral strip and a mounting flange according to the results of electrical design, as shown in Fig. 9.2. In structural design, a dielectric support cone is added for better stiffness of the spiral belt; a dielectric material is filled between the inner and outer conductors to ensure the stiffness of the inner and outer conductors, and a dielectric support sheet is used to connect the inner and outer conductors with the dielectric support cone for effectively improving the transmission path. The structural design model is shown in Fig. 9.3. With a series of design measures, the overall stiffness of the antenna will be greatly improved, and its strength will also be enhanced.

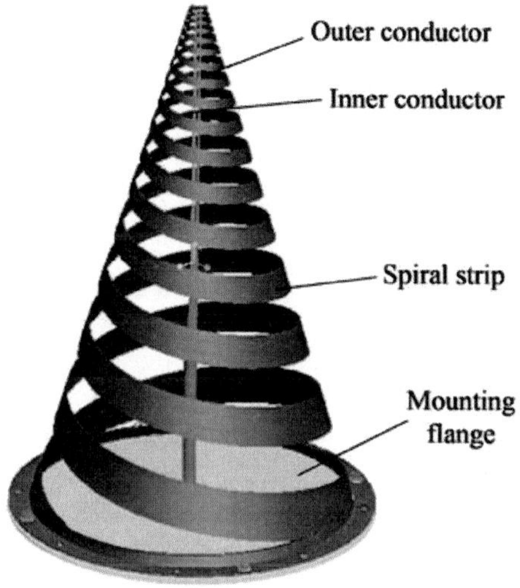

Fig. 9.2 Principle model of conical equiangular spiral strip antenna

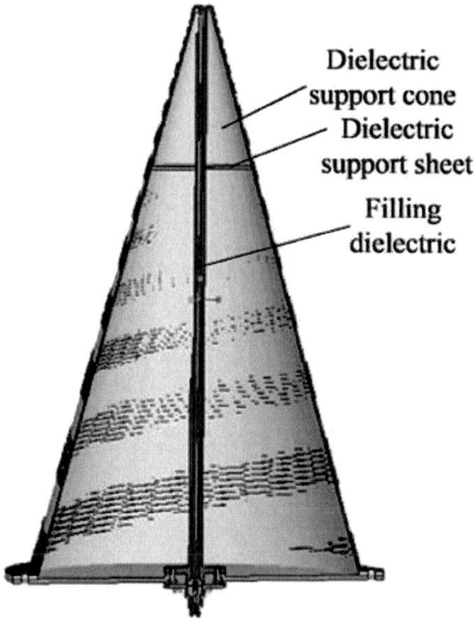

Fig. 9.3 Structure model of conical equiangular spiral strip antenna

9.1.2.2 Rational Design for Structural Connection Processes

The common connection processes of space-borne antennas include screw joint, glue joint, welding, etc. Among them, screw joint is the most commonly used connection mode, which is convenient for disassembly and assembly. Of course, all screw joints must have corresponding loosening prevention measures—with gluing or soldering for loosening prevention. The structural adhesives, conductive adhesives and anti-loosening adhesives are commonly used for antennas. As their name implies, by choosing adhesives, corresponding adhesives can be selected according to different uses. The common welding methods of antennas include soldering, silver brazing, electron beam welding, laser welding, etc. They are selected according to different materials and applications. Taking a bifilar helical data transmission antenna as an example (Fig. 9.4), beryllium bronze is used for helix and outer conductor, and silver brazing is used for the connection between helix and outer conductor.

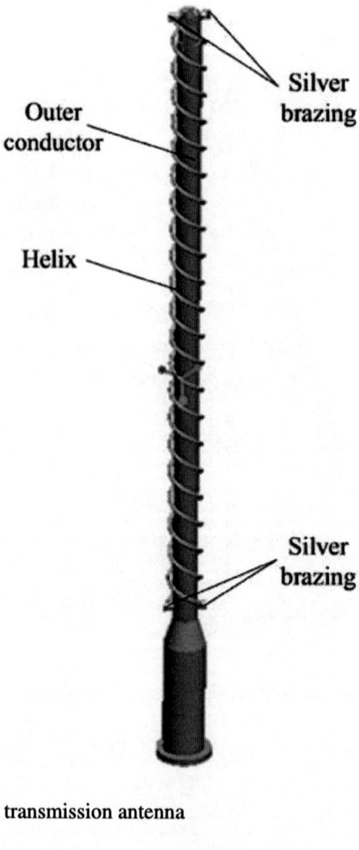

Fig. 9.4 Bifilar helical data transmission antenna

9.1.2.3 Influence of Thermal Deformation and Stress

Since spaceborne antennas undergo complex space environment, thermal deformation and stress should be considered in structural design. The main structure of the antenna should choose the same material or the material with similar thermal expansion coefficient as far as possible. Taking the bifilar helical data transmission antenna shown in Fig. 9.4 as an example, its outer conductor, inner conductor and helix are made of beryllium bronze. In addition, elastic alloy materials with relatively small thermal expansion coefficient can be used for parts requiring high-position accuracy, such as the feed of reflector antenna.

For the typical antennas which consist of inner and outer conductors, its inner conductors are usually slender rods, whose two ends shall be not fixed in the structural design, so as to avoid that the thermal deformation of the inner conductors cannot be released in the case of alternating heat and cold, and then the bending deformation occurs, which ultimately affects the electrical performance of the antenna. In this case, the inner conductor is generally fixed at one end, and the other end is designed with anti-deformation structure.

9.1.2.4 Venting Hole Design

Since spaceborne antennas undergo alternative load caused from ground to vacuum environment of space, ventilation should be provided in the structure. For the closed cavity structure, venting holes can be designed on the side wall, or venting grooves can be arranged at the socket position.

Generally, the number of venting holes should meet the requirements of the following formulas:

$$N \geq \frac{0.0017D}{S_{\mathrm{A}}} \tag{9.1}$$

where N is the number of venting holes to be designed, D is the internal volume of components, in m^3, S_{A} is the area of vents, in m^2.

9.1.3 Antenna Mechanical Analysis

After the structural design of the antenna product is completed, the structural strength and stiffness are verified by mechanical analysis to meet the requirements of the satellite design and construction specifications. The typical mechanical analysis process of antennas is shown in Fig. 9.5. The first is to ensure the accuracy of mechanical analysis model. A modal analysis, sinusoidal vibration analysis, random vibration

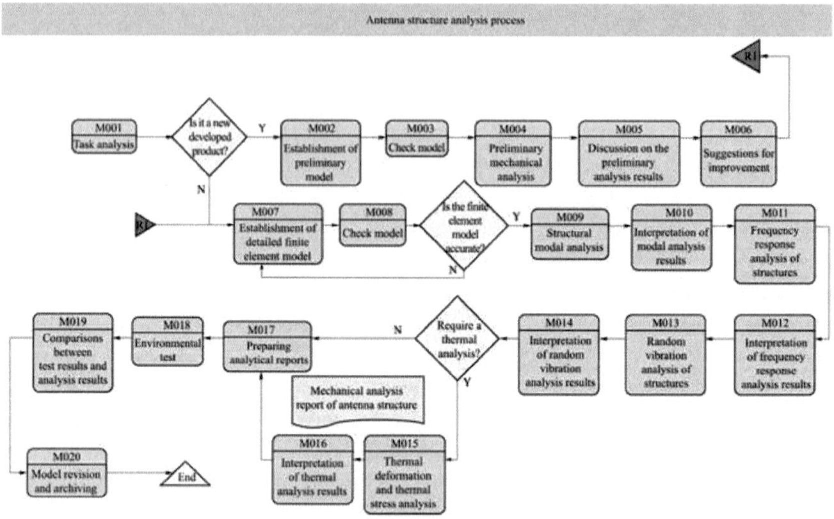

Fig. 9.5 Structural mechanics analysis flow of typical antennas

analysis, thermal deformation and thermal stress analysis are generally performed for antennas. The results of the analysis can provide guidance for the experiment, and can be used to verify the accuracy of the analysis model and guide the model modification.

9.1.3.1 Analytical Model Modeling

The establishment of structural analysis model refers to the transformation of actual structure into a mathematical model for analysis and calculation. Generally, the analysis model will simplify the actual structure in varying degrees, but must still retain the main mechanical characteristics of the structure. The establishment of analytical model is a very important part in structural analysis, which determines the rationality and correctness of the analysis results to a large extent.

1. Determination of structural geometric features

In the modeling of mechanical analysis, the geometric characteristics of the structure are first to be determined. The geometrical feature of the structure means that the component has several features such as rod, beam, surface, and body. In the process of structural simplification, it can be abstracted into lines, surfaces or bodies, which are simplified into rod, beam, shell and solid elements.

For example, a helix can be simplified to a beam element, as shown in Fig. 9.6a; a bracket can be simplified to a plate-shell element, as shown in Fig. 9.6b.

In addition, the bracket parts with complex structure can be simplified to solid elements, as shown in Fig. 9.7.

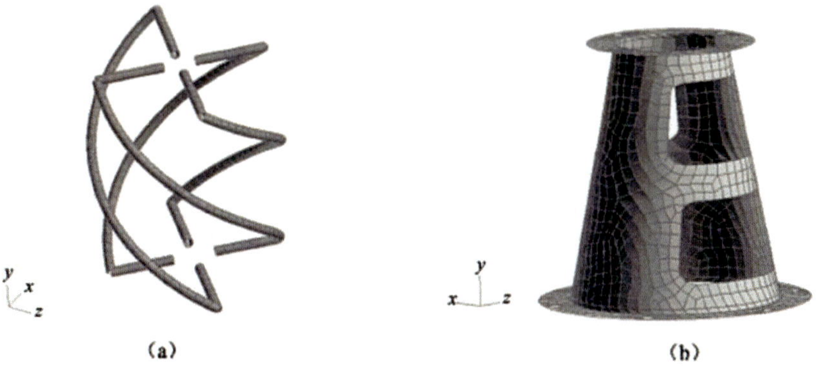

Fig. 9.6 A simplified example of structural modeling. a The application of beam element in the structural mechanics analysis of antenna; b the application of plate-shell element in the structural mechanics analysis of antenna

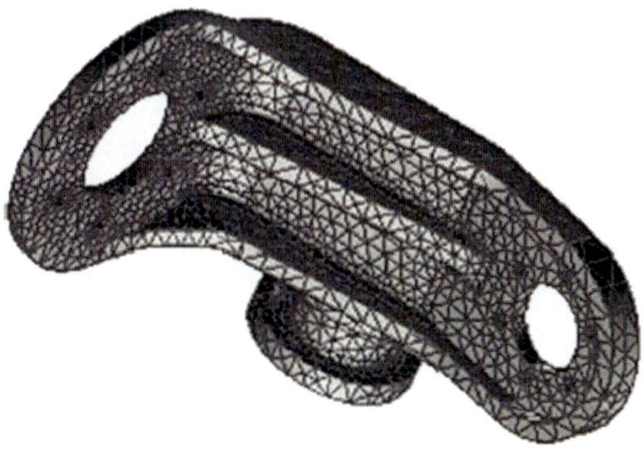

Fig. 9.7 Application of solid element in antenna structural mechanics analysis

2. Determination of connection and boundary conditions

The connection modes of welding and screw connection in antenna structure can be simplified to rigid elements, as shown in Fig. 9.8.

For the screw connection in the antenna structure, the screws can be simplified to beam elements, and then connect to the connected pieces through a rigid element, as shown in Fig. 9.9.

In terms of the boundary condition, by comparing and improving the results of the analysis and the test, it is found that the analysis results are in good agreement with the test results when the following methods are used: drawing the screw connection holes in the modeling, proper local dense gridding of hole edge, and then fixing and supporting the node at the screw connection holes, as shown in Fig. 9.10.

3. Definition of element characteristics

The definition of element characteristics is to assign material and cross-section characteristics to elements.

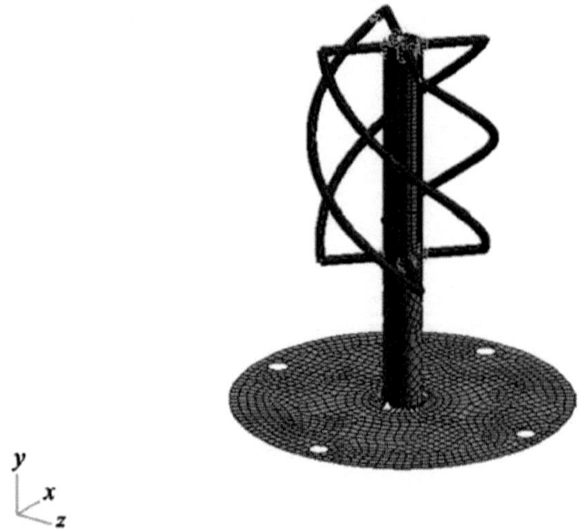

Fig. 9.8 Simplification of welding in mechanical analysis of antenna structures

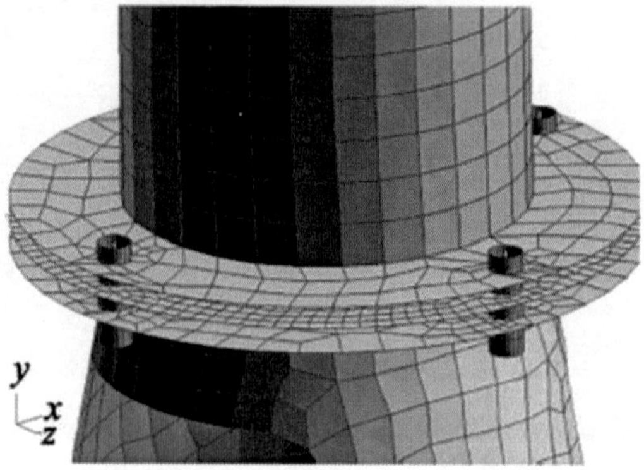

Fig. 9.9 Simplification of screw connection in mechanical analysis of antenna structures

The mechanical properties of materials are defined according to the used structural materials and the analyzed contents. In general, metal materials are defined as isotropic materials, and composite materials and honeycomb cores are defined as anisotropic materials.

In the application of shell element, only the thickness of common metal thin wall is required to be defined. The sandwich plate or laminated composite material can be defined according to laminated plate (i.e., define according to the characteristics of each single layer).

Shell element can be divided into different regions according to different thicknesses, and the section attributes can be assigned to them, respectively, as shown

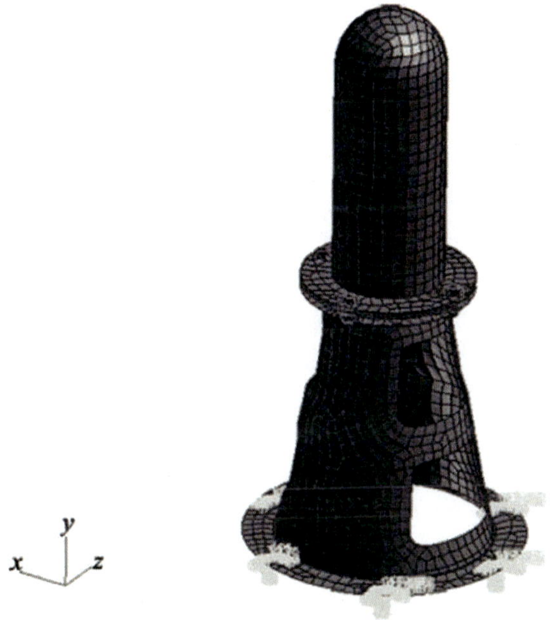

Fig. 9.10 Simplification of boundary conditions

in Fig. 9.11; beam elements can also be divided into different sections according to the section area, and the section attributes can be assigned to them, respectively, as shown in Fig. 9.12.

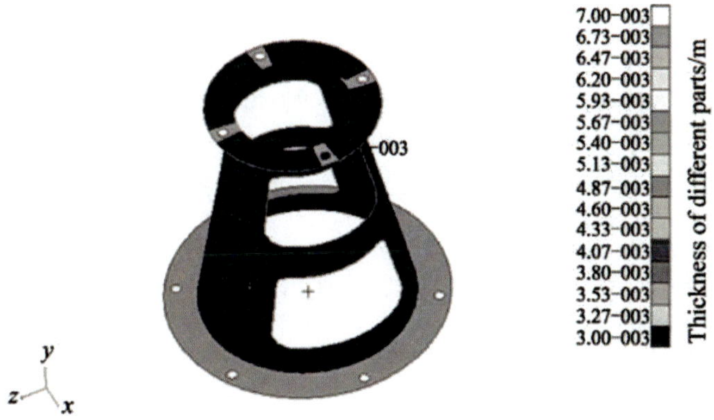

Fig. 9.11 Section properties of surface element

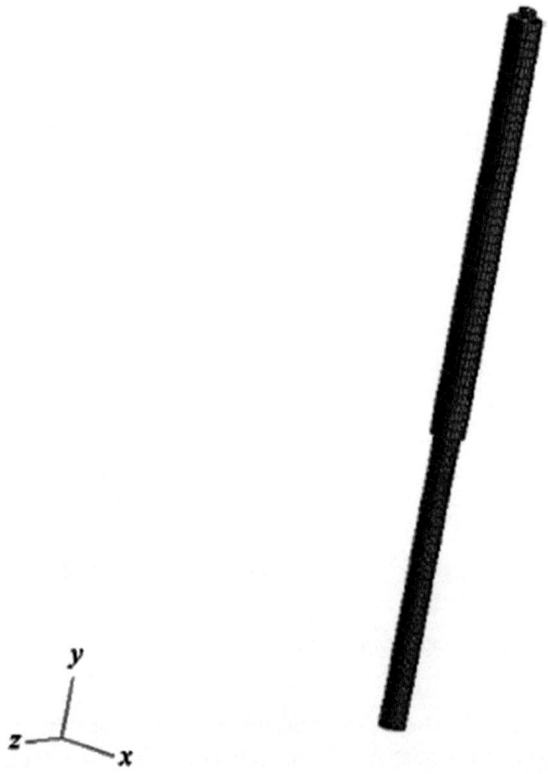

Fig. 9.12 Section properties of beam element

4. Check of computational analysis model

In order to establish an effective structural analysis model, it is necessary to check the analysis model to ensure the correctness of the analysis model. Commonly used analytical model checking methods include quality characteristic, free mode, stiffness test, etc.

(a) In **quality characteristic test**, the mass, moment of inertia and center of mass of the model are checked.
(b) In **free mode test**, the boundary constraints are relaxed. A structure without boundary constraints should have six rigid body modes with zero inherent frequency. If the results of analysis are less than six zero modes, the model has redundant constraints. Otherwise, the modal is in "mechanism" state, that is, there is no sufficient constraint on the structure.
(c) **Stiffness test** is used to check for improper connections between elements and unnecessary boundary conditions. For an unconstrained structure, a unit displacement is applied at a node along each translation and rotation direction, and the resulting unit load or strain should be close to zero.

9.1.3.2 Modal Analysis

Structural modal analysis aims to analyze the modes of the whole spacecraft or its main components, including inherent frequency and corresponding mode shape.

The characteristic equation of free vibration is

$$(\boldsymbol{K} - \omega^2 \boldsymbol{M})\vec{\varphi} = 0 \tag{9.2}$$

Modal analysis is based on the known stiffness matrix $\boldsymbol{K}$ and mass matrix $\boldsymbol{M}$ to solve the characteristic equation for getting the mode frequency ω and the corresponding mode shape $\boldsymbol{\varphi}$, and to analyze the characteristics of mode shape $\boldsymbol{\varphi}$.

In addition, the effective mass of a modal can be calculated to determine whether a mode is an integral mode (the main mode) or a local mode, and the direction of the mode.

9.1.3.3 Frequency Response Analysis

Frequency response analysis of structures is the response analysis of structures under basic sinusoidal excitation.

The distribution of sinusoidal excitation with frequency is generally represented by displacement amplitude and acceleration amplitude in the section, which should be transformed into a unified unit in the analysis, so it is generally expressed by the acceleration amplitude. Table 9.3 shows the general test conditions for sinusoidal vibration of component in qualification level.

In the analysis of sinusoidal vibration of antenna structure, the test magnitude corresponding to 5–20 Hz should be converted from displacement to acceleration. The conversion process is as follows.

The relationship between sinusoidal vibration, acceleration amplitude and displacement amplitude is

$$x = d \sin \omega t \tag{9.3}$$

$$\frac{\partial^2 x}{\partial^2 t} = -d\omega^2 \sin \omega t = a \sin(\omega t + \pi) \tag{9.4}$$

Table 9.3 General test conditions for sinusoidal vibration of components in qualification stage

Component characteristics	Frequency (Hz)	Test magnitude (0 ~ p)
First-order frequency > 100 Hz Mass < 23 kg	5–20	11 mm
	20–100	20 g

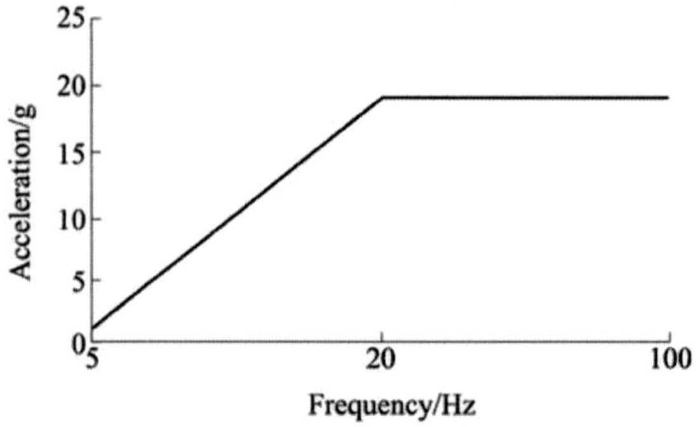

Fig. 9.13 Sinusoidal vibration analysis input

where d is the displacement amplitude, a is the acceleration amplitude, ω is the angular frequency, $\omega = 2\pi f$, then the relationship between the acceleration amplitude a and the displacement amplitude d is

$$a = (2\pi f)^2 d \tag{9.5}$$

Take the gravitational acceleration g (= 9800 mm/s^2) as a unit, then

$$a = \frac{df^2}{250} g \tag{9.6}$$

Then, the acceleration value corresponding to frequency of 5 Hz is

$$a = \frac{df^2}{250} g = \frac{11 \times 5^2}{250} g = 1.1g \tag{9.7}$$

Therefore, the input curve of the sinusoidal vibration of the antenna structure is shown in Fig. 9.13.

In addition, the selection of damping ratio has a great influence on frequency response analysis. The damping ratio, denotes as ξ, is a system parameter to represent the damping capability that the vibration energy consumed by system. If the inherent frequencies of two vibration systems are the same, the free vibration attenuation of the system with larger damping ratio is faster. If the damping ratios of the two systems are equivalent, the free vibration attenuation of the systems with higher inherent frequency will be faster. δ represents the natural logarithm of the ratio of adjacent amplitudes, which is only related to the damping ratio, as shown in Eq. (9.8).

$$\xi = \frac{\delta}{\sqrt{(2\pi)^2 + \delta^2}} \tag{9.8}$$

When $\xi \ll 1$, the approximate formula is shown in the common Formula (9.9)

$$\xi = \delta / 2\pi \tag{9.9}$$

In the vibration test, the response of the system damping under free vibration can be measured, then the logarithmic attenuation rate can be calculated, and finally the damping ratio of the system can be obtained. In random vibration analysis, the results satisfying the accuracy requirements can be obtained as long as the damping ratio at the resonant peak is input accurately, and the damping accuracy of other frequency bands can be relatively low. The damping ratio of metal structure is 2–3% and that of composite structure is 5–8%. In the random vibration analysis of the antenna structure, by comparing the analysis results with the experimental results, the analysis results are close to the experimental results when the critical damping ratio is taken as 2%.

9.1.3.4 Random Vibration Analysis

Random vibration is generally expressed by Power Spectral Density (PSD) function in frequency domain, and loaded by acceleration PSD. The main parameters describing random vibration include PSD, acceleration Root Mean Square (RMS), and frequency range. PSD is the distribution of the power contained in each frequency component of a random signal in the frequency domain, it is usually represented in W. For random signals, the acceleration should be calculated by the total RMS because random signal contains abundant frequency components. Practically, all kinds of conditions of random vibration tests adopt double logarithmic coordinates to calibrate the vibration value of the test. Figures 9.14 and 9.15 show the ascending and descending spectrum calibrated by double logarithmic coordinates, respectively.

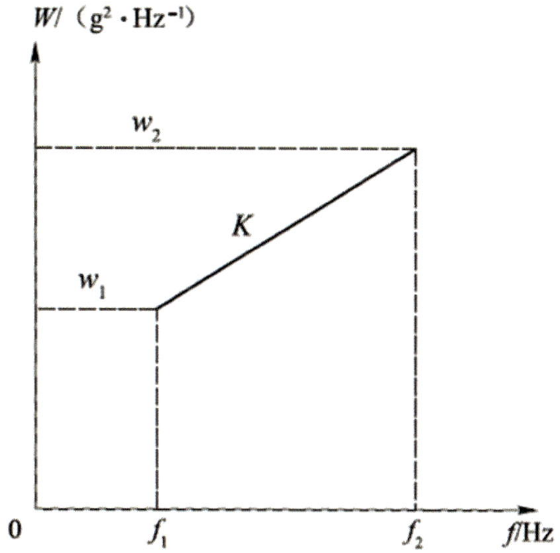

Fig. 9.14 Ascending spectrum

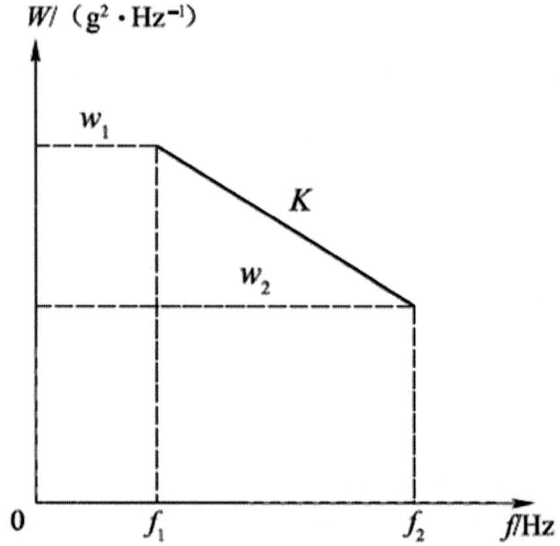

Fig. 9.15 Descending spectrum

The functional expressions of ascending and descending spectra are shown in Eqs. (9.10) and (9.11), respectively.

$$W(f) = W_2 \cdot (f_2/f)^{-k_1/3} \quad (9.10)$$

$$W(f) = W_2 \cdot (f_2/f)^{k_2/3} \quad (9.11)$$

In the following, the application of Eqs. (9.10) and (9.11) in practical will be illustrated with examples. The component-level random vibration loads of a satellite are shown in Table 9.4.

When the frequency is 10 Hz, the corresponding PSD of the qualification level is

$$W(10) = 0.2 \times (100 \div 10)^{-3/3} = 0.02 \quad (g^2/\text{Hz}) \quad (9.12)$$

Table 9.4 Component-level random vibration test conditions of a satellite

Item		Qualification level	Remarks
		PSD	
Frequency range	10–100 Hz	0.02–0.2 g^2/Hz	Ascending slope +3 dB/oct (Octave)
	100–600 Hz	0.2 g^2/Hz	
	600–2000 Hz	0.2–0.054 g^2/Hz	Descending slope −9 dB/oct (Octave)
RMS acceleration		12.81 g_{rms}	
Time		2 min	
Direction		x, y, z	

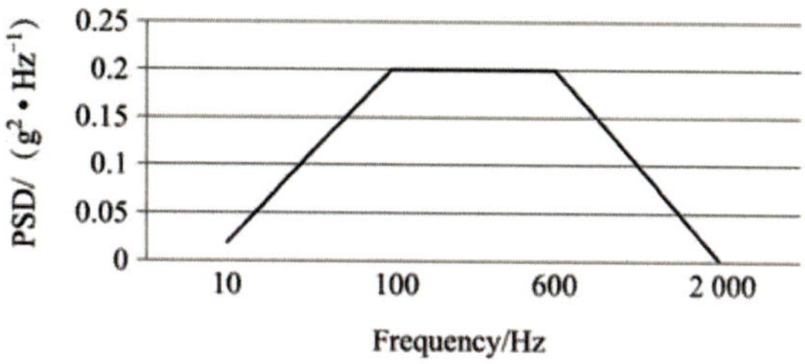

Fig. 9.16 Random vibration curve in qualification level

The PSD in the qualification level corresponding to the frequency of 2000 Hz is

$$W(2000) = 0.2 \times (2000 \div 600)^{-9/3} = 0.0054 \quad (\text{g}^2/\text{Hz}) \tag{9.13}$$

Therefore, the random vibration conditions in the qualification level provided in Table 9.4 can be transformed into the curves shown in Fig. 9.16, which can be input into the finite element software for calculation.

9.1.3.5 Safety Margin

In the design and construction specifications for satellites, the structural components of the equipment should have a positive safety margin under extreme loading conditions for yield limit and strength limit:

$$\text{M.S.} = S_a/S_e - 1 \tag{9.14}$$

where M.S. is the safety margin, S_a is the allowable stress (yield limit and strength limit), S_e is the stress generated by the product of the qualified load and the corresponding safety factor.

As shown in Table 9.5, the safety margin should meet the requirements for different materials.

Based on mechanical analysis, the safety margin of the antenna structure corresponding to frequency response analysis and random vibration analysis can be calculated.

In the frequency response analysis of the antenna structure, the analysis conditions are input according to the qualification level load conditions, and the maximum stress

Table 9.5 Safety margin

Metallic material safety margin		Non-metallic material safety margin	
Yield strength	≥0	First layer failure	≥0.25
Ultimate strength	≥0.12	Bearing strength	≥0.25
Stability	≥0.25	Stability	≥0.30

of different components can be obtained from the analysis results. In calculating the safety margin, S_e in Eq. (9.14) is the stress generated by the product of the maximum stress and the safety factor.

In the random vibration analysis of the antenna structure, whether the analysis results meet the requirements is also reflected by the safety margin. A large number of random vibrations in engineering can be considered as normal stochastic processes. A normal stochastic process means that the probability density function of a stochastic process is normal. For a normal process $X(t)$ with a mean μ_x of zero, the probability of taking a value outside the interval $(-3\sigma_x, \sigma_x)$ is 0.27%, and σ_x is the RMS value of X. Therefore, the maximum value of X can be taken as $3\sigma_x$.

In the random vibration analysis of the antenna structure, the 3σ criterion is used to check the safety margin of the structure, i.e., the probability that the stress does not exceed 3σ is 99.73%. Select the maximum stress RMS value $\boldsymbol{\sigma}$ of each component in the antenna structure and multiply by 3 to obtain the maximum 3σ stress of each component. S_e is the stress generated by the product of the 3σ stress and the safety factor for calculating the safety margin.

9.1.3.6 Analysis Example

For mechanical analysis, we take a directional antenna mechanical analysis as an example. The directional antenna consists of a reflector, a feed, a reflector bracket, a double-axis driving mechanism, a compress-release mechanism, an inter-axle bracket and a base, a cube mirror and mounting accessories for precision measuring at the edge of the reflector. In the modeling process, the reflector, the reflector bracket, the inter-axle bracket and the base can be simulated by the shell element; the feed is simulated by the shell element and the beam element; the casing of the double-axis driving mechanism is by the shell element, the stiffness is by the Bush element, and the uniaxial torsional stiffness and bending stiffness are assigned; the casing of the compression mechanism is simulated by the shell element, and the compression rod is by the beam element; the cube mirror and the accessory are by the mass element. The established directional antenna mechanics analysis model is shown in Fig. 9.17.

In the modal analysis, the natural frequency value and the corresponding vibration mode can be calculated by fixing and supporting the base mounting point. The first-order frequency is 84.25 Hz, and the corresponding vibration mode is shown in Fig. 9.18.

The frequency response analysis results show that the stress on the root of the directional antenna reflector and feed is large, as shown in Figs. 9.19 and 9.20, respectively. The safety margin of reflector and feed, which are calculated according to Formula (9.14), is 0.51 and 1.67, respectively, and meeting the safety margin requirements shown in Table 9.5.

The results of random vibration analysis show that the maximum stress of the reflector and the feed are 17.7 MPa and 56.5 MPa, respectively, and the safety margin is calculated according to Formula (9.14). The safety margins of the reflector and

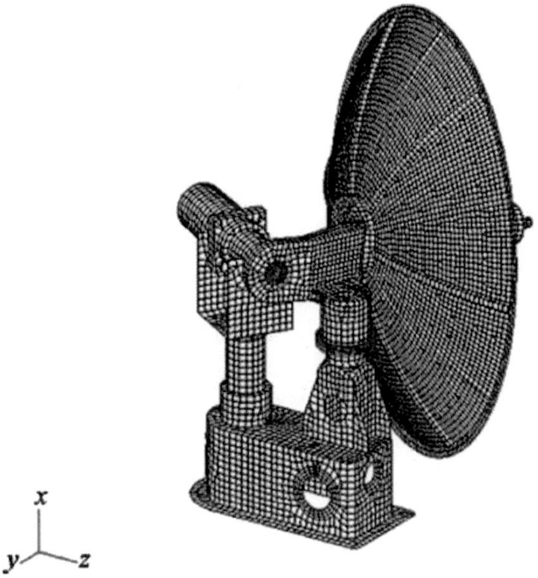

Fig. 9.17 Finite element model of directional antenna

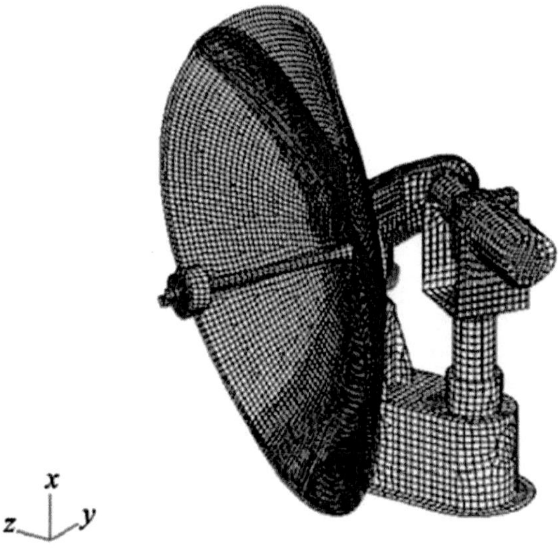

Fig. 9.18 First mode of vibration

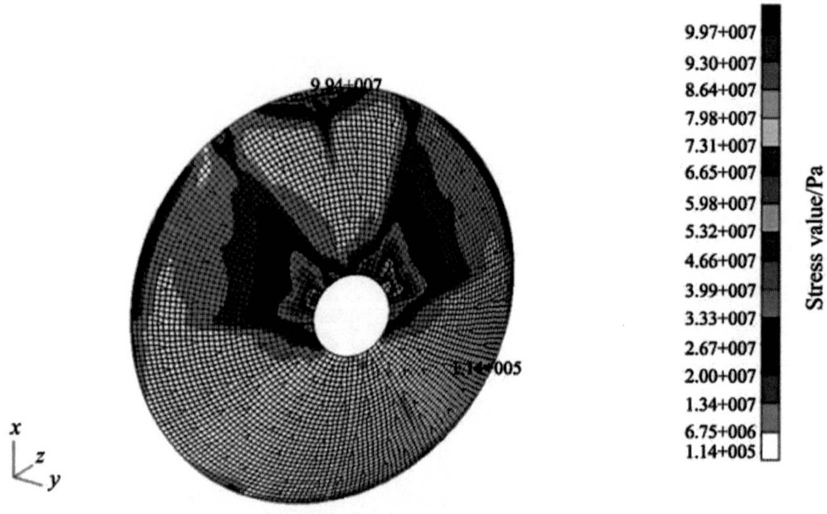

Fig. 9.19 Maximum stress nephogram of reflector

Fig. 9.20 Maximum stress nephogram of feed

the feed are 1.84 and 1.36, respectively, which meets the safety margin requirements shown in Table 9.5.

As can be seen from the mechanical analysis and modal analysis of the directional antenna, the fundamental frequency of the antenna is 84.25 Hz, and the corresponding mode shape is local warping. The results of frequency response analysis and random vibration analysis show that the safety margin of each component meets the design requirements.

9.2 Spacecraft Antenna Thermal Design

The spacecraft antenna is located on the surface of the spacecraft, works directly in the thermal environment of the orbital space, and subjects to greater thermal stress loads than that on the equipment in the cabin. Table 9.6 lists the operating temperature environment requirements for spacecraft antennas in different orbital environments.

9.2.1 Thermal Design Principles and Methods

In the antenna thermal design, passive thermal control measures should be priority taken for adjusting the heat exchange between the antenna and the space environment through multi-layer binding and thermal control coating, so as to improve the antenna temperature environment on-orbit. Active temperature control measures are applied to the components that cannot withstand the extreme temperature on-orbit, such as using a heating plate to raise the temperature of the components operating in the lower part of low-temperature operating range.

Based on clear technical parameters and a large number of on-orbit applications from the mature thermal control technology and products validated by flight, the results of design have high-confidence level.

The antenna should be insulated with the satellite as much as possible to minimize the thermal coupling and to reduce the mutual thermal coupling effect.

Table 9.6 Environmental requirements for spacecraft antenna operating temperature

Orbit	Temperature range (°C)
Medium- and low-sun synchronous orbit	−90 to +90
Medium-high orbit, geosynchronous orbit	−160 to +140
Deep space exploration	−160 to +135
Shenzhou spacecraft returning to orbit	−60 to +1100
Deep space exploration reentry orbit	−160 to +1400

9.2.2 Main Thermal Control Products of Antenna

9.2.2.1 Multi-layer Insulation Component

The multi-layer insulation component commonly used in antenna thermal design consists of reflective screen and spacer layer. The reflective screen is a 6 μm double-sided aluminized polyester film, the spacer layer is a polyester mesh, and both form a unit. The outermost layer of the component is a single-sided aluminized polyimide film with the conductive film facing outward. The aluminized film is with thickness of 25 μm, a solar absorption ratio $\alpha_s = 0.36$ (0.65, in the end of five years), and an infrared hemispherical emissivity $\varepsilon_h = 0.69$. The multi-layer insulation component of the antenna consists of 15 units (for the outside of the antenna bracket) with a thickness of 3–5 mm. The effective emissivity of the multi-layer insulation component ε_{eff} ranges from 0.02 to 0.12, and the infrared emissivity of the aluminized surface of the reflective screen is $\varepsilon_h = 0.1$.

Multi-layer insulation components are mostly fixed with polyimide pins, and a few are also fixed with nylon buckles. This not only facilitates the control of the tightness of multi-layer insulation components, but also is easy to be installed and removed.

Multi-layer insulation materials are generally used in antenna bracket, mechanism components, and back surfaces of reflectors to reduce the heat exchange between the antenna and the outside world.

9.2.2.2 White Thermal Control Paint

The white thermal control paint is generally applied to the surface of the antenna, and its solar absorption ratio α_s is 0.18 (0.36, in the end), and the infrared hemispherical emissivity ε_h is 0.87. It helps the antenna to radiate heat to the space under high temperature, and avoid the adverse effect of high temperature on the reliability of antenna structure.

9.2.2.3 Anti-static White Paint

Anti-static white paint is generally coated on the surface of the antenna, and its solar absorption ratio α_s is 0.21, and the infrared hemispherical emissivity ε_h is 0.85, which can promote the heat radiation of the antenna to the space under high-temperature conditions, and avoid the adverse effect of high temperature on the reliability of antenna structure, and prevent static electricity.

9.2.2.4 Black Thermal Control Paint

Black thermal control paint is mostly used on the surface of equipment in the cabin with solar absorption ratio α_s of 0.93 and the infrared hemispherical emissivity ε_h of 0.88. In some cases, antennas installed outside the cabin will also be painted with black paint, which should be considered based on the overall satellite demand.

9.2.3 *Thermal Design of a Satellite DCS Antenna*

The receiving antenna of space-borne Data Collection System (DCS) adopts a self-fed phase four-arm helical antenna and its structure is shown in Fig. 9.21.

The DCS antenna thermal design measures are as follows:

1. Insulation to the satellite body. A 5-mm-thick fiberglass insulation pad is placed between the antenna bracket and the satellite to reduce the thermal conduction coupling and the interaction between each other.
2. The part exposed to the space of the antenna bracket and the flange is partially covered with multi-layer insulation.
3. Spray the white paint SR107 ZK on the outer surface of the antenna feed tube.
4. No surface treatment on the antenna plate and the support piece.

9.2.4 *Thermal Analysis of Antenna On-Orbit*

After completing the antenna thermal design, the antenna on-orbit thermal analysis is carried out to verify whether the antenna thermal design meets the requirements of operating temperature range on-orbit. Still taking a satellite DCS antenna as an example, the influence of satellite configuration and layout, orbit, and the heat consumption of the antenna is analyzed on the antenna temperature.

9.2.4.1 Thermal Analysis Model

The DCS antenna itself has no heat loss, and its temperature distribution is mainly determined by the external heat fluxes, which is affected by the satellite orbit and flight attitude, and changes drastically in one cycle. Therefore, a transient model must be established for the thermal analysis of the DCS receiving antenna. The radiant heat environment and temperature field of the antenna on orbit are calculated by Thermal Desktop. The antenna grid diagram and model diagram are shown in Fig. 9.22, in which the satellite deck is not included. The whole antenna is divided into 452 nodes.

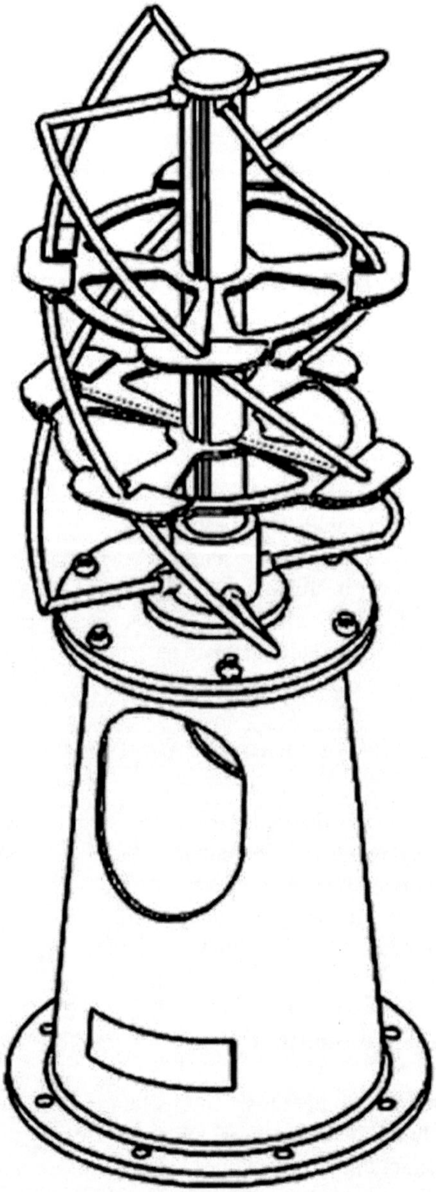

Fig. 9.21 DCS receiving antenna

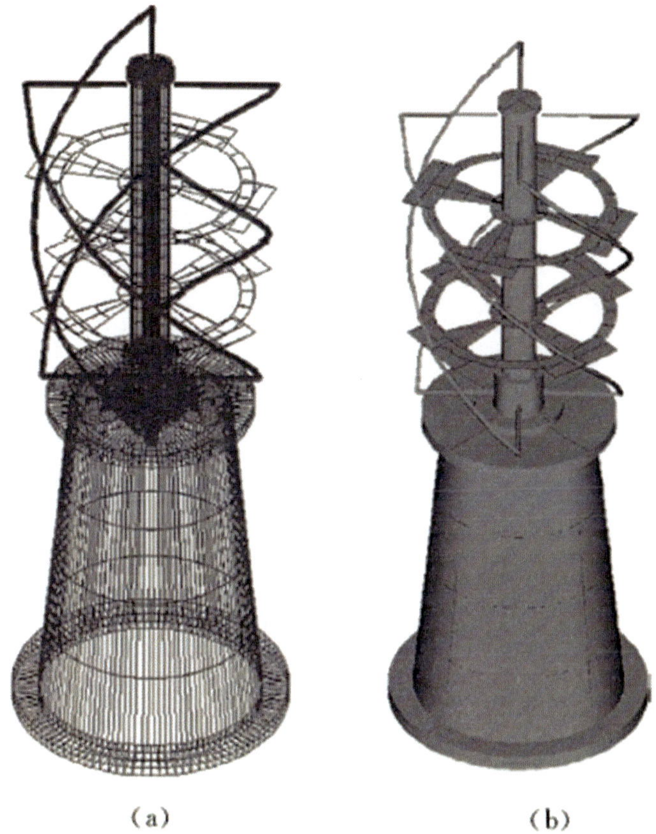

Fig. 9.22 Antenna thermal analysis model. **a** Grid map; **b** model map

9.2.4.2 External Heat Flux Analysis

The sun is the main heat source for the spacecraft on orbit, and the key determinant of the antenna's temperature condition on orbit. The change of the satellite β angle (the angle between the Sun and the Earth connection line and the satellite orbital plane, which characterizes the intensity variation of the satellite in the orbit by sunlight) is calculated, as shown in Fig. 9.23.

Based on the above calculation results of β angle, in combination with the variation law of solar constant within one year, the typical time of the satellite orbital heat fluxes can be determined as follows: ① near summer solstice, June 21, β angle is the smallest, all surfaces except the $-y$ plane have the largest heat fluxes by sunlight; ② the β angle is the largest on the February 26 or the October 15, which both are about 86.8°, and the solar radiation is nearly perpendicular to the satellite orbital plane. In this case, the solar radiation heat flux on the $-y$ plane is the largest, while that

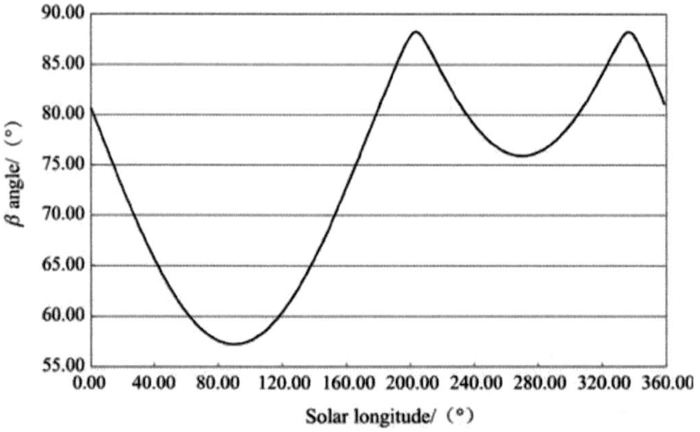

Fig. 9.23 Change of β angle in one year

on the other surfaces are the smallest, and the satellite operates relatively in low-temperature operating condition. Since the β angles are equal, the orbital heat fluxes are difference mainly due to the difference in solar constant. The solar constant on October 15 is 1374 W/m^2, which is smaller than that on February 26, so October 15 is another typical time; ③ near winter solstice, the solar constant is 1412 W/m^2, which is the largest.

The heat fluxes on the satellite surface are calculated in six directions. The average external heat fluxes in a typical season are shown in Table 9.7.

In order to determine the changes in the external heat fluxes of the on-orbit satellite at different times, transient calculation is also performed. From the ascending node, an orbital period is divided into 72 points by angle. For each point, arrived external heat fluxes are calculated in each direction according to the relative position of the satellite and the Earth, and the heated condition of the satellite is obtained at different times in a period. Figures 9.24, 9.25, 9.26 and 9.27 show the heat flux curves on an orbit.

The change rule of external heat fluxes in different seasons is complicated. It can be seen from the transient heat fluxes curve that the variation amplitude of the external heat fluxes on the $+z$ plane varies under different operating conditions. It is the flattest on October 15, and has large fluctuations in one cycle on the summer solstice and winter solstice.

The external heat flux of satellite is very different in six directions. The spiral, the support and the feed tube of the DCS antenna are greatly affected by the heat fluxes in the $\pm y$ direction. The greatly changed heat fluxes in the $\pm y$ direction result in a certain temperature difference among the antenna spiral, the bracket and the feed tube in the $\pm y$ direction. The spiral support and the flange of the antenna are greatly affected by the external heat fluxes in the $+z$ direction.

Table 9.7 The arrived external heat fluxes on the satellite surface in all directions in W/m^2

Orientation	Arrived external heat fluxes								
	Winter solstice (December 23)			Near the summer solstice (June 21)			October 15		
	Sun	Earth albedo	Earth infrared radiation	Sun	Earth albedo	Earth infrared radiation	Sun	Earth albedo	Earth infrared radiation
$+x$	105.92	7.26	48.37	211.08	13.52	45.45	8.2	3.9	47.9
$-x$	105.92	7.27	48.49	211.08	13.55	45.35	8.2	3.9	47.6
$+y$	0	2.13	48.6	0	8.84	45.1	0	0	47.7
$-y$	1 356.91	14.18	48.47	955.34	18.55	45.46	1 373.7	12.2	47.5
$+z$	105.92	27.59	184.7	131.98	51.5	172.93	8.2	14.4	181.4
$-z$	105.92	0	0	220.32	0	0	8.2	0	0

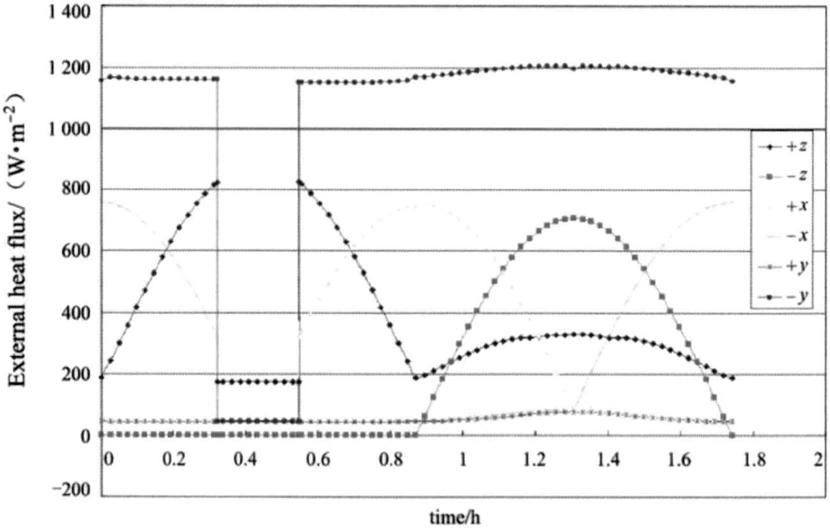

Fig. 9.24 Transient curve of the arrived external heat flux on each surface in summer solstice

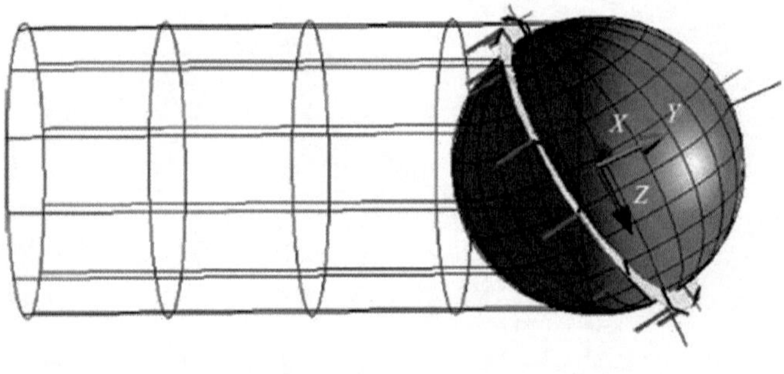

Fig. 9.25 Sketch of position of summer solstice orbit relative to the Earth and the illumination

9.2.4.3 Operating Condition Selection

Based on the above analysis of the β angle and the external heat fluxes, the typical time of the satellite's orbital external heat flux is determined. Therefore, the operating conditions are set according to the typical time period, as shown in Table 9.8.

The thermophysical parameters of the main materials in the calculation are shown in Table 9.9.

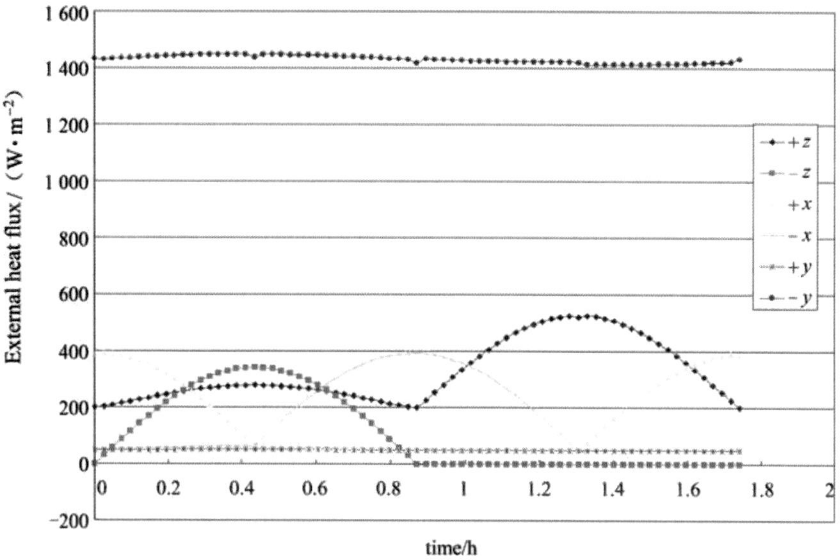

Fig. 9.26 Transient curve of the arrived external heat flux on each surface in winter solstice

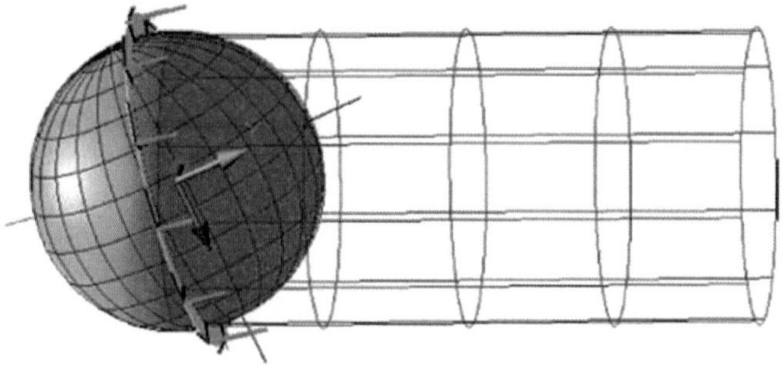

Fig. 9.27 Sketch of position of winter solstice orbit relative to the Earth and the illumination

Table 9.8 Operating condition selection

No	Working condition type	Orbital moment	Solar constant (W m^{-2})	Coating degradation	Boundary temperature (°C)
Operating condition 1	Initial stage at low temperature	October 15	1374	No	−10
Operating condition 2	End stage of the summer solstice at high-temperature 1	Summer solstice	1323	Yes	+45
Operating condition 3	End stage of the winter solstice at high-temperature 2	Winter solstice	1412	Yes	+45

Table 9.9 Selection of thermophysical parameters

Material name	α_s		ε_h	Thermal conductivity [W (m °C)$^{-1}$]	Specific heat capacity [J (kg °C)$^{-1}$]
	Initial stage	End stage			
Polyimide secondary surface mirror	0.36	0.65	0.69	–	–
Beryllium bronze gold-plated surface	0.24	0.30	0.05	–	–
Epoxy phenolic laminated glass cloth	0.82	0.82	0.8	0.2	1400
PTFE	–	–	0.90	0.244	1050
Duralumin 2A12	0.34	0.40	0.13	117.2	921
Brass H62	0.4	0.4	0.15	108.86	400
Beryllium bronze	–	–	–	104.67	440
Glass fiber composite	0.82	0.82	0.8	0.25	1150

9.2.4.4 Thermal Analysis Result

Based on the selected operating conditions of the previous section, a transient thermal analysis model is established, and the calculation results are given in the form of a temperature graph. In the temperature graph of Fig. 9.28, the abscissa is time in seconds and the ordinate is temperature in °C. Figure 9.28 shows the temperature of several consecutive cycles after the antenna injection and stable.

As can be seen from the above results that the temperature of each part of the antenna is as follows.

Under the initial low-temperature condition on October 15, the DCS antenna is at the lowest temperature, and the minimum temperatures of the spiral, spiral support, antenna support and antenna platen reach −12.48 °C, −29.18 °C, −14.22 °C and −15.11 °C, respectively. This is because the satellite orbit is almost perpendicular to the sun incident light on October 15. The sunlight is directed from the −y direction to the antenna, the antenna +*y* plane is blocked by the antenna itself, and the external heat fluxes are low. The temperature of the deck where the antenna is located is low, and heat is leaked to the deck. However, it seems that all equipments are in the range of −90 to +90 °C, thus meeting the temperature specification.

Under the high-temperature operating condition at the end of the summer solstice, the DCS antenna spiral and spiral support are at the highest temperature, up to 79.49 °C and 70.79 °C, respectively, meeting the temperature requirements of the antenna. Under the high-temperature operating condition at the end of the winter solstice, the temperature of DCS antenna spiral made of beryllium bronze is the

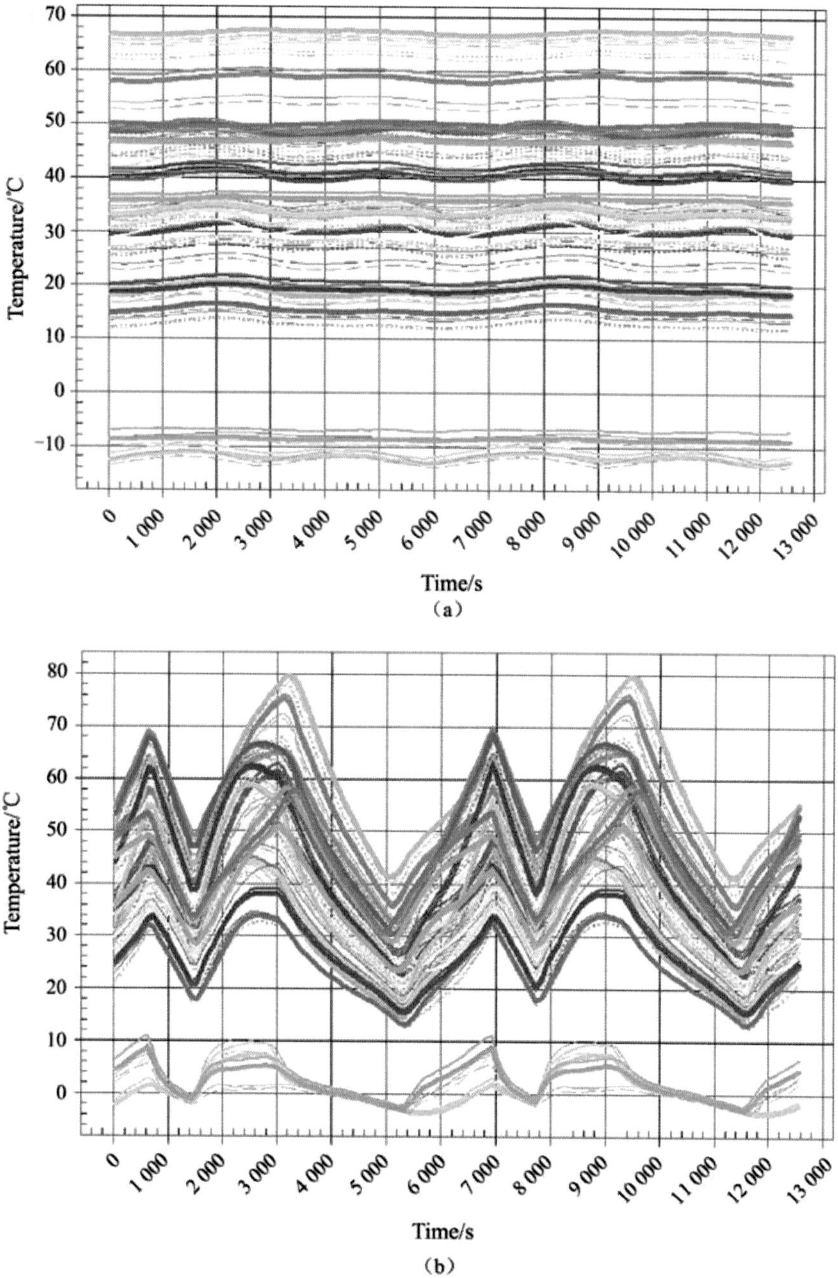

Fig. 9.28 Antenna injection temperature (different curves represent temperature changes at different locations). **a** Helix (working condition 1); **b** helix (working condition 2); **c** helix (working condition 3)

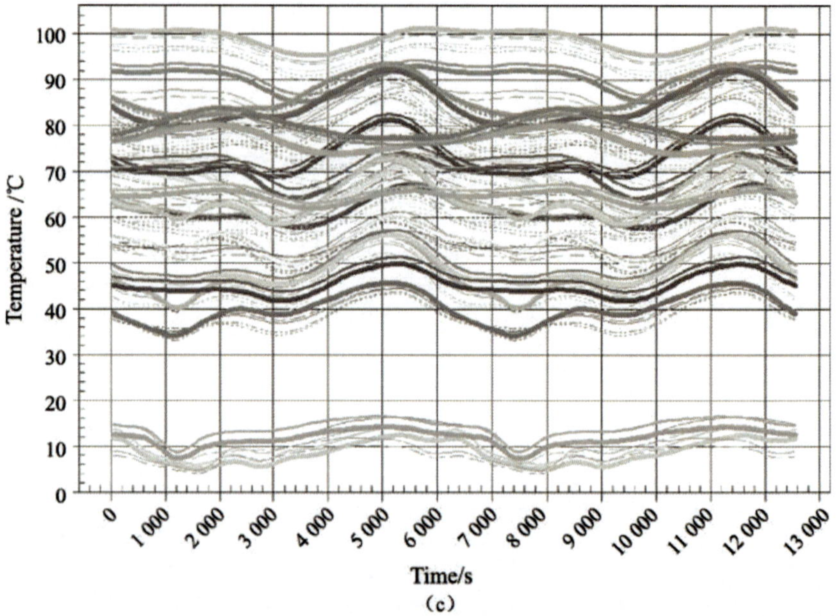

Fig. 9.28 (continued)

highest, reaching 101.42 °C. The temperatures of other components are all between 5 and 61 °C, which meets the requirements of antenna temperature specification.

9.2.5 Thermal Deformation and Thermal Stress Analysis

For the reflector antenna and the array antenna with high shape accuracy requirements, it is necessary to consider the influence of the thermal deformation of the antenna structure under the on-orbit temperature condition on the product surface accuracy. The temperature field distribution of the antenna is obtained through the on-orbit thermal analysis, the temperature field data is mapped into the finite element model of the antenna structure to further implement the thermal deformation analysis, thus obtaining the thermal deformation parameters of the antenna under the on-orbit condition, and checking the structural strength under the temperature load in parallel.

Figures 9.29 and 9.30 depict the temperature field distributions of an antenna array under the high-temperature operating condition. The temperature ranges from −11.3 to +76.8 °C. The temperature field distribution is mapped to the corresponding structural analysis model, and the linear expansion coefficient of each material is supplemented in the material parameters. The deformation and strength of the structure are calculated by using room temperature of 20 °C as initial temperature and temperature

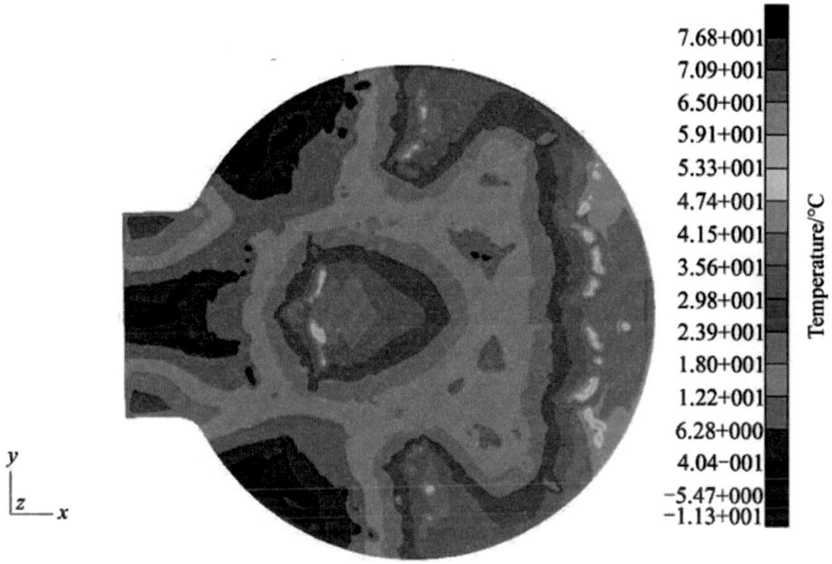

Fig. 9.29 Surface temperature field distribution under high temperature

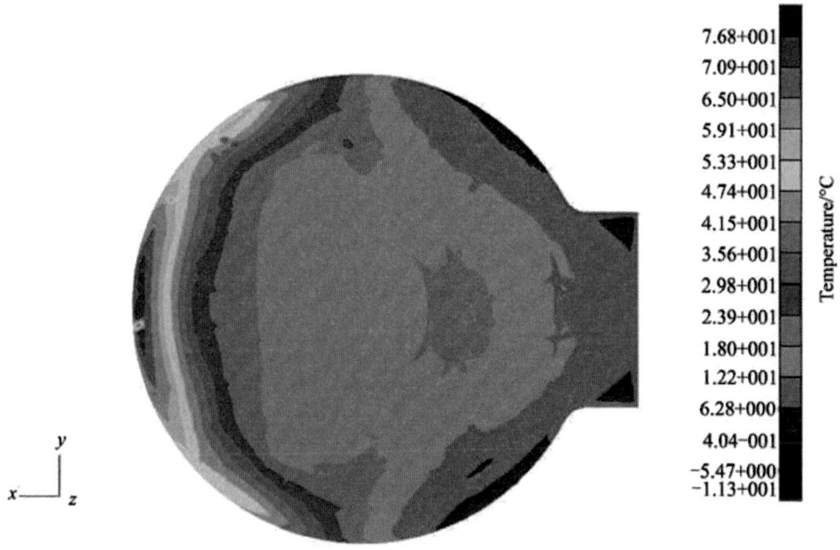

Fig. 9.30 Surface temperature field distribution under high temperature

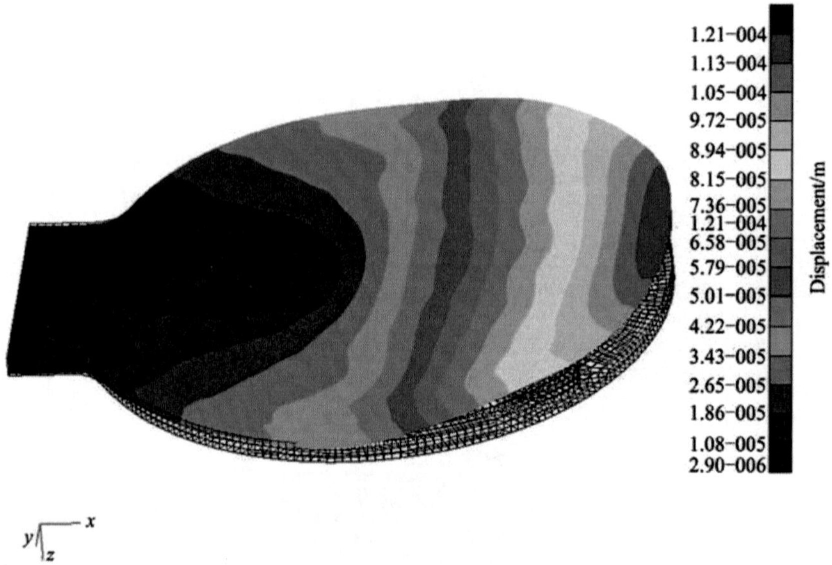

Fig. 9.31 Surface displacement area under high temperature

field distribution as load. The deformation is shown in Fig. 9.31. As can be seen from Fig. 9.31, the maximum displacement of the upper and lower surfaces of the antenna array under high temperature operating condition is 0.12 mm, which occurs at the mounting edge. Then the influence of thermal deformation on the performance of the antenna is analyzed based on the thermal stress shown in Fig. 9.32. Besides, the structural strength under the temperature load is checked according to the thermal stress.

9.3 Design of Spacecraft Antenna Multipaction Protection

Multipaction (also known as secondary electron multiplication) is a phenomenon of secondary electron multiplication between metal surfaces generated by electrons under the acceleration of a strong microwave electric field under vacuum conditions, that is, a radio frequency breakdown in passive components that transmit high-power microwave.

The multipaction effect is a vacuum resonant discharge phenomenon that occurs between two metal surfaces or on a single dielectric surface. It is usually excited by a radio frequency electric field transmitted in a component, and formed by a secondary electron, that is generated by an electron impacts on the surface, and then is accelerated in a radio frequency electric field to obtain energy. The conditions for its occurrence vary with the type of multipaction. For a multipaction that occurs on

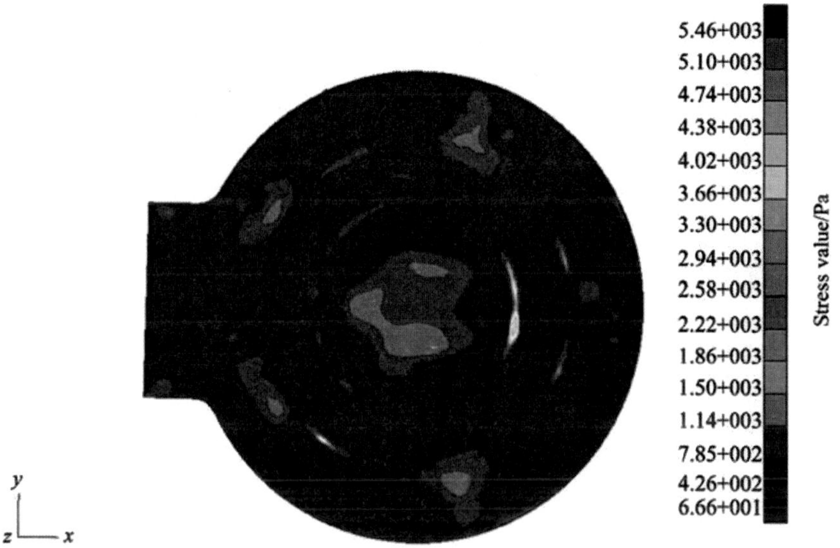

Fig. 9.32 Calculation of thermal stress caused by temperature gradient under high-temperature working condition

a single surface of a medium, the DC electric field generated by its surface charge must be capable of accelerating electrons back to the surface of the medium, thus enabling the generation of secondary electrons. For the multipaction between the metal surfaces, the conditions for its occurrence are that the electron mean free path must be greater than the clearance distance between the two metal surfaces, and the average electron transit time between the two surfaces must be an odd times of the RF electric field half-period.

For example, the multipaction effect between two plates must meet the preconditions that the electrons entering the plates are accelerated toward one plate in the positive half circle of the RF field, the electrons just impact the polar plate when the electric field passes through the zero point, then secondary electrons will be generated and accelerated back to the other plate during the negative half cycle. This process will be repeated. At each impact, the electron secondary emission releases more free electrons until in a steady equilibrium state, resulting in a multipaction effect. The specific process is given in Fig. 9.33 [1].

If a microwave passive component working in high-power state is not designed properly, multipaction is easily occurred when the power, frequency and internal structure gap size of the component satisfy a certain condition. This depends on various reasons such as processing technology, surface treatment, material properties and pollution. Once the multipaction phenomenon occurs, it will cause serious consequences, such as an increase in the standing wave ratio of the microwave transmission system; fluctuations in cavity tuning, coupling parameters, waveguide loss,

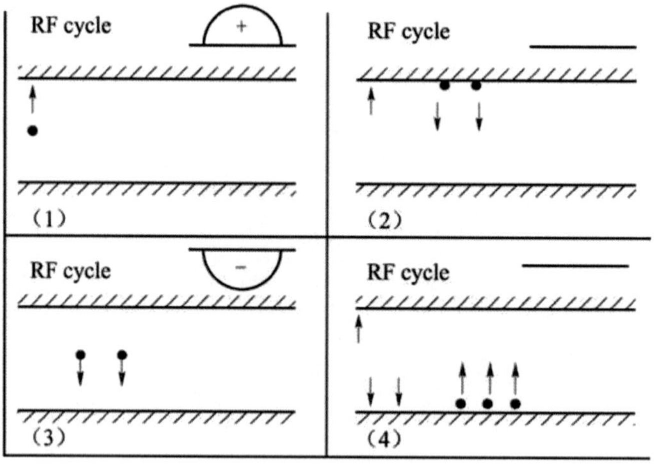

Fig. 9.33 Plate multipaction process

and phase constant caused by an increase in the reflected power; slow erosion of cables, connectors, and components; and an increased additional noise of the system.

When the permeability of the air passage inside the microwave element is not very satisfied, the multipaction effect in the individual component first induces a gas discharge that produces a stronger energy than the multipaction effect itself, and finally results in ultimate damage of the component.

The multipaction phenomenon is caused by secondary electron emission from the surface of the component. In a vacuum environment, the electron transit time can be calculated by the average free path length of the electron (compared to the size of the structure gap), the gap size (d) and the microwave frequency (period T) in the component:

$$t_\tau = (2n - 1)T/2, \quad n = 1, 2, 3, \ldots \tag{9.15}$$

If the surface secondary electron emission coefficient is greater than 1, the electron produces a phenomenon of secondary electron multipaction by electron avalanche resonance phenomenon under the acceleration of the strong microwave electric field.

For a sinusoidal varying RF field with a peak value of E, the motion equation of the electron with mass m is

$$m\frac{\mathrm{d}^2x}{\mathrm{d}t^2} = eE\sin(\omega t + \varphi) \tag{9.16}$$

where e is the electron charge, ω is the angular frequency and φ is the initial phase.

After integration, it is obtained

$$U = \frac{4\pi^2(fd)}{(e/m)\Phi} \tag{9.17}$$

Equation (9.17) gives the relationship between the breakdown voltage and the frequency f and the gap size d, where Φ is a function related to phase φ.

Based on the above formula is combined with experimental test data, ESA established a computer simulation model to obtain a multipaction sensitivity engineering curve between the parallel plates, which can be used both as a waveguide component design reference material and for other microwave transmission system design. Figure 9.34 shows the multipaction sensitive areas of different materials for the waveguide structure.

In terms of the numerical simulation of multipaction effect of microwave components with complex structures, a multipaction electromagnetic particle joint simulation method is established based on electromagnetic Finite Difference Time Domain method (FDTD) and particle simulation technology in combination with secondary electron emission simulation [3]. Under the consideration of the Coulomb force among real electrons and the influence of charge and current changes generated by electron motion on the electromagnetic field, the three-dimensional numerical simulation technology problem of multipaction in microwave components in complex structures is solved; the electromagnetic field value evolution is calculated in the uniform three-dimensional mesh and time stepping; electronic motion state change propulsion and secondary electron yield and energy distribution are calculated with the numerical model; the multipaction threshold is predicted based on the trend

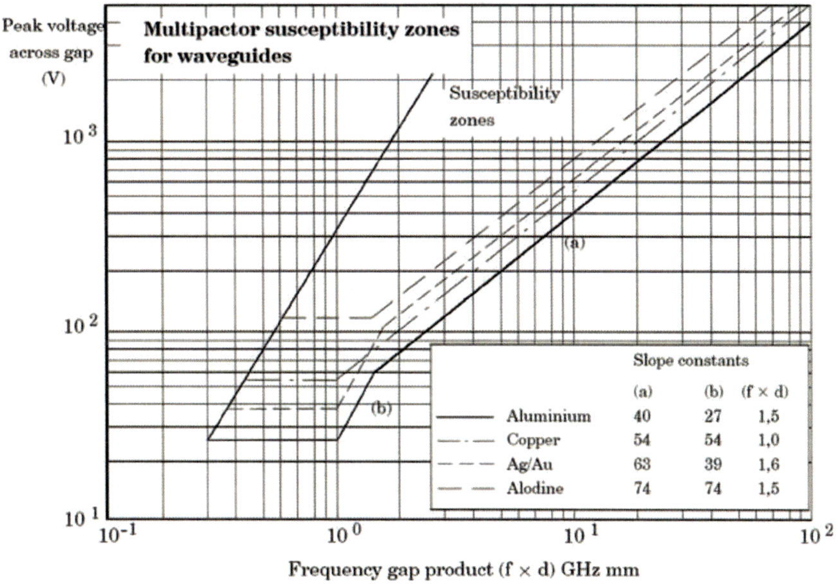

Fig. 9.34 The multipaction sensitive areas for waveguide structure [2]

of the obtained secondary electron number with time; and the specific physical images and discharge locations during the multipaction process are obtained through multipaction electrons evolution with time.

The numerical simulation records the trend of all secondary electron motion state changes with time, and shows the physical image of the multipaction effect formation process in three dimension. During the formation of multipaction, it can be observed that electrons undergo physical processes such as secondary electron emission, phase selection, phase focusing, and secondary electron resonance multiplication. The discharge phenomenon induced in a local region of the metal surface of the microwave component provides a valid basis for the study of a microscopic mechanism of multipaction.

The three-dimensional electromagnetic particle numerical simulation method can describe the physical process of the multipaction effect of high-power microwave components and the specific discharge position in three dimensions. Compared to the predicted threshold value with the actual device multipaction experiment, it is found that the coincidence is good and the error is less than 1.2 dB, thus verifying the validity and accuracy of the method.

The results show that if the multipaction effect is to be excited in the microwave component, the electromagnetic field must satisfy certain phase and frequency conditions according to the physical structure of the microwave component, and the input power must be higher than a certain threshold power, that is, the multipaction effect initialization condition should be satisfied.

The measures to suppress multipaction in engineering:

1. Control the product of frequency and gap size ($f \times d$) with the large gap size design method, so that it is outside the sensitive area of multipaction discharge.
2. Use a suitable surface treatment process to reduce the secondary electron emission coefficient of the surface.
3. Strictly control the process, avoid processing burrs and filaments, and keep the surface continuous and smooth.
4. Prevent surface contamination and maintain a high-multipaction power threshold.
5. Appropriate vent hole should be designed to exhaust the gas remaining inside the component.
6. Fill the media as much as possible to increase its discharge threshold, as conditions permit.

9.4 Spacecraft Antenna Passive Inter-modulation Protection Design

In the early research of high-power multi-channel communication equipment for mobile communication, a new electromagnetic interference source, Passive Inter-Modulation (PIM), was discovered, which refers to the inter-modulation caused by the inherent nonlinearity of passive components.

PIM means that two or more frequencies are mixed in a passive device to create a new spurious signal. In high-power, multi-channel spacecraft communication systems, the nonlinearity of passive components produces higher harmonics relative to the operating frequency, which mix with the operating frequency to produce a new set of frequencies. Therefore, in nature, PIM can be considered as a kind of mixing product. The inter-modulation frequency generated by a dual-frequency system is shown in Eq. (9.18).

$$f_{IM} = (f_1 + f_2) + (2f_2 - f_1) + (2f_1 - f_2) + (3f_2 - 2f_1) + (3f_1 - 2f_2) + \cdots \tag{9.18}$$

The inter-modulation frequency product is usually represented by its order, for example, $(2f_2 - f_1)$, $(2f_1 - f_2)$ is third order, and $(3f_2 - 2f_1)$, $(3f_1 - 2f_2)$ is fifth order, where $(2f_2 - f_1)$ and $(2f_1 - f_2)$ are inter-modulation products with the most power, which is also known as third-order inter-modulation distortion. It is generated by the cross-modulation of the second harmonic of a dominant frequency signal and the fundamental wave of another dominant frequency signal. The power of the inter-modulation component is directly related to the power of the two signals participating in the mixing. Theoretically, the third-order inter-modulation level changes by 3 dB, i.e., 2 dBc, when the carrier frequency power changes by 1 dB.

The basic PIM phenomenon is caused by current flowing through nonlinear components. Passive components, such as filters, coaxial cables and connectors, metal connection surfaces, antenna feeds, and antennas, can have inherent nonlinearities for various reasons. The microscopic mechanism that causes nonlinearity of passive components is very complex, not only related to material properties and structural forms, but also related to the channel loading and process quality of system assembly.

Inter-modulation interference is the useless signal generated by the combination of signals in a nonlinear device or a propagation medium to modulate the presence signal. Inter-modulation interference is difficult to perform more accurate quantitative EMC prediction. Its mechanism is like the electron tunneling phenomenon of the Metal-Insulator-Metal (MIM) connection surface and a thin insulating coating. In electronic systems, this has happened in L-band, S-band, C-band antennas and multiplexers. So far, studies have shown that there are several potential reasons for the PIM phenomenon:

1. In hardware systems (such as connectors) with ferromagnetic materials, the PIM phenomenon is caused by the nonlinearity of the ferromagnetic material itself.
2. In the test with corroded samples, a high-PIM level is found.
3. PIM can also occur in the coaxial connector, which is not tightly connected.
4. PIM may be caused by tiny cracks, tiny whiskers, and blisters in metal structures.
5. The presence of dirt at the metal joint or the presence of a metal-insulator-metal joint due to coating causes nonlinearity and leads to PIM.
6. Research shows that PIM will be affected by mechanical load change due to thermal expansion and contraction under different temperatures. Coaxial cable

braids and fill media also affect the PIM level. PIM is also related to the length of the coaxial cable. The longer the cable, the higher the PIM level. The PIM level is also related to the operating frequency. The higher the frequency, the higher the PIM level. The PIM level is related to the transmit power level and sometimes to time. PIM is extremely sensitive to physical motion, temperature cycling processes, and temperature changes, which often have a threshold effect.

Due to the instability of the PIM phenomenon, it often shows unpredictability, so the PIM level is too difficult to predict by analysis. Although its spectrum can be calculated and the calculation method is the same as the active inter-modulation spectrum calculation, the PIM discrete spectrum increases sharply with the increase of the number of carriers in the system. When the number of carriers is large, PIM interference is no different from broadband noise.

In a sense, PIM interference is mainly detected through multiple measurements under long-term and multiple conditions.

The investment of PIM measurement system in engineering is huge. It is now practical to take measures in design to prevent the occurrence of PIM. The precautions for PIM are as follows:

1. Avoid using ferromagnet materials in high-frequency connectors and antennas, feeds, and all connectors.
2. Anti-rust treatment of high-frequency connectors and antennas, feeds and all connectors.
3. The coaxial connector is tightened and the various connectors are prevented from loosening.
4. Ensure the quality of the processing technology of the antenna and feed, and avoid micro-cracks, whiskers or blisters in metal structures.
5. Avoid using braids as fillers for coaxial cables.
6. Multi-carrier communication system does not adopt transmitting and receiving common antenna schemes as much as possible.

For example, the PIM level of a reflector antenna is effectively reduced from −80 to −130 dBm by removing the gap structure in the inner conductor of the feed structure, and the measurement results are compared as shown in Figs. 9.35 and 9.36.

Although the PIM is very complicated and there are still many unclear points, based on the current research progress, it is entirely possible to ensure lower PIM level of the satellite by taking special measures for design, production, measurement and final assembly. Many satellite companies have developed a rigorous and complete PIM plan. In the satellite design, strict requirements have been put forward in terms of frequency allocation, materials, mechanical dimensions, purchased devices, processes, etc., and the allocation PIM budget have been carried out. The system PIM index is guaranteed by the combination of measurement and simulation.

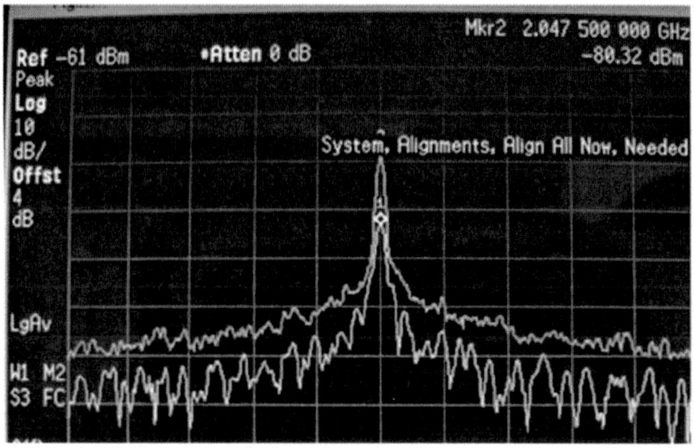

Fig. 9.35 The PIM test result before changing

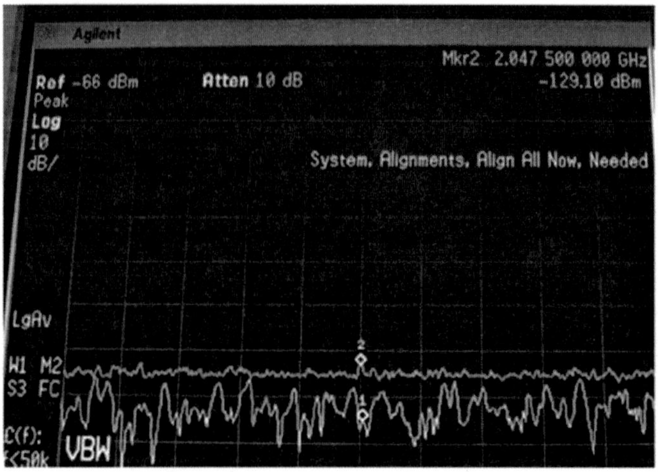

Fig. 9.36 The PIM test result after changing

9.4.1 PIM of Antennas in Communication Satellites

Antennas are an indispensable part of high-power communication satellites. Satellite communication systems usually have multiple antennas (for commands, beacons, communications, etc.), which are the primary part for high-power signal transmission and high-sensitivity receiving channels. In practice, the antenna system is very prone to producing PIM, and the antenna reflector itself is an important source of PIM, so the PIM of the antenna system is very crucial.

In general, there are two types of nonlinearities in antenna systems, namely, contact nonlinearity and material nonlinearity. Contact nonlinearity is caused by metal contact with nonlinear current and voltage, which is characterized by looseness, oxidation and contamination. Material nonlinearity is associated with the inelastic behavior of a component or system. Here, it refers to the materials used, such as ferromagnetic, carbon fiber, iron, nickel, and aluminum, which exhibits nonlinear characteristics.

Since multi-carrier satellite system uses reflector antenna as communication antenna, there are some design principles and guidelines for reflector antenna:

1. The production of reflector support structures is critical to the generation of PIM. The riveted aluminum support structure is most used, and each rivet is a potential factor for the generation of PIM. This problem is particularly acute in large-scale antenna and can be solved by using glue technology in the manufacturing process.
2. The place where the high-current density exists is easy to generate inter-modulation products. The design of the feed channel is crucial. In principle, the whole feed should be used, but it is almost impossible to use in practice. Therefore, many flanges with high-contact pressures that have been developed so far can be used. The influence of the secondary reflector cannot be ignored, especially in the axisymmetric system, the entire secondary reflector and its supporting structure are very relevant to PIM. The apex of the secondary reflector is very sensitive to PIM. The research shows that the metal-to-metal contact is not suitable for the secondary reflector supporting structure and its attachments. The following techniques can be used in the design of the feed channel:
    (a) The contact potential difference caused by dissimilar metal contact and uneven expansion can be eliminated by using a silver-plated aluminum structure.
    (b) The electron beam bonding is used in polarizer processing.
    (c) The whole manufacturing technique with spark erosion is used for other feed components.
    (d) The waveguide flange joint with high contact pressure is used.
    (e) There is no dielectric and therefore no electrostriction.
3. Due to the large temperature range experienced by satellite antennas in space, the influence of temperature on various contact levels shall be considered. The direct contact between multiple layers of thermal coating arranged near the antenna must be avoided, especially the feed horn. Moreover, the balance between RF design and thermal control must be fully considered.

9.4.2 *PIM of Coaxial Cable in Communication Satellite*

A large number of coaxial cables are used in satellite systems. As a signal transmission path and a short jump connection between the antenna and the feed system in the communication cabin, the coaxial cable should ensure minimal group delay and loss. In high-power multi-channel systems, the coaxial cable should also ensure that it does not cause fatal passive inter-modulation interference. The coaxial cable is one of the important sources of PIM. Especially in communication satellites or direct broadcast satellites, multiple transmitters and receivers share a feed, forming a common antenna working mode, a large amount of inter-modulation products can be generated when signals with multiple different frequencies and occupying a certain bandwidth pass through the same cable.

The factors affecting the coaxial cable PIM are summarized as follows:

1. **Length of coaxial cable**. The longer the coaxial cable, the higher the level of the inter-modulation products produced. The size of the inter-modulation products produced by a certain length of cable is determined by the total current flowing through the contact point of the braid. The total inter-modulation product produced by the cable is the vector sum of the inter-modulation products of each contact point in the cable. Therefore, with the increase of the length and contact point, the amplitude of the inter-modulation product will increase.
2. **Multi-core stranded wire and single-core inner conductor wire**. The inter-modulation products produced by these two cables are basically the same, which may be due to the fact that the smaller amount of inter-modulation products produced by the multi-core stranded wire is completely masked by the higher inter-modulation products produced by the outer braided fabric.
3. **Braid fill factor**. The inter-modulation product generated by a braided cable with a high-fill factor is lower than that generated by a braided cable with a low-fill factor, because the current intensity flowing through a single weave wire in the braided cable with a low-fill factor under the same input signal power is large.
4. **Ribbon and line form braid**. The inter-modulation products produced by these two braids have essentially the same amplitude level. Although the two braids differ in their expression, there is no difference in the nature of the structure, the ribbon braided cable has much less metal contact points per unit length, and this just compensates for the effects of braid contact sparseness.
5. **Corrosion of braid**. The corroded copper wire is measured in the experiment. The surface of copper generally produces two types of oxides, namely, cuprous oxide and copper oxide. The properties of the two are poles apart. The cuprous oxide has high conductivity, while copper oxide is a highly insulating medium, one of which will increase the voltage amplitude of the inter-modulation product and the other will reduce the voltage amplitude of the inter-modulation product. Under normal temperature and pressure, only cuprous oxide is generated when copper is exposed in air, so the voltage amplitude of inter-modulation products will increase.
6. **Braid contact pressure**. When the pressure at the contact point of the ribbon and line form metal with the braid is reduced, the metal contact points are dispersed, and the dispersed contact points produce a high-inter-modulation product, so the breakage or peeling-off of the outer protective layer will increase the voltage amplitude of the inter-modulation product.

9.4.3 PIM of Connectors in Communication Satellites

Sub-systems should be cascaded with various connectors in satellite systems, while metal-to-metal connectors are an important source of PIM generation. The voltage-current nonlinear relationships in connector contacts can be caused by many factors, such as semiconductor, electronic tunneling effect, voltage hysteresis, saturation of current and geometric structure of contacts.

Since contact points are an important factor in causing PIM, a "contactless" connector solution is proposed. The physical object is a two-port network with low-transmission loss and air media between the center conductors, without physical contact. For the designed frequency, the impedance transformation is performed by a quarter-wavelength impedance characteristic. New connector includes a pair of axially arranged coaxial wires, whose inner and outer conductors are electrically connected by a low-resistance rubber joint to achieve a "contactless" design, see related literature for details [4].

References

1. Woode A, Petit J (1989) Diagnostic investigations into the multipactor effect, susceptibility zone measurements and parameters affecting a discharge. ESTEC working paper no. 1556
2. Space engineering-multipaction design and test. ECSS-E-20-01A Rev.1, 1 Mar 2013
3. Li Y, Cui W, Zhang H et al (2017) A novel simulation method of multipactor in complex component for satellite application. Chin Space Sci Technol 37(2):73–80
4. Graßl M, Zißler W (2015) Novel components for a fully automated PIM and S-parameter test system. Microw Meas Conf (ARFTG) 85:1–5

Chapter 10
Spacecraft Antenna Test Technology

Abstract The adequate essential ground measurement and verification tests are crucial for the development of the spacecraft antenna for ensuring its high reliability. This chapter discusses test methods for electrical performance including radiation, impedance, PIM, multipaction and facilities configuration for spaceborne antenna. Dynamic environment and thermal environment test methods and facilities configuration are also provided in this chapter.

10.1 Spacecraft Antenna Electrical Performance Testing Technology

10.1.1 Electrical Performance Test Task

During or after the development process of the spacecraft antenna, the antenna design, processing and assembly are verified by the electrical performance measurements, and the electrical performance specifications of the antenna are confirmed to comply with the design requirements. For spacecraft antenna with low mature or complicated technology, the antenna is used on the satellite only after that all the electrical prototype, qualification prototype and flight prototype of the antenna have been tested.

The basic purpose of the spacecraft antenna electrical performance test is to experimentally measure and verify the antenna circuit and radiation characteristics. In the actual measurement, the spacecraft antenna electrical performance test items can be summarized into three aspects, namely, circuit performance parameter test, radiation performance parameter test and system performance parameter test, as shown in Fig. 10.1.

10.1.1.1 Circuit Performance Parameter Test

Irrespective of the antenna's radiation characteristics, the antenna is equivalent to a circuit or microwave network, or even a terminal load. In order to ensure efficient

R. You et al., *Technologies for Spacecraft Antenna Engineering Design*, Space Science and Technologies, https://doi.org/10.1007/978-981-15-5833-7_10

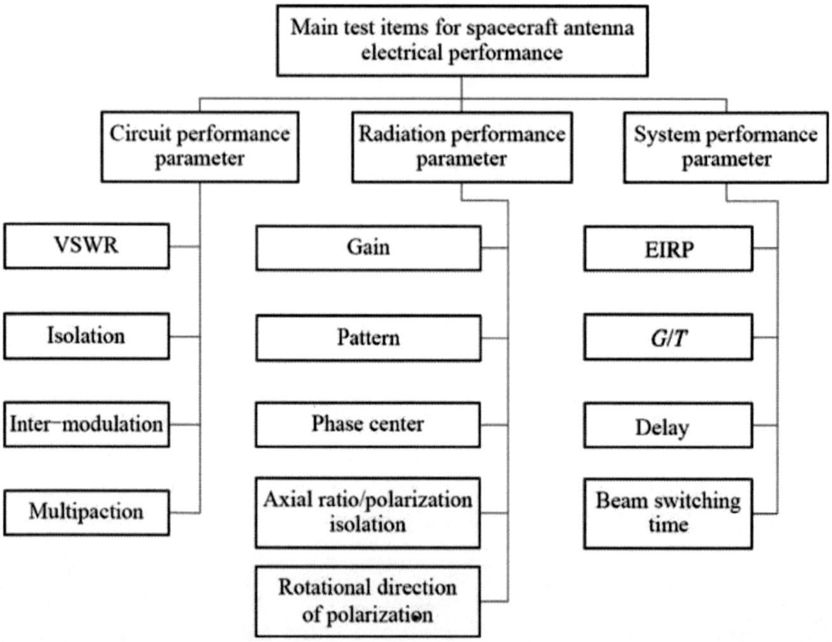

Fig. 10.1 Main test items of the spacecraft antenna

transmission of energy or signals, the antenna shall match with the transmission line, while avoiding distortion of the signal. For this purpose, the parameters related to circuit characteristics should be tested, mainly including network scattering parameter test and inter-modulation parameter test. The former includes measurements such as VSWR, isolation, efficiency, transmission phase, and delay; the latter includes tests such as reflection inter-modulation and transmission inter-modulation.

The spacecraft antenna on-orbit operating environment is significantly different from that of the ground antenna. The antenna or RF device is prone to low-pressure discharge or vacuum multipaction (i.e., electron secondary multiplication effect) under the high power, resulting in increased antenna reflection coefficient, enhanced inter-modulation interference, the malfunction of entire communication system or even burning of RF device. To this end, the special inspection items, such as low-pressure discharge, vacuum power tolerance and multipaction test, shall be performed on the ground for the spacecraft antenna (described in Sect. 10.1.5).

10.1.1.2 Radiation Performance Parameter Test

The electromagnetic wave radiated by the antenna shall have a certain directionality or spatial shape to improve the radiation efficiency, avoid harmful interference to satellite-ground and inter-satellite communication of other spacecraft, or suppress harmful interference from specific geographical coordinates. For this, the space radiation performances are tested, mainly including pattern, beam width, side lobe level, gain, direction coefficient, polarization, phase center and phase center stability.

10.1.1.3 System Performance Parameter Test

The spacecraft antenna is an important part of the satellite-ground and inter-satellite communication system or scientific measure equipment. The overall performance parameters of the system related to the antenna are tested for inspection of the overall technical specification or tactical specification of the system. Theoretically, the system performance parameter test can replace almost all the radiation parameter test, but it is generally only used for the indicators related to the key performance or function of the system or the whole satellite, such as EIRP, G/T and delay, because of its high test cost and high risk. Due to space limitations, this section will introduce the system performance parameter test of a phased array antenna.

10.1.2 Circuit Performance Parameter Test

The antenna can be seen as an "impedance converter" between the transceiver and space, which can convert the high-frequency current generated by the transmitter into electromagnetic waves and send it to a specified spatial region, or convert electromagnetic waves from a specified spatial region into the high-frequency current and send them to the receiver. As an important part of the wireless communication system, the antenna circuit performance parameter test mainly includes the VSWR, isolation and inter-modulation test. The main concerns of each performance parameter test are described in Fig. 10.2.

10.1.2.1 VSWR Test

VSWR is defined as the ratio of the maximum to minimum voltage along the transmission line, which is also called the standing wave coefficient. In the communication system, in order to ensure long-term stable operation of the system and obtain minimum transmission loss, the input and output impedances of all devices (including transceiver, feeder line, antenna, etc.) in the communication system match each other as much as possible, so that the entire transmission system can always be in the traveling wave state, avoiding the generation of standing waves.

The VSWR and isolation test becomes very simple with the introduction of vector network analyzer. From the perspective of the output port of the transmitter to the antenna, the antenna can be regarded as a single-port network. When testing the VSWR of a spacecraft antenna port, the Antenna Under Test (AUT) should

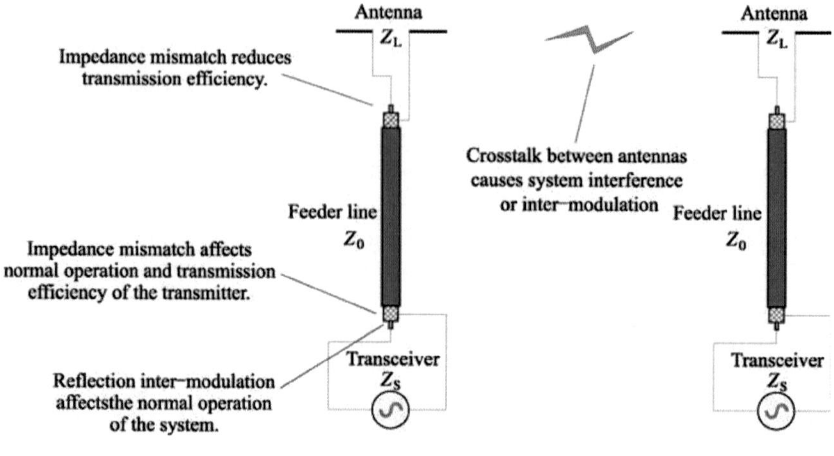

Fig. 10.2 Concerns of circuit performance parameter test

be installed or placed in a free space or in an anechoic chamber that is relatively unreflected and far from the test equipment and test personnel.

10.1.2.2 Isolation Test

When two or more communication systems are installed on the same carrier platform, the crosstalk among antennas will occur (Fig. 10.2). Even if the two transmitters operate in different frequency bands, the harmonics or inter-modulation interference may be caused. A variety of different functional antennas are densely placed in a small space on the surface of the spacecraft or satellite, which may introduce severe electromagnetic interference. In engineering, the isolation is used to quantitatively characterize the degree of coupling between two antennas on the same platform. It is defined as the ratio of the transmit power of one antenna to the power received by the other antenna, expressed in logarithm (dB). In this case, the two antennas to be tested shall be regarded as a two-port network. From the perspective of the antenna port, the isolation between the antennas can be described by the two parameters S_{21} and S_{12}. The equivalent circuit is shown in Fig. 10.3.

The spacecraft antenna isolation test is similar to the VSWR test. The AUT is installed or placed in a free space or in an anechoic chamber that is relatively unreflected and far from the test equipment and test personnel. The two antennas are installed at each predetermined location on the satellite surface and are typically tested using a dual-port vector network analyzer. The insertion loss between the two test ports of the vector network analyzer is the isolation between the two antennas. If the isolations of more than two antennas are to be tested, the multi-port vector network analyzer can be used. By the current available four-port or eight-port vector

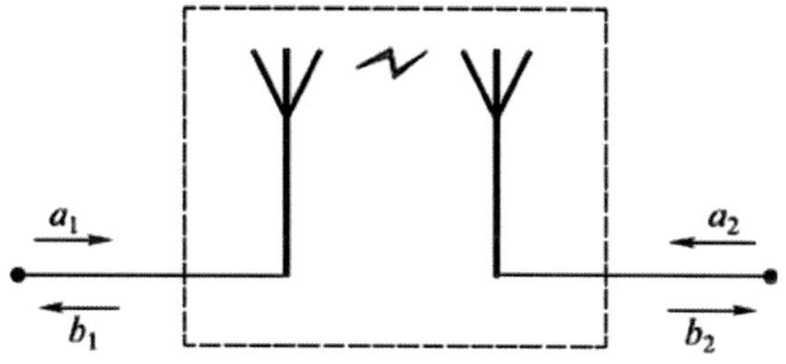

Fig. 10.3 Antenna isolation equivalent circuit

network analyzer, the number of connections can be reduced and test efficiency can be improved obviously.

10.1.2.3 Inter-Modulation Test

1. Inter-modulation classification

For the spacecraft antenna, the measurement can be divided into transmission inter-modulation and reflection inter-modulation type according to the inter-modulation interference objects. In the transmission inter-modulation, an antenna is considered as a two-port network. If two carrier frequencies are simultaneously input to the antenna port, the inter-modulation is generated in the radiated electromagnetic wave (see Fig. 10.4). In the reflection inter-modulation, an antenna is considered as a single-port network. If two carrier frequencies are simultaneously input to the port, the inter-modulation is reflected to the port (Fig. 10.5). The transmission inter-modulation focuses on the interference of the antenna system to other onboard systems, while the reflection inter-modulation focuses on the interference of the antenna to its own receiving system.

2. Inter-modulation test

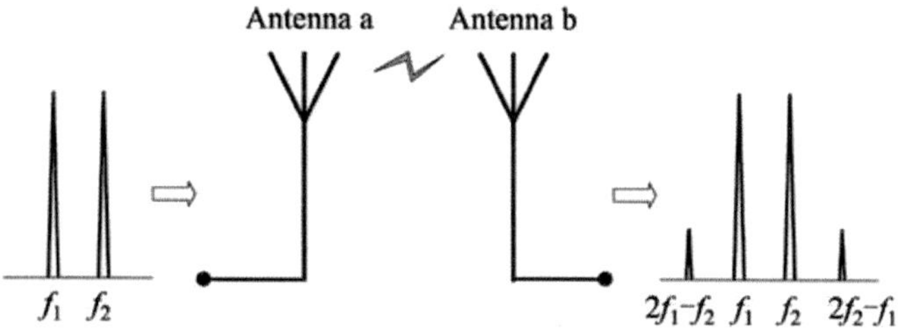

Fig. 10.4 Transmission inter-modulation

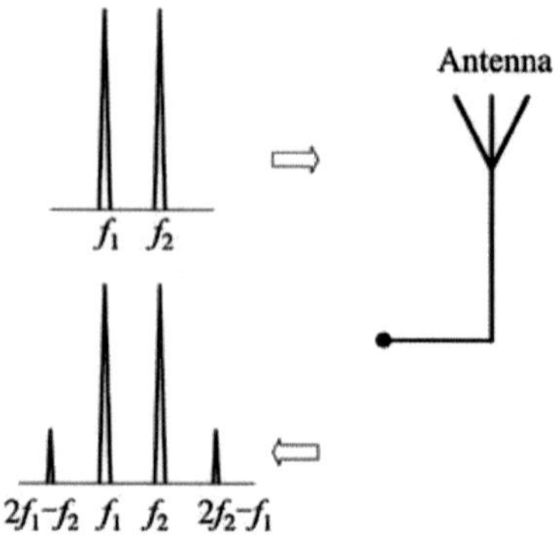

Fig. 10.5 Reflection inter-modulation

Spacecraft antenna inter-modulation is typically tested in a low inter-modulation anechoic chamber or in a free space. Before testing, it should be ensured that the inter-modulation level of anechoic chamber environment is lower than that of the AUT for more than 10 dB and that the inter-modulation level of the receiving subsystem is lower. The diplexer or filter used in the test system should be designed with low inter-modulation, and the coaxial and waveguide adapters should be used as little as possible or out of use.

Figure 10.6 shows a typical block diagram of the antenna transmission inter-modulation test. The two high-power carrier frequency signals f_1 and f_2 are, respectively, amplified and filtered into the P_1 and P_2 ports of the combiner. After synthesis

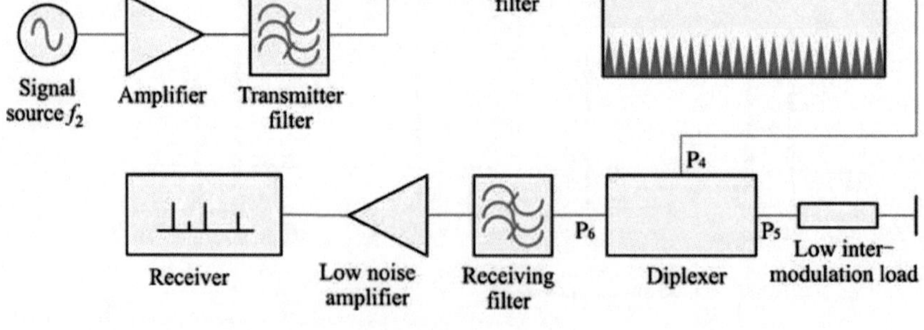

Fig. 10.6 Antenna transmission inter-modulation test method

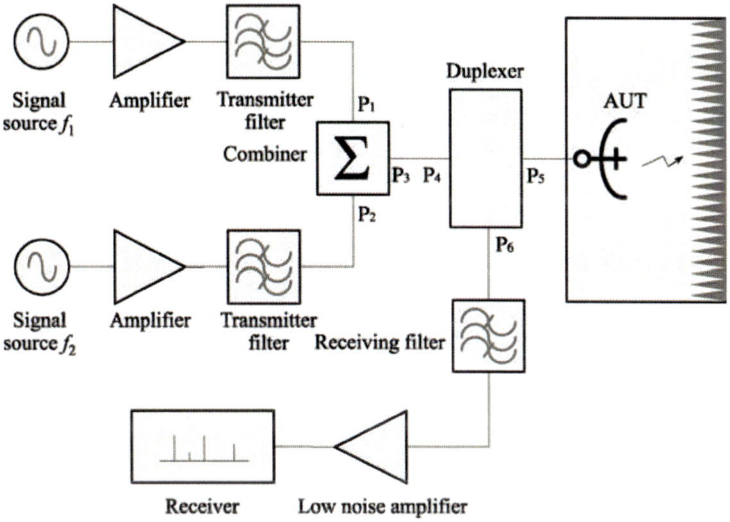

Fig. 10.7 Antenna reflection inter-modulation test method

and filtering, two pure high-power carrier frequency signals are obtained and sent to the feeding port of the AUT. The inter-modulation products generated by the antenna are sent into the *P*4 port of the diplexer connected to the auxiliary antenna, output by the P6 and passed through the receiving filter into a high-sensitivity receiver or spectrum analyzer, and then converted to obtain the antenna transmission inter-modulation test result.

The test of antenna reflection inter-modulation is similar to that of the transmission inter-modulation, as shown in Fig. 10.7. The synthesized high-power carrier frequency signal enters the duplexer port P_4 and is fed into the antenna via P_5. The inter-modulation products generated by the AUT are reflected back to the port P_5 of the duplexer, and the inter-modulation signal is guided from the port P_6 to the receiving filter, and then sent to the high-sensitivity receiver or spectrum analyzer, thus obtaining the antenna reflection inter-modulation test results through conversion.

In general, spacecraft antenna inter-modulation values are low, very close to the test system noise inter-modulation. Before the test, the test system noise inter-modulation value is generally verified. A low inter-modulation antenna like the radiation characteristics of the AUT is usually used to perform system verification according to the same test power, frequency point, and cable connection state, and to check the sensitivity of cables and joints to bending, mechanical stress and structure. After the system inter-modulation verification, the AUT can be tested by using the same settings.

Antenna inter-modulation testing is complex and sometimes hard to implement. Since only the interference frequency band is often cared in the engineering, the inter-modulation interference test is performed for the spacecraft antenna. For example, to verify the passive interference of the transmitting device to the receiving device, it is first calculated that two frequencies in the transmitting device are most likely to generate passive inter-modulation in the receiving band of the system, which are transmitted by the power source of the transmitting device, and then the high-sensitivity reception is performed by the receiver of the system, observing whether passive inter-modulation products can be received.

10.1.3 Radiation Performance Parameter Test

10.1.3.1 Spacecraft Antenna Test Site Selection

In order to obtain the antenna radiation pattern, test sampling must be performed on a certain surface around the AUT. According to different sampling positions, it can be divided into two test methods: far-field test and near-field test. The far-field test includes traditional outdoor far-field test, indoor far-field test and compact field test in an anechoic chamber. According to the selected scanning surface shape, the near-field test includes planar near-field test, cylindrical near-field test and spherical near-field test.

For different payloads, platforms and mission requirements, the required frequency, form and dimension of the spacecraft antennas are much different. According to the operating frequency band, antennas can be divided into various types, e.g., shortwave, ultra-shortwave, L-band, S-band, C-band, X-band, Ku-band, Ka or even millimeter wave and sub-millimeter wave. By the radiation form, there are weak directional antennas with only a few centimeters in size, and deployable reflector antennas and large array antennas with tens of meters in size. The spacecraft antenna test basically covers all current antenna testing technologies and is an important driving force for the development of antenna testing technology.

For the various types of spacecraft antennas described in this book, the common test sites for radiation performance testing during development and production are listed in Table 10.1.

10.1.3.2 Introduction to Spacecraft Antenna Test Site

As can be seen from Table 10.1, the current spacecraft antenna radiation performance is tested mainly in the indoor far field, compact field or near field.

1. Indoor far field

The indoor far-field test refers to the far-field test in the microwave anechoic chamber. The microwave anechoic chamber with the inner walls shielded by the absorbing material can reduce the reflection of the inner wall, so that a non-echo zone close to "free space", a quiet zone, is formed in the inner region. The indoor far field has many advantages such as all weather testing, good confidentiality, anti-electromagnetic interference, etc. Due to the limitation of indoor space and construction costs, the

Table 10.1 Spacecraft antenna corresponding test site

No	Antenna type	Test site
1	TT & C antenna	Indoor far field
2	Data transmission antenna	Indoor far field, planar near field
3	GNSS receiving antenna	Indoor far field, spherical near field
4	Inter-satellite link antenna	Planar near field, indoor far field, compact field
5	Antenna for return capsule	Indoor far field, spherical near field
6	Communication payload antenna	Planar near field, compact field
7	Microwave remote sensing payload antenna	Planar near field, indoor far field, compact field
8	Low-frequency antenna	Indoor far field, spherical near field

antenna diameter that usually be tested in the indoor far field is small. The common anechoic chamber for indoor far-field test includes tapered and rectangular anechoic chambers, as shown in Fig. 10.8.

The divergent geometry of the tapered anechoic chamber avoids large-angle specular reflections from the side walls, floors and ceilings is with good low-frequency performance. Both ends of the rectangular anechoic chamber can be used for pattern

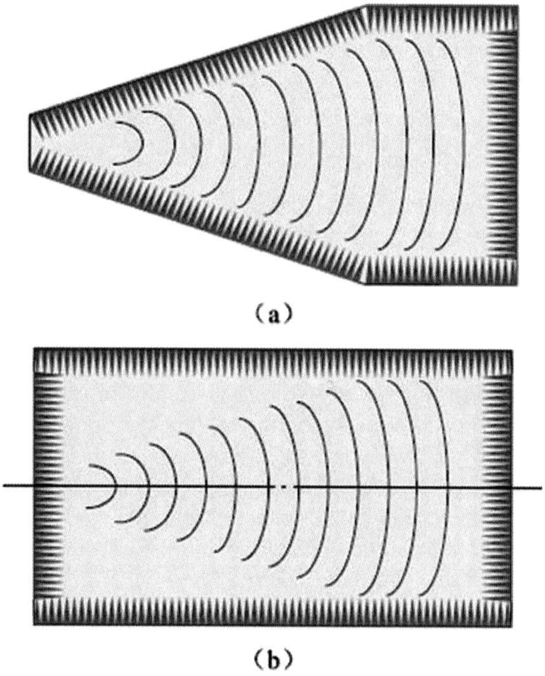

Fig. 10.8 Shape of a common microwave anechoic chamber. **a** Tapered anechoic chamber; **b** rectangular anechoic chamber

Fig. 10.9 Indoor far-field anechoic chamber (photo provided by ISSE of CAST)

test or system parameter test, with their good versatility. In the development of spacecraft antennas, indoor far field is widely used in the gain and pattern test of spacecraft platform antennas such as TT&C, GNSS, data transmission and antenna of return capsule. Figure 10.9 shows a typical indoor far-field test anechoic chamber.

2. Compact range

As the aperture of the spacecraft antenna and the operating frequency increase, the test distance is required to reach several kilometers or even tens of thousands of meters to meet the far-field conditions. Such a large transmission and reception distance makes the far-field test very difficult, and this problem can be solved by the compact range technology. The compact range technology is to form a quasi-plane wave environment in a relatively short distance after calibrating the spherical wave emitted by feed source using a lens or reflector according to the concept of geometrical optics. According to the calibration method, the compact field can be divided into two types: the transmission compact field and the reflective compact field. According to the number of reflectors used, the reflective compact field can be classified into single-reflector, double-reflector and multi-reflector compacting field. At present, the Cassegrain double-reflector compact field is used in the most of the spacecraft antenna tests. Figure 10.10 shows the layout of the Cassegrain double-reflector compact field.

The sub-reflector of the Cassegrain compact field is a rotating hyperboloid, and the main reflector is a rotating paraboloid. The sub-reflector calibrates the spherical

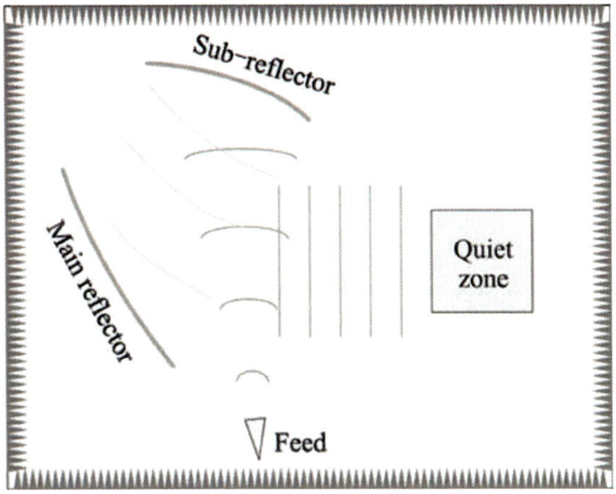

Fig. 10.10 Layout of Cassegrain compact range

wave from the feed to a spherical wave emitted by the virtual source, and the main reflector calibrates it to a plane wave. In order to obtain the maximum equivalent focal length ratio, the feed and reflectors layout generally adopts a four-shaped structure. Theoretically, the cross-polarization current is zero when the Cassegrain compact field satisfies the circularly symmetric condition and is fed forwardly by the line-polarized feed. Due to the influences of uncertainties such as machining accuracy, edge diffraction and actual anechoic chamber reflection, cross-polarization current is not zero, but usually less than −40 dB, which is very suitable for high-polarization isolation communication satellite antenna testing. Figure 10.11 shows the common Cassegrain compact field.

3. Near field

In antenna near-field testing technology, the near-field region at several wavelengths from the antenna aperture to be measured is scanned using a probe with known characteristics, the amplitude and phase distribution of the antenna on the scanning surface are measured at discrete points, and then the far-field radiation characteristics of the AUT are calculated by the strict mathematical transformation. The basic idea is to expand the spatial radiation field of the AUT into the sum of the spatial spectrum functions. The weighting function in the expansion contains the complete far-field information. With the near-field test data, the weighting function is calculated to obtain the far-field pattern of the antenna. According to the Huggens-Kirchhoff principle and the equivalent principle, the test can be performed on any closed surface. The near-field test mainly has planar, cylindrical and spherical surface scanning method under the consideration of the achievability of the scanning mechanism and the convenience of data processing, as shown in Fig. 10.12. At present, most of the spacecraft antenna tests are used in the planar near field and spherical near field.

By planar near-field scanning, a sufficiently large plane is scanned by a probe at an appropriate distance from the AUT and is expanded by plane spectrum, thus obtaining far-field pattern of the antenna. Obviously, the complete closed surface cannot be

Fig. 10.11 Cassegrain compact range (photo provided by ISSE of CAST)

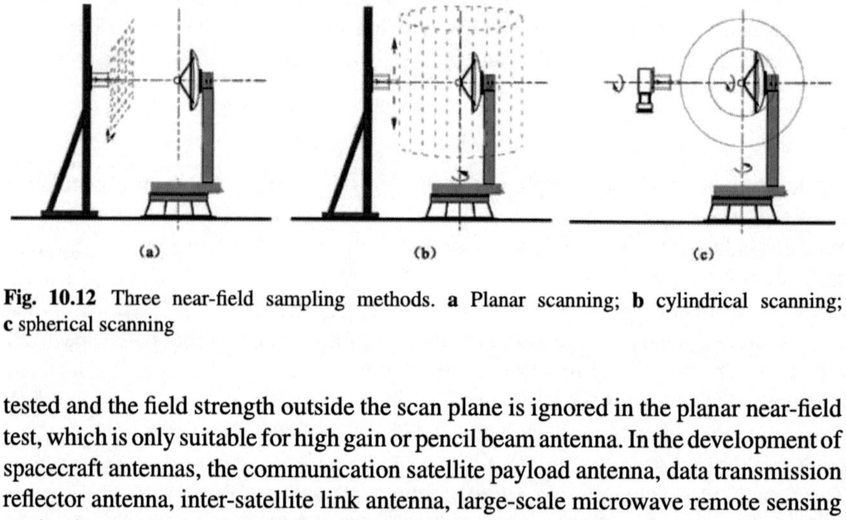

Fig. 10.12 Three near-field sampling methods. **a** Planar scanning; **b** cylindrical scanning; **c** spherical scanning

tested and the field strength outside the scan plane is ignored in the planar near-field test, which is only suitable for high gain or pencil beam antenna. In the development of spacecraft antennas, the communication satellite payload antenna, data transmission reflector antenna, inter-satellite link antenna, large-scale microwave remote sensing payload antenna are mainly tested by the planar near-field tests. The planar near field plays another important role in the amplitude phase correction and near-field diagnosis of phased array antennas. Figure 10.13a shows a photo of a planar near field.

In the spherical near-field test, the spherical field distribution of the AUT is tested by a probe at an appropriate distance from the AUT, and expanded by the spherical

Fig. 10.13 Near-field lab. **a** Planar near field; **b** multi-probe spherical near field (photo provided by the ISSE of CAST)

spectrum to obtain the far-field pattern of the antenna. Depending on the number of sampling probes, the spherical near field can be categorized into single-probe spherical near field and multi-probe spherical near field. Single-probe spherical near-field test uses a single probe to obtain a spherical field distribution surrounding the AUT through multiple positioner. In multi-probe spherical near-field test, several multi-band low-reflection probes are placed in one dimension (usually θ) in the spherical coordinate system, and electronic scanning is used instead of mechanical scanning to quickly obtain a spherical field distribution. The spherical near-field test technique can obtain the complete pattern and gain of the AUT and is an ideal antenna test method. The TT&C antenna, GNSS receiving antenna, data transmission antenna and some low-frequency spacecraft antennas are mostly tested by the spherical near field. Figure 10.13b shows a multi-probe spherical near-field photo.

10.1.3.3 Full Size Radiation Model Satellite Test

The radiation model satellite, referred to as RM satellite, is a simplified model of the satellite with electromagnetic radiation performance equivalent to that of real satellite. The dimension, device layout and boundary conditions of the RM satellite are consistent with that of real satellites, which can realistically simulate the electromagnetic environment of the antenna. With the RM satellite, whether the antenna meets the system design requirements will be checked by testing the radiation characteristics of the antenna mounted on the satellite. Therefore, the RM satellite should be designed as close as possible to the real satellite to ensure that the test results truly reflect the true operating state of the antenna.

In the actual test, the RM satellite test is generally carried out in the initial stage of model development, after the radiation performance of the single antenna meets the design requirements, and the spacecraft configuration size and the layout of the device on the satellite surface are initially determined. When the RM satellite is processed, the space-borne equipment can be simplified according to its occlusion or influence on the antenna. For example, the solar wing generally has less impact on the RM satellite test results for a conventional platform antenna or a communication payload antenna, so its model is usually omitted in the test. The RM satellite test is exactly the same as the antenna stand-alone test, except that the demand for the plane wave quiet zone is increased. At present, the RM satellite test for spacecraft antenna is mainly carried out in the compact field or the spherical near field.

10.1.4 System Performance Parameter Test

10.1.4.1 Test Content

In a satellite-ground or inter-satellite communication system consisting of antenna, transceiver or transponder, some parameters such as the directional characteristics of the receiving or transmitting beam, beam pointing accuracy, polarization isolation, phase center are determined almost entirely by the antenna itself. The test results of the antenna can fully represent the performance of the system, and these parameters are tested only at the antenna level. Some properties such as Equivalent Isotropic Radiated Power (EIRP), Gain-to-noise Temperature ratio (G/T) and Saturated Power Flux Density (SPFD) are determined by the antenna and the active part, which shall be connected to a system before testing. Moreover, with the gradual application of space-borne active phased array antennas represented by distributed T/R components, the antennas are developing toward integration and some system performance parameters must be tested at the system level instead of by segmentation.

System performance parameter test involves a very wide range, and new projects are constantly being added as communication technology is evolved. The system performance parameter test in this section focuses on the test of the space-borne phased array antenna, and the important system performance parameter tests directly related to the antenna performance are mainly introduced. Theoretically, the system performance parameter test can contain almost all antenna radiation performance. But it is generally only used for testing the key performance indicators of the whole satellite or subsystem because of its high test cost.

When performing the system performance parameter test related to the spacecraft antenna, in order to reduce the test workload and the test risk, only the system performance parameters of individual cut (or pointing), wave position and frequency point are usually tested on the basis of ensuring test coverage, and then combined with the antenna radiation pattern to obtain or evaluate the overall system performance of the three-dimensional space, all wave positions and full frequency bands of the communication system.

10.1.4.2 Testing Method

1. EIRP

EIRP is usually tested with the absolute power method. The test block diagram is shown in Fig. 10.14. Taking a circularly polarized antenna test as an example, an AUT operates in a transmitting state and directs to the auxiliary receiving antenna in the angle to be measured. The receive antenna, a standard gain antenna, receives two orthogonal direction power levels. The power of the antenna input signal is gradually changed and the received power of the standard gain antenna is recorded. According to the test system parameters, such as the standard antenna gain and link loss, the phased array antenna EIRP in transmitting state is obtained. According to the relationship between the EIRP of the AUT and the input power, the compression point or the saturation point position of the output power is calibrated.

EIRP is calculated as follows:

$$\mathrm{EIRP} = 10\lg\left(10^{0.1P_1} + 10^{0.1P_2}\right) - G_S - 20\lg\left(\frac{\lambda}{4\pi R}\right), \tag{10.1}$$

where P_1 and P_2 are the receiving power of the standard gain antenna, in dBm; G_S is the standard antenna gain, in dBi; R is the spatial distance between the receiving and transmitting antennas, in m; λ is the antenna operating wavelength, in m.

2. *G*/*T* value test

The *G*/*T* value test is usually performed in the continuous wave method, and the test block diagram is shown in Fig. 10.15. During the test, the matching load is connected on the input port of the spectrum analyzer first, the spectrum analyzer

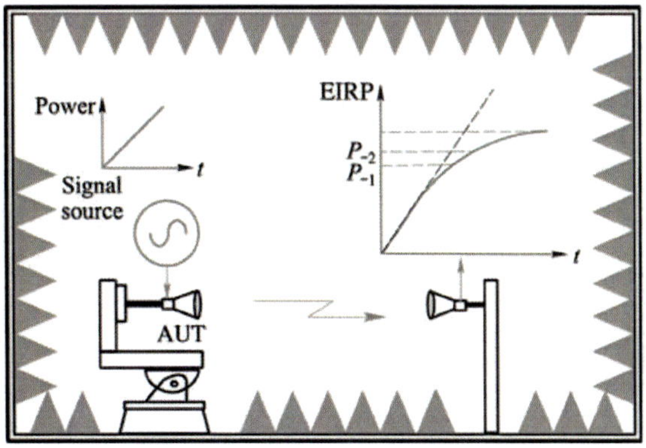

Fig. 10.14 Block diagram of EIRP value test

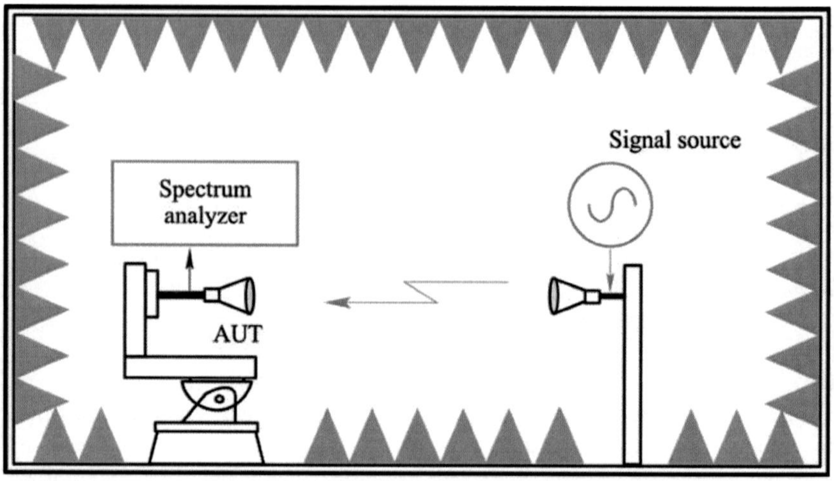

Fig. 10.15 Block diagram of G/T value test

own noise power is recorded; then the AUT is powered on and set in the receiving state. At this time, the output of the signal source is turned off, and the noise power received by the spectrum analyzer is recorded. The transmit signal is turned on, and the transmit signal level is so adjusted that the signal power received by the spectrum analyzer is at least 3 dB higher than the noise power. Finally, the standard antenna input power and the signal power received by the spectrum analyzer are recorded.

The *G/T* value is calculated as follows:

$$G/T = 10 \times \lg \frac{k \times B \times (Y_2 - 1) \times Y_1}{Y_1 - 1} + 20\lg\left(\frac{4\pi R}{\lambda}\right) - P_{Tx} - G_{FEED} \quad (10.2)$$

$$Y_1 = \frac{P_2}{P_1} \quad (10.3)$$

$$Y_2 = \frac{P_3}{P_2} \quad (10.4)$$

, where k is the Boltzmann constant, in J/K; B is the spectrum test bandwidth, in Hz; λ is the antenna operating wavelength, in m; R is the distance between the receiving and transmitting antennas, in m; P_{Tx} is the standard gain antenna input power, in dBm; G_{FEED} is the standard gain antenna gain, in dBi; P_1 is the spectrum analyzer noise power, in W; P_2 is the received noise power from AUT, in W; P_3 is the received signal power from AUT, in W.

3. Absolute delay

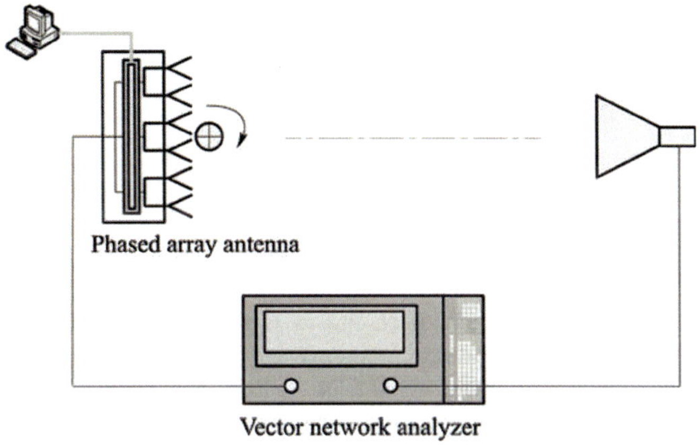

Fig. 10.16 Sketch of antenna delay test

The delay characteristic test system consists of anechoic chamber, vector network analyzer, test cable, standard horn and AUT. Figure 10.16 shows the test system connection.

Frequency sweep test with vector network analyzer includes transmission parameter phase $\varphi(\omega)$ of complete test system consisting of test cable, standard horn, space transmission section and AUT, which is obtained according to the following equation:

$$\tau = -\frac{d\phi}{d\omega}. \tag{10.5}$$

The delay of the entire test system is automatically calculated by vector network analyzer. The sectionalization test is performed for cable and the standard horn separately, and absolute delay of the phased array antenna itself is obtained by subtracting the delays of each section from the total delay.

10.1.5 Multipaction Test

The spacecraft antenna is mounted outside the spacecraft, and the most significant difference from the ground antenna is its special operating environment. Therefore, spacecraft antenna test also includes RF performance test under ambient simulation conditions. The RF test under spacecraft antenna environment simulation conditions mainly includes three types: low-pressure discharge test, multipaction test and power tolerance test.

When the air pressure is in the range of 1 kPa–1 Pa, the low-pressure environment causes gas ionization. The low-pressure discharge test is mainly used to detect whether a low-pressure discharge occurs in the ascent segment of the spacecraft antenna.

When the air pressure drops to 10^{-2} Pa or lower, electron multiplication effect may occur on the surface of the RF device under the action of high-power microwave signal. The multipaction test is used to detect whether the multipaction phenomenon occurs in the spacecraft antenna when operating on orbit.

The power tolerance test, mainly used to test the ability of the spacecraft antenna to withstand high-power signals, is usually performed in both atmospheric and vacuum environments.

The three tests of low-pressure discharge, multipaction and power tolerance are the state assessments of the spacecraft antenna during high-power operation. Their test equipment and conditions are similar. Generally, the same test system is used to implement three tests. This section takes the multipaction test as an example to briefly introduce the multipaction test system and test conditions.

10.1.5.1 Multipaction Test System

The spacecraft antenna multipaction test system mainly includes the following five basic components:

1. Vacuum subsystem

The vacuum subsystem consists of wave-transparent or metal vacuum tank, vacuuming equipment and vacuum testing equipment for simulating the on-orbit vacuum environment to provide test space for the DUT.

2. High and low-temperature subsystem

High and low-temperature subsystem generally includes heat sink equipment and temperature test equipment for simulating the temperature change environment on orbit, achieving high- and low-temperature control and detection in the vacuum tank.

3. Signal generating device

The signal generating device comprises signal source, microwave signal modulating unit and microwave power amplifier for generating high-power pulses or continuous wave test signals.

4. Free electron generation equipment

In multipaction test, free electrons are added to induce multipaction. The free electron generating source includes radioactive beta source, UV light source, electron gun, power-on probe, etc.

5. Multipaction detection subsystem

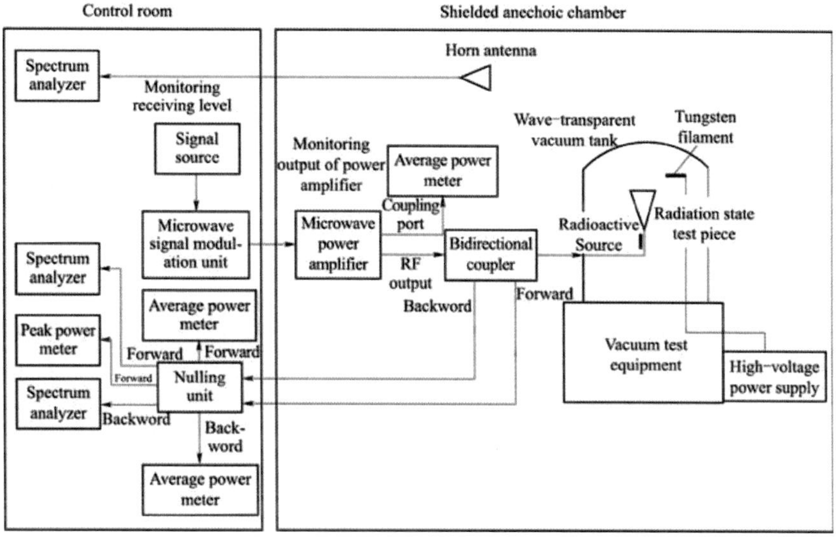

Fig. 10.17 Block diagram of multipaction test system

In the multipaction test, detection methods use different devices. The main detection methods include nulling detection method, incident and reflected power detection method, second or third harmonic detection method, near-carrier noise time-frequency conversion detection method, near-carrier noise detection method, optical detection method, electron probe detection method, etc.

Figure 10.17. shows a block diagram of a multipaction test system with a wave-transparent vacuum tank.

10.1.5.2 Test Conditions

1. Vacuum degree

Usually, the vacuum in the tank shall be less than 1.5×10^{-3} Pa and maintained for a period before the multipaction test. For the antenna with fully open structure, the vacuum holding time is usually 2 h, while for the antenna with the closed waveguide is generally 4–6 h and for the antenna with the coaxial structure is 24 h. For RF components and equipment consisting of a large number of composite materials, the vacuum degree requirement can be appropriately relaxed and the vacuum holding time should be greater than 24 h.

2. Free electron

In continuous wave test, no electron source is required, while in pulse test, an electron source is required. For the multipaction test, it should be ensured that the critical region of the multipaction of the device contains enough free electrons, if necessary, the existence of electrons should be verified. The free electron emission source should be placed as close as possible to Device Under Test (DUT). If a radioactive source is used to generate free electrons, the radioactive source can be applied on the surface of DUT. In order to ensure sufficient low-energy free electron uniformity, the multipaction test should be performed after the free electron emission source has been operated for a period of time.

3. Temperature

The multipaction tests should be, respectively, carried out at low, normal and high temperature. The thermal environment during the test shall simulate the actual operating environment. The low and high test temperatures shall be determined by the lowest and highest operating temperatures that the RF components and equipment can reach on orbit.

4. Pulse setting

The multipaction test is generally carried out by adding a microwave pulse signal, and the duty ratio of the microwave pulse signal is adjustable from 1 to 10%. The continuous wave method can also be used for test if the test conditions permit. In order to simulate the actual hot operating state, the general pulse signal shall be maintained at a certain pulse bottom level, so that the pulse average power reaches the actual rated operating power, or the pulse signal width is increased, so that the pulse average power reaches the actual rated operating power.

5. Test frequency

If the critical frequency point of DUT cannot be determined by analysis, the non-resonant device should be tested at the lowest operating frequency, and the resonant device should be tested at the center frequency and the side frequency.

10.1.5.3 Detection Method

The multipaction detection method is divided into two categories: the overall detection method and the local detection method. The overall detection method mainly includes near-carrier noise detection method, third harmonic method (sometimes also second harmonic method), forward and backward power detection method (nulling), and residue detection method. Local detection methods mainly include charge detection and optical detection, each of which has its own advantages and disadvantages and is widely used globally, and basically satisfies the requirements of multipaction detection. Multipaction test can be selected according to actual equipment conditions and test requirements. In order to avoid misjudgment of the occurrence of multipaction, in general, at least two detection methods are used at the same time.

10.1.6 Spacecraft Antenna Assisted Test Technology

10.1.6.1 Pulse Pattern Test Technology

When the traditional antenna test system is used for radiation performance test, continuous wave signals are used in the tests. With the development of space-borne antenna testing technology, especially after the emergence of active T/R components, pulse pattern testing has become one of the test methods to be considered. If a specific antenna or test system has one or more of the following conditions, the pulse pattern test should be considered:

1. Different characteristics under the continuous wave and pulse excitation

Since many devices used inside the antenna for pulsed environments exhibit different characteristics under continuous wave and pulse excitation conditions, the pattern can only be tested under pulse conditions.

2. Power tolerance limitation of the AUT

Many active antennas cannot be tested with continuous wave signals as an excitation at the desired power level. For example, the internal active device is not designed to withstand the power dissipation during continuous operation, or the antenna thermal control system is not perfect, which lacks sufficient heat dissipation capability for continuous wave signal excitation. Testing with pulse signals not only keeps the test power of these devices consistent with the actual operating power but also avoids overheating to damage the antenna, thus obtaining more realistic characteristics of the antenna.

3. Power tolerance limitation of test system

For example, due to the limitation of the power tolerance capacity of the absorbing material in the anechoic chamber, the AUT cannot work at full power during the test. If the true antenna radiation characteristics consistent with the actual operating power are to be measured in the anechoic chamber, the test is carried out only in the pulse mode.

4. Reduce the reflection effect of test site

When the pattern of the low-frequency spacecraft antenna (200–1000 MHz) is tested, the multipath interference from the ground and side walls of the chamber is serious due to the poor performance of the absorbing material in the low-frequency range. If the pulse mode is used for the gating test, the multipath effect in the chamber can be effectively eliminated and the test accuracy of the antenna pattern can be improved by reasonable setting of the pulse width.

The test signal in the pulse pattern test is a pulse signal. When a single carrier signal is continuously turned on and off (pulse modulation) in the time domain, the spectrum of this signal in the frequency domain exhibits a sin$x$/x response (Fig. 10.18). The width of each lobe in the frequency domain is inversely proportional to the Pulse Width (PW). The spectrum zero appears at $f_0 \pm n/\text{PW}$ ($n = 1, 2, \ldots$), that is, the shorter

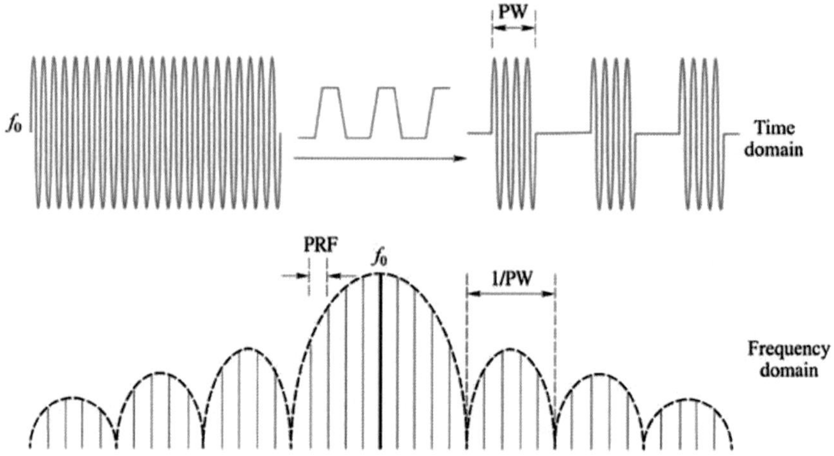

Fig. 10.18 Pulse spectrum curve

the duration of the pulse, the wider the frequency band occupied by the spectrum energy. The spacing between the different spectral components is equal to the Pulse Repetition Frequency (PRF).

Antenna pattern test in pulse mode is more complex than that in traditional Continuous Wave (CW) mode. In the CW test state, the receiver's Intermediate Frequency Band Width (IFBW) is mainly used to reduce the noise level of the receiver, thus improving the dynamic range of the test system. In pulse mode, the wide pulse or narrow pulse test mode should be selected according to the maximum sustainable operating time of the T/R component, and then the IFBW is set by the selected test mode.

According to the current dynamic requirements on the microwave receiver and antenna test system, the wide pulse test mode should be preferred for the antenna test with T/R component continuous operating time greater than 1 μs. The advantage of wide pulse detection is the fast test speed and no duty cycle loss. Since the wider IF bandwidth is used in the detection process, the noise floor of the receiver is high. In the wide pulse test mode, the receiver must be synchronized with the pulse stream to ensure that data acquisition occurs only when a pulse signal is present. This means that the receiver must be provided with a pulse trigger signal with the same frequency as the PRF, and the delay relationship of this pulse trigger signal with respect to the pulse stream shall be accurately set. Therefore, the wide pulse test mode is also referred to as a synchronous data acquisition mode. Figure 10.19 shows a block diagram of a pulse pattern test using a vector network analyzer.

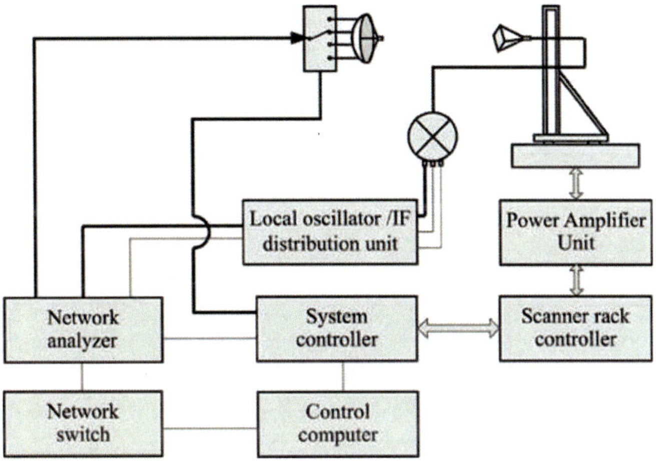

Fig. 10.19 Block diagram of pulse pattern test system

10.1.6.2 Multi-beam Parallel Test

With the application of phased array antenna technology, the higher requirements for improving the near-field test efficiency of phased array antennas have been put forward in engineering. Taking the space-borne SAR payload antenna as an example, there are usually many imaging modes for obtaining the Earth remote sensing data in different environments. Thousands of wave positions are designed for each mode, and the parameters of the directivity coefficient, beam width, side lobe level, dispersion, polarization mode, beam forming and broadening of each beam position must be tested and optimized on the ground. To reduce test time, several beams are tested in parallel.

The antenna operates in a fixed beam position state, and only one beam position pattern data can be obtained in a single near-field scan in the conventional near-field pattern test. During the multi-beam near-field pattern test, the antenna controller and the antenna test system are coordinated and synchronized, and multiple phased array beam position pattern data can be obtained by a single near-field scan. The core issue of multi-beam test is the synchronous of the test system with the phased array antenna beam controller. For multi-beam parallel test, current commercial near-field test systems typically provide a standard interface from the test system controller to output fixed-format beam position control codes and trigger signals. For different types of phased array antennas, in order to complete multi-beam parallel testing, a matching switching device is usually developed to convert the beam position code and trigger signal outputted by the test system into a control signal that the beam controller can recognize. Figure 10.20 shows a block diagram of a multi-beam parallel test system.

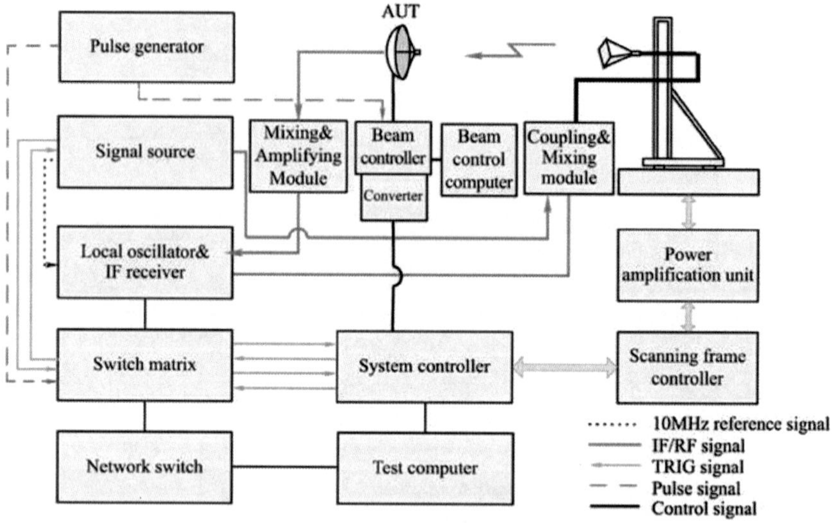

Fig. 10.20 Block diagram of multi-beam parallel pattern test system

When the phased array antenna is tested in multi-beam and multi-frequency mode, the test state is more complicated. Each sampling test of the receiver is accompanied by loop nesting such as position triggering, beam position switching and frequency switching. Figure 10.21 shows the timing relationship of an antenna for multitasking test. During the test, the probe reaches the sampling position according to the preset sampling parameters, and the beam position control interface in the test system controller outputs the beam position switching pulse and the beam code to the converter, and the converter converts the information into identifiable beam position information and sends it into the beam controller after receiving the information, then the beam position is switched. After the switching is completed, the feedback

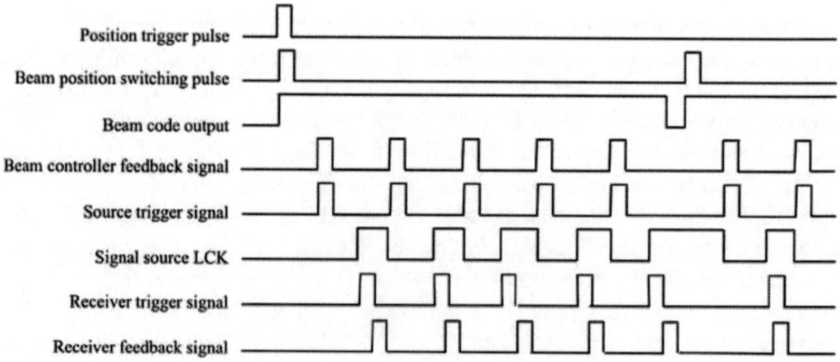

Fig. 10.21 Timing relationship of multi-beam parallel test

signal is output to the test system controller, and the near-field test system performs multi-frequency data collection.

10.1.6.3 Near-Field Diagnostic Technology for Phased Array Antenna

Antenna testing also includes debugging and diagnostics of the antenna. With the application and development of space-borne phased array antenna, antenna fault diagnosis techniques are becoming more and more important. Phased array antenna test and diagnosis technology have always been a research hotspot in the field of phased array testing.

Phased array antenna fault diagnosis relies on the available test data, i.e., the basic characteristic parameters, such as antenna gain, beam width, beam pointing, pattern zero depth, side lobe level, side lobe position, polarization, scanning range, are firstly obtained through antenna performance tests, and then it is found that the parameter is abnormal through the research and comparison. Based on the abnormal information, analyze the amplitude and phase distribution of the aperture to find out the illogical mutation point and the cause of the pattern distortion. Figure 10.22 shows the abnormal curve of the elevation profile of a satellite antenna.

In Fig. 10.22, the thick solid line indicates the fault pattern curve. The side lobe is higher and the zero depth is shallower than that of the normal side lobe. It can be inferred that there is an abnormality in the antenna operation, and the antenna shall be analyzed and diagnosed. The phased array antenna test diagnosis is usually performed in the planar near field. The outstanding advantage of the antenna near-field scan test is that the data collected during the test can be used to calculate the far-field pattern, the amplitude and phase distribution of aperture field. The aperture field distribution shown in Fig. 10.23 is the inversion result of the aperture field of

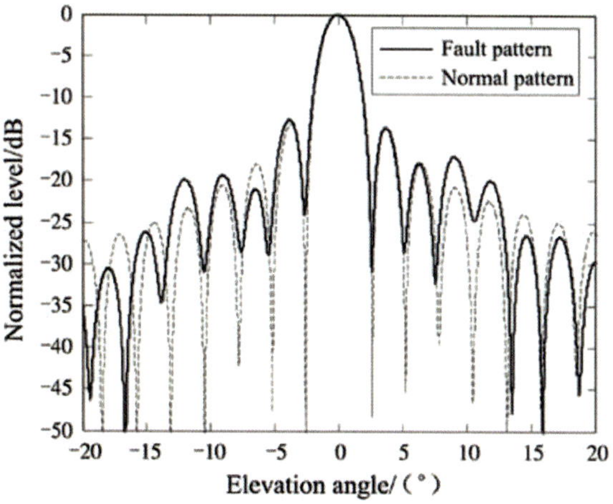

Fig. 10.22 Antenna fault pattern curve

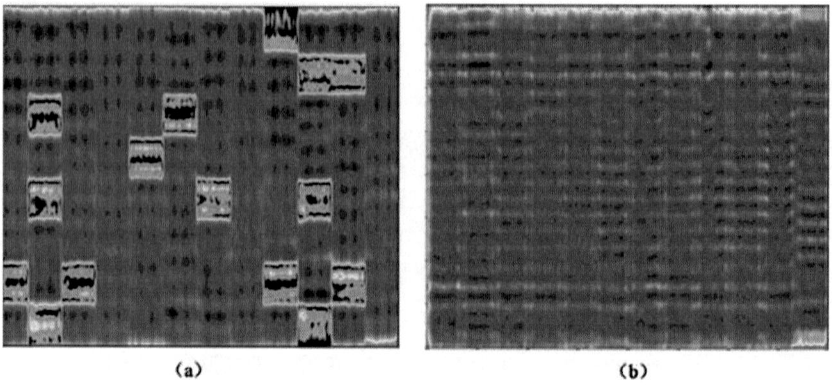

Fig. 10.23 Antenna aperture field distribution. **a** The aperture field with abnormal pattern; **b** the aperture field with normal pattern

the antenna in Fig. 10.22.

Figure 10.23a shows the two-dimensional aperture field distribution when the antenna is in an abnormal working state. Some radiating elements are not operating properly, so that the specific positions of these units in the array can be found. In certain circumstances, near-field scanning equipment can also be used to place the probe close to the radiation array and directly perform the aperture field test and diagnosis.

10.2 Spacecraft Antenna Environmental Test Technology

10.2.1 Dynamic Environment Test

The dynamic environment test of spacecraft antenna generally involves acceleration test, shock test, sinusoidal vibration test and random vibration test. The antenna test can be appropriately tailored according to the actual requirements of different missions, specifically, the test matrix given by the satellite system.

10.2.1.1 Test Objective

The main objective of the antenna participating in the dynamic environment test is to verify the correctness of the structural design scheme, assess whether the product structure can withstand the dynamic environment that may be encountered and expose the defects that may exist in the material, component selection and manufacturing assembly process, in order to minimize early failures and improve reliability.

10.2.1.2 Acceleration Environment

Acceleration refers to the acceleration of the spacecraft in launching by means of the thrust of the rocket engine, which changes slowly and approximates the steady state, and the low-frequency part of the response acceleration of the vibration environment, both constitute the acceleration environment of the spacecraft.

There are static load test and centrifuge test for satellite acceleration environment simulation.

1. Static load test

The several parts of the satellite structure are loaded by the load system to generate an appropriate stress distribution, thereby assessing the strength and stiffness of the satellite structure under simulated steady-state acceleration inertial loads. The whole satellite is generally tested by static load.

2. Centrifuge test

The centrifuge is essentially a long arm that rotates around a fixed axis in a plane. The test piece is mounted on one end of the arm and the other end is weighted to maintain balance. When the arm is rotated at a certain angular velocity ω, the test piece is subjected to an inertial load in the direction away from the axial center, to maintain the acceleration $r\omega^2$ directed to the axis, where r is the distance from the centroid of the test piece to the axis of the centrifuge. Generally, onboard equipments are tested by centrifuge.

Both above methods have their own characteristics and can be used for different purposes. Centrifuge test is generally used in the antenna acceleration environment simulation test.

10.2.1.3 Shock Environment

The shock environment mainly refers to the extremely short-term impact load generated by the pyrotechnic device during the separation of the spacecraft and the launch vehicle, the fairing throwing, the cabin separation, the unlocking, releasing and unfolding of various extension components. In addition, the spacecraft is also subjected to an impact environment during ground transportation and hoisting, and when the return capsule returns to the ground.

Since the shock environment of a spacecraft can be divided into simple impact (i.e., a typical impact) environment and complex impact (i.e., an explosion shock) environment, the impact simulation can be categorized into simple impact simulation and complex impact simulation. The test equipment for simulating simple impact mainly includes a drop impact machine and an electric vibration table. The equipment for simulating the explosion impact mainly includes an electric vibration table, a resonance plate type impact simulation device and a pendulum type horizontal impact test machine.

10.2.1.4 Vibration Environment

In the process of rocket propulsion, pulse thrust and jet noise, etc. will generate vibration excitation. These excitations will be directly transmitted to the satellite main bearing structure through the satellite-rocket adapter and act on each secondary structure step by step, thus affecting the performance and reliability of the equipment.

The traditional mechanical environment vibration test mainly includes sinusoidal scanning vibration test, random vibration test and noise test. In the sinusoidal scanning vibration test, only one frequency is applied to the test piece at a certain moment. During the whole test period, the excitation frequency is smoothly increased by linear or logarithmic sweep frequency, and the magnitude of the vibration at each frequency is controlled according to the test conditions. The random vibration test is to apply the energy contained in all frequency components to the test piece at the same time. The frequency range of random vibration is usually 20–2000 Hz, and the excitation condition is generally expressed by the trapezoidal acceleration power spectrum along the double logarithmic coordinate distribution. The noise test is usually done in the reverberation room, where the energy contained in all frequency components is simultaneously applied to the test piece. The frequency range of the noise test is usually 31.5–8000 Hz, and the excitation condition is usually expressed by the sound pressure level in the 1 octave or 1/3 octave bandwidth.

Vibration environment test simulation is divided into low-, medium- and high-frequency vibration. Low-frequency vibration (generally 5–100 Hz) is tested with the low-frequency sinusoidal scanning test, which is generally applied to both satellite and onboard equipment. In medium and high-frequency acoustic vibration environment, small satellites and small components are subjected to random vibration test (generally 10–2000 Hz) on the vibration test-bed, and large satellite and large components are subjected to noise test (generally 31.5–8000 Hz) in the reverberation room.

10.2.1.5 Test Program

The test program, provided by the test proposer, aims to clarify the input requirements of the test implementation, generally including the test purpose, test piece technical status, test conditions and allowable deviation, test control requirements, test measurement requirements, test procedures, environmental requirements, supporting facilities requirements, success criterion, test report requirements, etc.

The test purposes, test conditions, allowable deviations, environmental requirements, etc. can refer to the environmental test specification given by the satellite system. For the technical state of the test piece, the number of products, the product interface and the test loading direction are clarified. The single-point control or multi-point control should be specified in the test control requirements. Single-point control is used in the shock test. The vibration test can select multi-point control or single-point control according to the specific conditions of the product under test. The control points are generally located at the interface between the test fixture and the product. For the test measurement, measurement points shall be reasonably set according to the specific conditions of different products. The typical antenna testing process is as follows:

Pre-test VSWR test, product appearance inspection—pre-test product photographing—*x*-direction shock test—*y*-direction shock test—*z*-direction shock test—VSWR test, antenna appearance inspection—photographing of the tested product—the *x*-direction feature-level scanning—the *x*-direction sinusoidal vibration, the random vibration test—the *x*-direction feature-level scanning—the *y*-direction feature-level scanning—the *y*-direction sinusoidal vibration, random vibration test—*y*-direction feature-level scanning—*z*-direction feature-level scanning—*z*-sinusoidal vibration, random vibration test—*z*-direction feature-level scanning—post-test VSWR test, product appearance inspection—photographing of the tested product.

Success criteria are generally divided into test success criteria and product pass test criteria. Check whether the test is carried out in accordance with the provisions of the test program, if yes, the test is qualified. The test criteria for the product generally include the following: After the test, if no damage or other abnormal phenomena have been found in the product appearance inspection, the test is passed; if the feature-level scan results of the vibration test before and after the test are consistent, the test is passed; if the results of the VSWR test before and after are consistent, the test is passed. All the three are passed, then the product passes the test.

10.2.1.6 Test Fixture

Before the test, it is necessary to communicate with the test executant to confirm whether the test bench meets the test conditions. The test fixture is designed according to the interface of the test bench. When the product is small, the shape of the test fixture is recommended to be "L", which is convenient for the three-direction conversion of the antenna and reserves the necessary operation space. When the product is large, the test fixture is recommended to be flat and connection point is designed at the center to avoid the effect of the fixture on the test. It is recommended that the same fixture can realize simultaneous testing of multiple products to save resources and improve efficiency on the premise of the same test conditions, equipment capacity and product safety. A typical antenna vibration test is shown in Fig. 10.24.

10.2.1.7 Test Announcements

Before the shock test, the test fixture shall be calibrated, and the control curve is calibrated within the tolerance range; if the conditions permit, the test fixture can be calibrated with the product structural parts to improve the accuracy of the control curve; the measuring point is pasted at the corresponding position to measure its response curve.

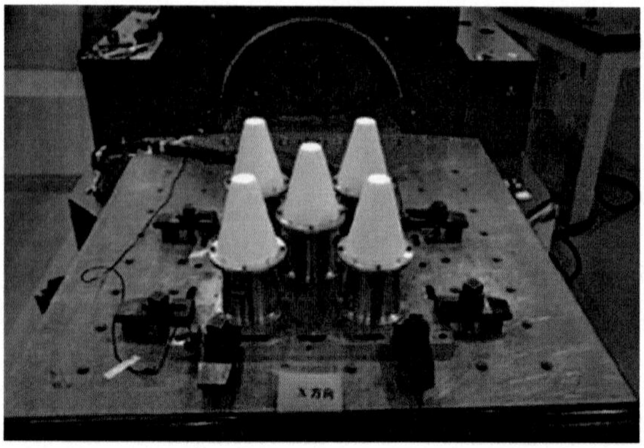

Fig. 10.24 Typical antenna vibration test

Before and after sinusoidal vibration and random full-scale vibration test, feature-level test is performed for comparing the feature-level response curves before and after the full-scale test to determine whether the test piece has a structural modal frequency offset. If the base frequency value of the test piece changes greatly, it should be checked by visual inspection or non-destructive testing to detect the failures such as the broken structure, the loosen connecting bolt or the degumming. If necessary, transfer function measurements can be carried out to determine the exact location of the structural fault.

For products whose electrical performance is monitored during the test, special tooling must be designed or the test cable should be effectively bundled to minimize the impact of the measuring equipment on the product under test. After the test, the appearance of the antenna is checked, and the electrical performances of the antenna are tested and compared with the test data before the test.

During the test, the response curve of the measuring point is interpreted to obtain the peak frequency, response, etc., which are interpreted whether they are within the safe range, and compare them with the analysis results. In sinusoidal vibration test, it is sometimes encountered that the response of important measuring points in certain frequency bands may exceed the highest value predicted by the environment. In order to avoid the product failing due to the test, it is necessary to limit the amplitude of the response by the active notching control or the response limiting control, and the notching control is required to obtain the consent of the system design engineer.

10.2.2 Thermal Environment Test

The thermal environment test of spacecraft antenna generally involves thermal cycling test and thermal vacuum test. It can be appropriately tailored according to the actual requirements of different missions. The specific test matrix should be given according to the satellite system.

10.2.2.1 Test Objective

By the thermal cycling qualification and proto-flight tests, the ability of the product to work under the qualification-level and proto-flight-level thermal cycle stress environment are verified, which indicate that the flight product can withstand the thermal cycling environment during acceptance test.

The thermal cycling acceptance test is to find out potential defects in the quality of materials and process manufacturing of the product in a normal pressure thermal cycling environment.

In the thermal vacuum qualification and proto-flight tests, the operating ability of the product under the specified pressure and the qualification and proto-flight vacuum thermal cycle stress environment are verified, which indicate that the flight product can withstand the thermal vacuum environment during the acceptance test.

The thermal vacuum acceptance test is to find out potential defects in the material and process manufacturing quality of the component in a vacuum thermal cycling environment.

10.2.2.2 General Requirements

The test temperature range of the product can refer to Fig. 10.25.

The general principles for choosing temperature control points are as follows:

1. The position of the temperature control point should be consistent with that of the telemetry temperature point of the product as far as possible.
2. Temperature control points should be as few as possible, and only one if possible. Several temperature control points can be considered for complex products.
3. For the thermal vacuum test, the temperature control point should be selected at the representative of the test product where is not near heat source. That is usually near the mounting lug.
4. For the thermal cycle test, the temperature control point should be selected on the leeward side of the product, and where the heat radiation from the external heating source cannot be directly reached.
5. For the test of multiple products, the temperature measurement points should be so arranged on each product that the test temperature of each product meets the requirements.

The temperature retention time requirements are as follows:

1. When the temperature at the temperature control point of the test product reaches the limit of the specified test temperature, and the temperature change in the first

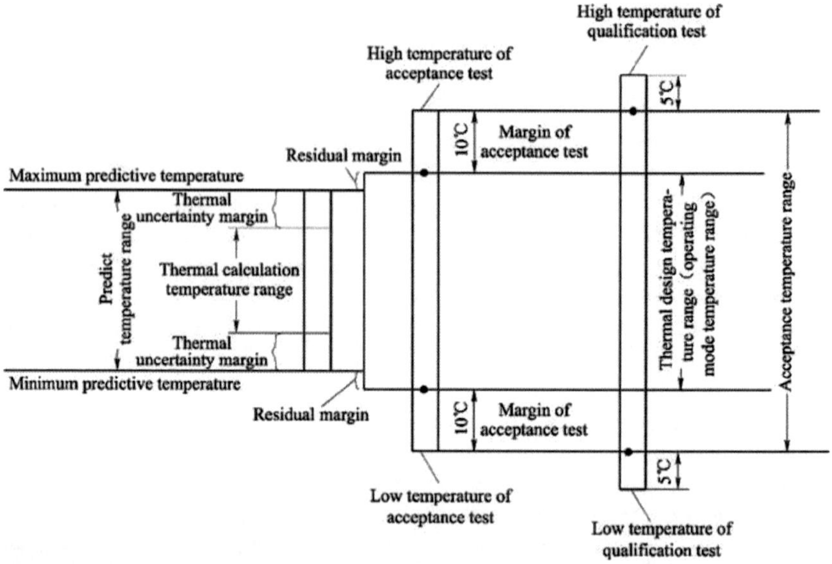

Fig. 10.25 Relationship among different temperatures. *Note 1* Temperature limit to meet performance requirements. *Note 2* Residual margin is determined by thermal design, which may be adopted or not

hour is less than 1 °C/h, then it is considered as that the test product is in a stable temperature state.

2. In order to achieve the temperature stability of the internal components and elements of the test product, the temperature should be kept for a certain time after the product is in a stable temperature state. Due to the different masses and the different thermal inertia of the products, the temperature holding time required for the internal temperature stability of the products differs too. For the implementation details, refer to the satellite environmental test specifications.

The test product should be inspected on the appearance, cleanliness and redundancy before entering and after leaving the test vessel; the test product should be subjected to detailed electrical and functional tests before, after and during the test. Pre-test results should be used as the basis for comparing the performance test results during and after the test.

The typical thermal vacuum test for the antenna is shown in Figs. 10.26 and 10.27.

10.2.2.3 Test Program

The test program is provided by the test proposer to clarify the input requirements of the test implementation, generally including the test objective, technical condition of the test piece, test conditions and allowable deviation, temperature holding time, test

Fig. 10.26 Installation status of the test piece on the octahedron tooling

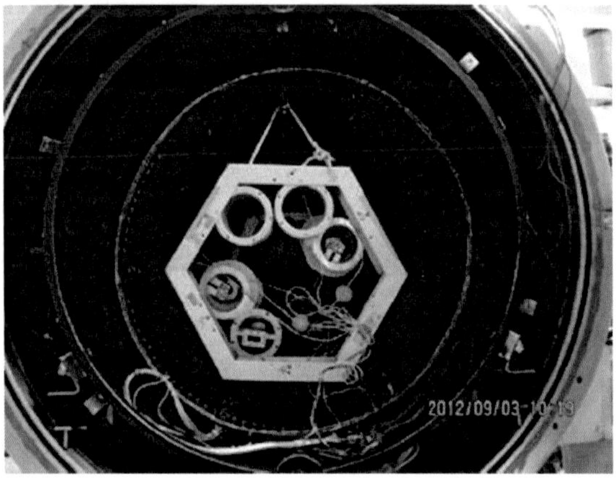

Fig. 10.27 Installation status of the test piece in a vacuum environment simulator

control requirements, test and measurement requirements, test procedures, thermal cycle test requirements, thermal vacuum test requirements, environmental requirements, supporting facility requirements, success criteria, test report requirements, etc.

Chapter 11
Design Case of Typical Spacecraft Antenna System

Abstract This chapter addresses several typical design schemes of spacecraft antenna system, which are based on the spacecraft development projects in China. These antenna system design cases include earth remote sensing satellite, meteorological satellite, space science satellite, communication satellite, recoverable satellite, Shenzhou space ship and Chang'e-5 lunar probe.

11.1 Design Case for Earth Remote Sensing Satellite Antenna System

In this section, a medium-resolution optical remote sensing satellite with an orbital height of 500 km is taken as an example. In order to broaden the imaging swath, two identical optical cameras are loaded for mosaic remote sensing measurement. In order to expand the imaging coverage, the function of left and right-side swing maneuvering imaging is available in the satellite. Stated thus, higher requirements are put forward for TT&C antenna and data transmission antenna.

Considering the satellite's on-orbit operation and side-swing working state, side-swing TT&C and data transmission are the hard nuts that must be cracked by the antenna system.

The satellite has two extreme working states: 35° from the land to the East and 35° from the sea to the West. In both extreme working states, the TT&C ground station and the data transmission ground station are always in the $+y$-direction of the satellite coordinate system.

The satellite has other two extreme working states: 35° from the land to the West and 35° from the inland to the East. In both extreme working states, the ground control station and the digital transmission ground station are always in the $-y$-direction of the satellite coordinate system.

Therefore, the layout of antenna system should be considered to meet the requirements of TT&C and data transmission of satellite side-swing and normal working conditions.

R. You et al., *Technologies for Spacecraft Antenna Engineering Design*, Space Science and Technologies, https://doi.org/10.1007/978-981-15-5833-7_11

The antenna system is composed of TT&C antenna, data transmission antenna and GPS antenna. The specifications of the antenna system in the whole satellite state are shown in Table 11.1.

Table 11.1 Antenna system performance specification

Subsystem	Item	Specification
TT&C antenna	Polarization mode	Dual orthogonal circularly polarized binary array
	Working frequency	2042, 2217 MHz
	Coverage gain	(1) The antenna array forms a spherical pattern approximating the full-space radiation (2) The net gain of radiation in the full-space is ≥ -10 dBi (including antenna matching network loss, high-frequency cable loss and polarization loss, i.e., normalized to the transponder port) (3) The pattern has a flat amplitude and phase characteristics in the non-interference region within $\pm 102°$, which meets the requirements of TT&C (4) The range of interference region is less than 15%, and each interference concave point is less than 2° at -10.0 dBi
GPS antenna	Polarization mode	RHCP
	Working frequency	1575.42 MHz
	Coverage gain	(1) The pattern is near hemisphere (2) Coverage net gain: $\pm 45°$, $G \geq -1$ dBi; $\pm 75°$, $G \geq -3.0$ dBi; $\pm 80°$, $G \geq -3.5$ dBi; $\pm 85°$, $G \geq -4.0$ dBi
Data transmission antenna	Polarization mode	RHCP
	Working frequency	(8222 ± 160) MHz
	Coverage gain	The radiation pattern is a saddle-shaped beam with satellite-to-Earth axis as rotation axis. Its maximum directional angle is $\theta_m = \pm 66.2°$, $G_m \geq 5.5$ dBi; 0° axis direction, with $G_0 \geq -1$ dBi The antenna must satisfy the requirement on the satellite with 35° side swing in the direction of $\pm y$ relative to the zero point below the satellite. The transmitting power should be $P_t \geq 50$ W

11.1.1 Composition of TT&C Antenna

Comprehensively considering the on-orbit operation and side-swing working state, that is, under the satellite side-swing state, the beam coverage area of the $+z$-antenna increases from $\pm 66.2°$ to $\pm 101.2°$. The layout of TT&C antenna array should meet the requirements of both satellite side-swing and normal working state.

The TT&C antenna consists of two LHC antennas, two RHC antennas, two equal power division networks and six radio frequency cable assemblies. Figure 11.1 shows the composition of the TT&C system. Two-element array layout in which antenna 1 (Earth-facing L) and antenna 3 (Earth-aft R) arrays are connected to transponder A through cable assembly 5; antenna 2 (Earth-facing R) and antenna 4 (Earth-aft L) arrays are connected to transponder B through cable assembly 6.

(1) In the case of satellite receiving uplink remote control signal

There are two independent TT&C uplink channels on the satellite with hot-backup working mode. Both transponders on the satellite can receive remote control signals when the ground sends any circular polarization uplink signals. For the whole TT&C system, the effective signal received by onboard receivers is one of the two signals selected according to the agreed mode. The total radiation characteristic of TT&C antenna binary array is the incoherent synthesis of two identically polarized TT&C antennas.

(2) In the case of ground receiving downlink remote telemetry signal

The ground receiving station adopts diversity reception mode and the satellite transmitter uses cold backup mode. At a time, only one transponder transmits the telemetry

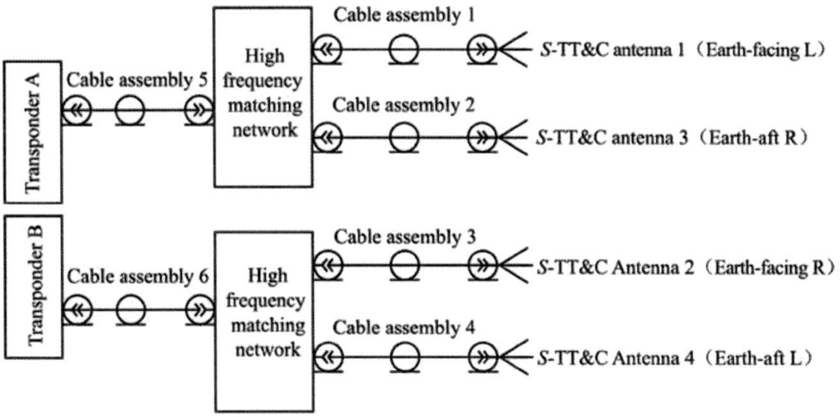

Fig. 11.1 Schematic diagram of TT&C system composition

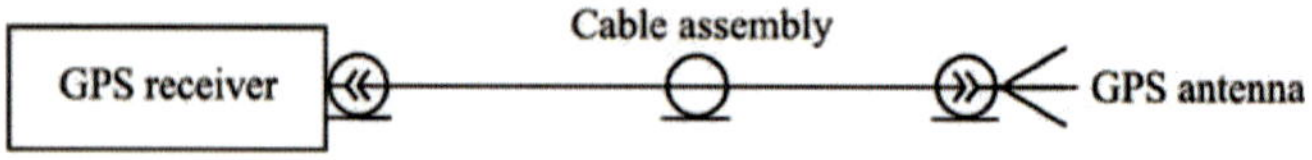

Fig. 11.2 Principle block diagram of GPS receiving system

signal. The telemetry signal is transmitted through dual-orthogonal circularly polarized binary array installed on the Earth-facing and Earth-aft side, and the two circularly polarization signals can be received by the ground station. The total radiation characteristic of TT&C antenna binary array is the coherent synthesis of two differently polarized TT&C antennas.

Only one GPS antenna is installed on the Earth-aft side of the satellite. The principle block diagram of the receiving system is shown in Fig. 11.2.

The four-arm spiral antenna in Chap. 2 is used as TT&C antenna element, and the self-feeding four-axis spiral GNSS antenna in Chap. 4 as GPS receiving system.

11.1.2 Composition of Data Transmission Antenna

The data transmission antenna consists of two groups of antenna switching arrays, each of which is, respectively, connected with two TWTAs that are cold backup to each other. Each group of the switching arrays consists of three X-band data transmission antennas, one SP3T and three sets of waveguide assemblies. The working principle block diagram is shown in Fig. 11.3. The data transmission antenna transmits the high-frequency signal from the data transmitter to the ground receiving station according to the requirements of switching connection of the backup TWTA.

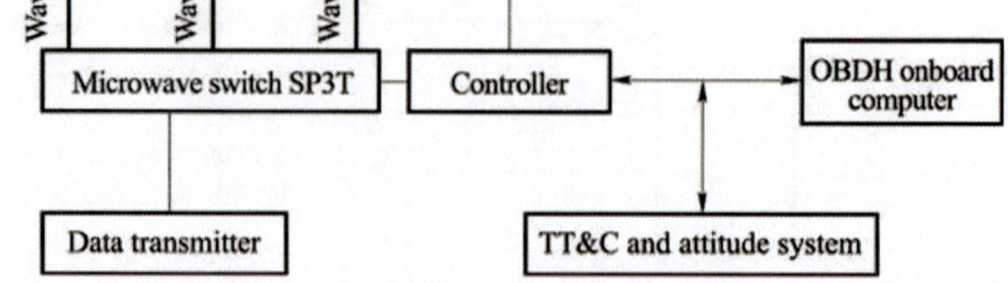

Fig. 11.3 Principle block diagram of single switching array for data transmission antenna

Besides, the radiation pattern of the antenna element should meet the overall requirements of the coverage gain. When the satellite is in the visible area of the ground under the side-swing mode within the range of ±35°, the data transmission antenna should be competent for data transmission to the ground station.

The X-band backfire bifilar helical antenna in Chap. 3 is selected as data transmission antenna.

11.1.3 Antenna System Layout

Considering the satellite's on-orbit operation and side-swing state, the TT&C antenna should meet the requirements of the side-swing and normal working state.

Through simulation analysis, the layout of TT&C antenna is as follows: TT&C antenna 01 and antenna 02 are arranged in satellite Earth-facing plate which in direction $+y$ and $-y$, respectively, antenna 03 and 04 are arranged in the Earth-aft plate, the specific positions are shown in Figs. 11.4 and 11.5, which can satisfy the

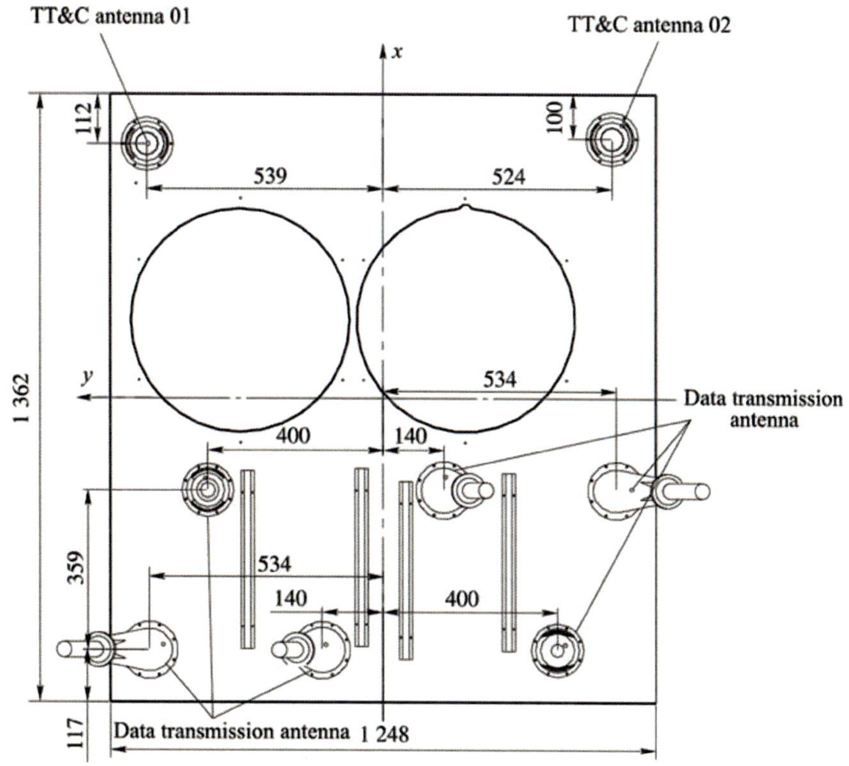

Fig. 11.4 Satellite $+z$-plate layout

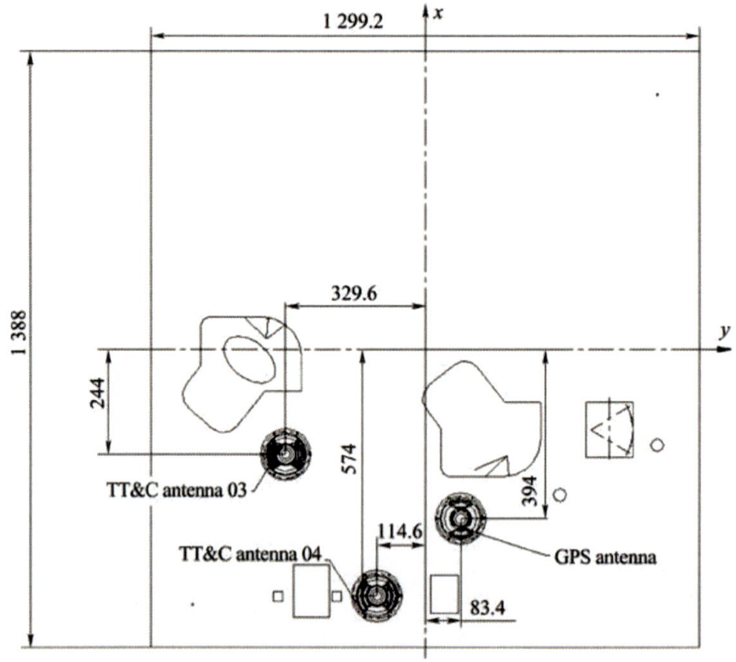

Fig. 11.5 Satellite-z-plate layout

TT&C requirement under satellite side-swing and normal working mode.

The layout of the data transmission antenna is as follows: six data transmission antennas are all arranged on the Earth-facing plate (+z-plane) in two rows (3 per row). Four data transmission antennas are inclined to install at a certain angle. The specific position shown in Fig. 11.4 can meet the data transmission requirements of the satellite side-swing and normal working state.

11.1.4 Performance of Antenna System

Figures 11.6, 11.7, 11.8 and 11.9 shows the corresponding polarization response patterns of two sets of TT&C antenna binary arrays in RM satellite state. The composition of TT&C antenna system is consistent with that on the satellite. The yOz-plane test results of the layout meet the requirements of satellite side-swing TT&C.

Figure 11.10 shows the test results of the GPS antenna RM satellite state, which meet the technical requirements.

Figure 11.11 shows the test results of the data transmission antenna RM satellite state (taking the satellite normal working state as an example), which meet the technical requirements.

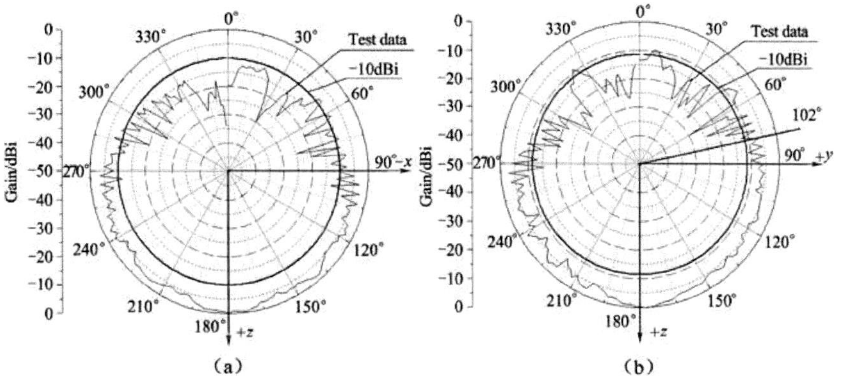

Fig. 11.6 TT&C antenna 01 LHCP uplink. **a** *xOz*-plane pattern; **b** *yOz*-plane pattern

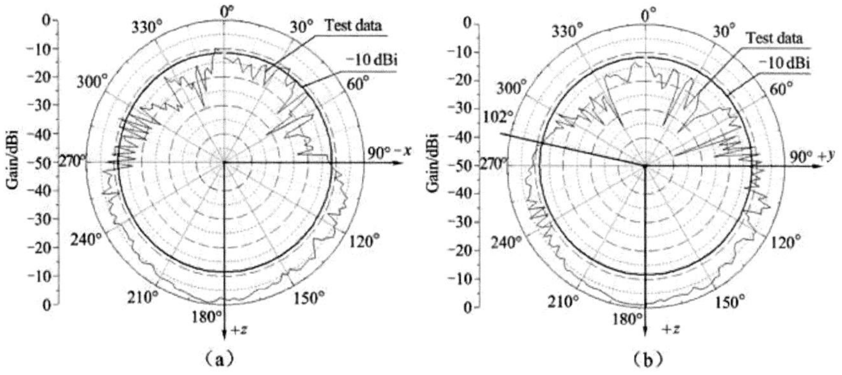

Fig. 11.7 TT&C antenna 02 RHCP downlink. **a** *xOz*-plane pattern; **b** *yOz*-plane pattern

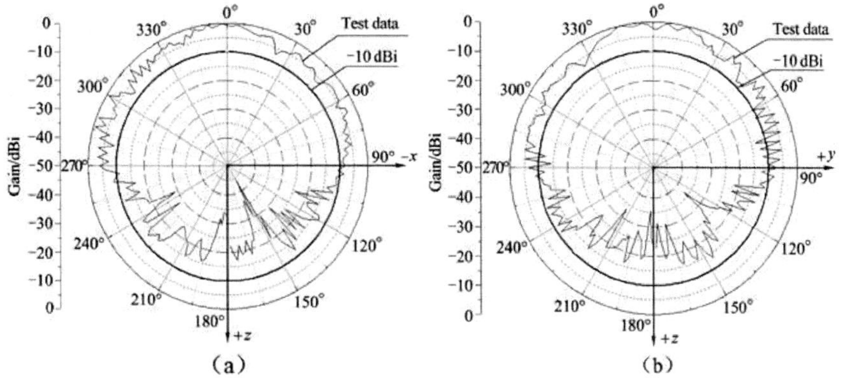

Fig. 11.8 TT&C antenna 03 RHCP uplink. **a** *xOz*-plane pattern; **b** *yOz*-plane pattern

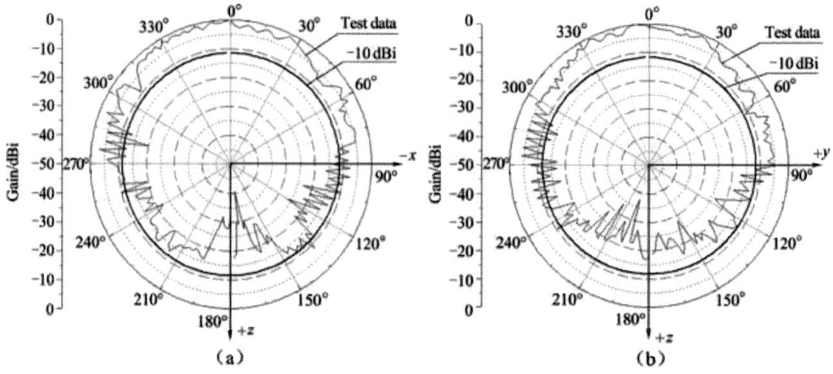

Fig. 11.9 TT&C antenna 04 LHCP downlink. **a** xOz-plane pattern; **b** yOz-plane pattern

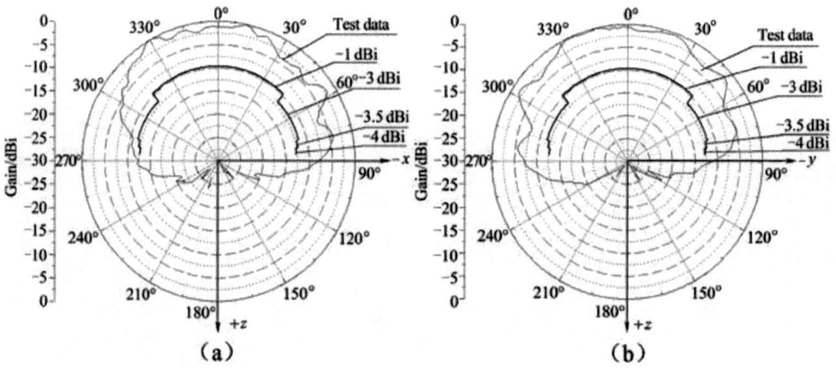

Fig. 11.10 GPS antenna RHCP. **a** xOz-plane pattern; **b** yOz-plane pattern

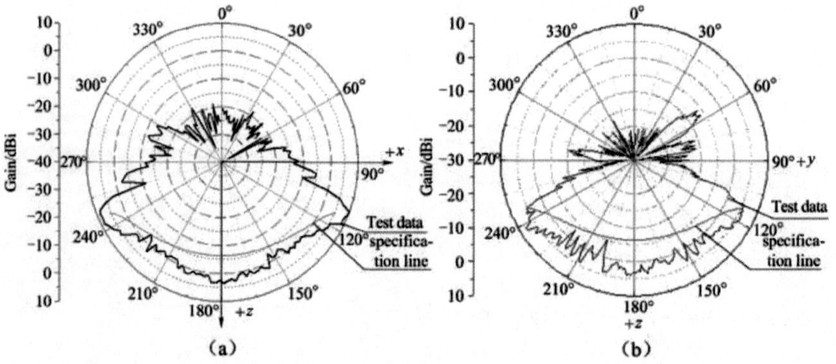

Fig. 11.11 Directional pattern of data transmission antenna in satellite environment. **a** xOz-plane pattern; **b** yOz-plane pattern

Gain test results include power division network loss, polarization mismatch and feed loss.

Six such satellites are operating on orbit, and the wireless channel is normal under the condition of side swing, which verifies the correctness of the antenna system design.

11.2 Design Case of Meteorological Satellite Antenna System

China's meteorological satellites include FY-1, FY-2, FY-3 and FY-4. China has so far successfully launched 15 FY meteorological satellites, of which 8 are operated on orbit now, and become an indispensable part of the world meteorological satellite observation system. These satellites are widely used in meteorological forecasting, disaster prevention and mitigation, scientific research and other fields. The satellite FY-2 is the first-generation geostationary orbit meteorological satellite in China, covering the Asia–Pacific region. The satellite was first developed in the mid-1980s. The first satellite was launched in June 1997. Later, six satellites were launched successively. At present, it is still the main means of short-term weather forecasting. The satellite is based on the spin stabilization platform of the communication satellite DFH-2 in China. In order to meet the needs of high stability image, the spin speed of the satellite is increased from 50 to 100 cycles/min. The payload of the satellite is visible light/infrared multi-channel scanning radiometer, which can monitor the cloud and water vapor distribution in the coverage area for 24 h. Since the communication satellite platform is adopted, C-band of communication is used for TT&C, image data is downloaded and forwarded through S-band, and data of ground meteorological stations across China is transmitted through the satellite UHF data collection system, so the satellite is composed of C-, S- and UHF-bands. Since the satellite is a spinning platform with an antenna weight of only 10 kg, the antenna must orient the Earth through a despun motor, so the whole antenna system also includes a three-channel microwave rotating joint. In addition, the antenna uses advanced materials and processes such as carbon fiber composite material and Nomex honeycomb composite material, to achieve lightweight design.

11.2.1 Satellite Function

FY-2 is a representative of the geostationary orbit spin stabilized satellites. The main functions of the satellite are to:

1. obtain visible light, infrared cloud images and water vapor distribution maps;
2. forward cloud images and low-rate information transmission;

3. transmit the environmental data collected by the ground data collection platform and.
4. monitor space environment.

FY-2 with three frequency bands, C-band TT&C channel, S-band cloud image forwarding and UHF-band data collection, communicates with the ground through the antenna of the corresponding frequency band.

11.2.2 Technical Specification

The most important characteristic of geostationary orbit spin stabilized satellite is that the antenna should be oriented to the ground using a despun motor. All radio frequency channels must be transmitted through the hollow axis of the despun motor whose diameter is 52 mm. Therefore, the microwave rotating joint is an important component of the satellite. Because the satellite antenna is in three bands, the rotating joint needs three channels, that is, the corresponding C-band antenna, S-band antenna and UHF-band antenna are connected with the equipment in the satellite through the rotating joint.

Because of the high requirement of attitude and stability of satellite, some technical specifications such as rotation accuracy, lateral inertia difference and dynamic unbalance are added to the antenna system performance.

Due to the special characteristics of FY-2, the antenna should meet not only the electrical performance requirements of TT&C and data transmission, but also the requirements of the satellite attitude control. This puts forward more constraints for antenna design, which must be considered in the design. The main performance indicators of the antenna system include as follows:

(1) Operating frequency range

The UHF-band uplink frequency ranges from 401 to 403 MHz.

The S-band uplink frequency ranges from 2042 to 2060 MHz, and the downlink frequency range is 1670–1712 MHz.

The range of the C-band uplink frequency range is 5925–5970 MHz, and that of the downlink frequency ranges from 4165 to 4200 MHz.

(2) Antenna beam

Using satellite coordinates, the sub-satellite point is in the positive direction of the x-axis, e.g., $\begin{cases}\varphi = 0^\circ \\ \theta = 90^\circ\end{cases}$

Conical beam of UHF-band antenna $\begin{cases}\varphi = 0^\circ \pm 9.2^\circ \\ \theta = 90^\circ \pm 9.2^\circ\end{cases}$

Conical beam of S-band antenna $\begin{cases} \varphi = 0^\circ \pm 9.2^\circ \\ \theta = 90^\circ \pm 9.2^\circ \end{cases}$

Ring-shaped beam of C-band antenna $\begin{cases} \varphi = 0^\circ \sim 360^\circ \\ \theta = 90^\circ \pm 30^\circ \end{cases}$

(3) Polarization mode and axial ratio

The RHCP of UHF antenna is $\leq$3 dB (within beam angle).

The vertical line polarization of the S-band antenna is $\geq$25 dB (within beam angle).

The uplink frequency range of the C-band antenna in RHCP is $\leq$5 dB (within $\theta = 90^\circ \pm 3\ 0^\circ$) and the downlink frequency range in LHCP is $\leq$5 dB (within $\theta = 90^\circ \pm 30^\circ$).

(4) System input standing wave ratio

UHF-band channel $\leq$1.5.

S-band channel $\leq$1.5.

C-band channel $\leq$1.65.

(5) Net gain

The beam peak value within the uplink frequency range of the UHF-band channel is $\geq$10 dBi and is $\geq$9.5 dBi at beam edge ($\pm$9.2°).

The beam peak value in the uplink frequency range of the S-band channel is $\geq$17 dBi and is $\geq$14 dBi at beam edge ($\pm$9.2°); the beam peak value in the downlink frequency range is $\geq$17.5 dBi and is $\geq$14.5 dBi at beam edge ($\pm$9.2°).

The uplink frequency range of the C-band channel is ≥ -4.5 dBi at beam edge ($\pm$30°), and the downlink frequency range is ≥ -4.0 dBi at beam edge ($\pm$30°).

(6) Isolation between channels

The isolation between the S-downlink frequency band and the UHF frequency band is $\geq$50 dB.

The isolation between the S-band and the C-band is $\geq$50 dB.

The isolation between the C-band and UHF-bands is $\geq$48 dB.

The isolation between the C-band uplink and downlink channels is $\geq$40 dB.

(7) Power capacity

S-band channel: $\geq$30 W.

C-band channel: $\geq$10 W.

(8) Insertion loss ripple of each channel by despun

UHF channel is $\leq$0.1 dB.

S-band channel is $\leq$0.1 dB.

C-band channel is $\leq$0.1 dB.

(9) Weight

The weight of despun section is less than 10 kg.

The weight of spin section is less than 2.5 kg.

(10) Mass characteristics of the despun section

Center of mass: $x = (0 \pm 0.3)$ mm, $y = (0 \pm 0.3)$ mm, $z \leq 165$ mm.

Lateral main inertia difference: $|I_{T\max} - I_{T\min}| < 0.72$ kg m^2.

Dynamic unbalance: the measured value is not more than 0.034 kg m^2 at 100 rpm (with wind resistance).

(11) Installation precision requirements

The radial jump of the rotating joint is not more than 0.02 mm.

The omnidirectional antenna radial jump is not more than 1.0 mm.

The parallelism between the mechanical axis of S-band antenna and UHF-band antenna is not more than 1 mm.

The parallelism between the mechanical axis of S-band antenna (UHF-band antenna) and the equatorial plane of the satellite is not more than 1 mm.

(12) Thermal control temperature range of antenna system

-180 to $+100$ °C.

(13) Antenna system reliability allocation index

>0.999 (end of 4 years).

11.2.3 Design of Antenna System

As shown in Fig. 11.12, the antenna system consists of two parts: spin section and despun section. The upper half of *y*-axis is despun section and the lower half is spin section.

The despun section includes S-band reflector antenna, UHF-band helical antenna, C-band omnidirectional antenna, omnidirectional antenna duplexer, antenna frame and support structure, despun part of three-channel microwave rotary joint and five-RF coaxial cable assembly for connection.

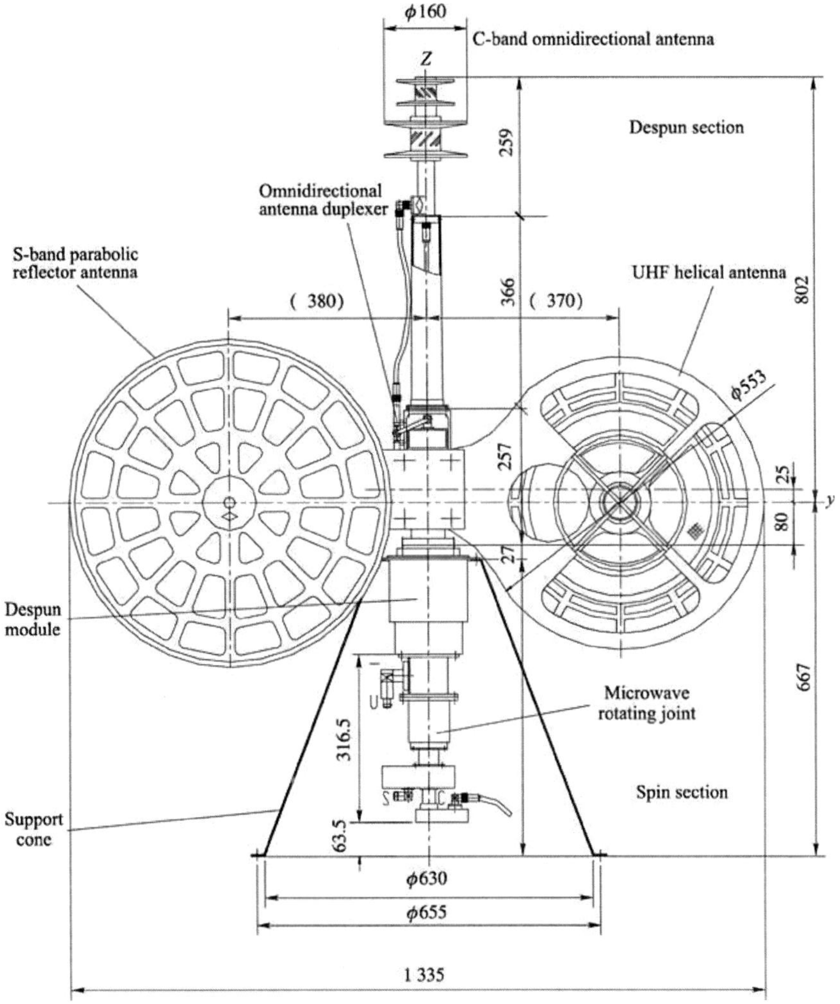

Fig. 11.12 Antenna system layout of FY-2 satellite

The spin section comprises the spin part of the three-channel microwave rotating joint, the transponder duplexer and a RF coaxial cable assembly.

The antenna is arranged in the shape of triangle according to the requirements of electrical performance and electromagnetic compatibility, as well as the overall assembly and structural layout of the satellite. S-band reflector antenna, UHF helical antenna, C-band TT&C antenna and duplexer are mounted on the antenna frame and assembled as a whole, and then mounted on the support of the microwave rotating joint despun section through the antenna frame. The S-band reflector antenna and the UHF-band helical antenna are basically symmetrical to the rotating axis of the rotating joint and can easily realize the dynamic unbalance. The C-band omnidirectional antenna is located on the rotating axis, which is the center of the omnidirectional plane of the antenna pattern, in order to minimize the radiation fluctuation when the satellite launches and spin-ups but does not yet despun. In order to reduce the weight, the main structures of the S-band reflector antenna and UHF helix adopt lightweight composite material, which results in the weight of the whole despun section less than 6 kg, and all antennas (including rotating joints, cables, thermal control materials) less than 10 kg. The physical photos of the antenna system are shown in Fig. 11.13.

C-band TT&C antenna uses coaxial slot and double-cone-disk to form circularly polarized eight-shaped pattern. The specific design is shown in Sect. 2.5.

The S-band antenna is an electrically small ($D/\lambda \sim 3.5$) antenna and the feed is a cup-reflected dipole. Moreover, the dual-frequency matching technology is adopted, which makes a good match at both 1.7 and 2.1 GHz.

Under the condition of limited size, the UHF-band antenna adopts a bifilar helical scheme to achieve high gain for a given length.

The microwave rotating joint adopts a three-layer coaxial struture, and the input and output of the UHF channel adopts dielectric materials filling mode to reduce the structural size; the input and output of the C- and S-band channels adopt a waveguide coupling design to realize low loss and high stability during rotation, as shown in Fig. 11.14.

C-band duplexer (omnidirectional antenna and transponder duplexer) uses air strip line feeding network structure to achieve the goal of small size, low loss and high isolation, as shown in Fig. 11.15.

The measured pattern of the antenna system in the satellite environment is shown in Figs. 11.16, 11.7 and 11.18. UHF-band antenna is asymmetric due to the plane

Fig. 11.13 Physical photos of FY-2 satellite antenna system

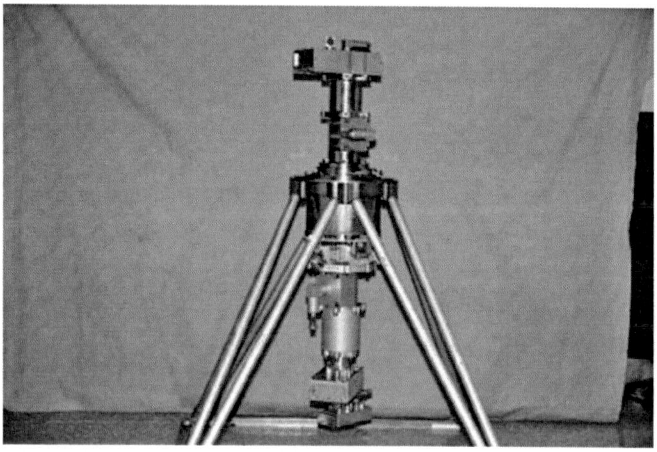

Fig. 11.14 Three-channel microwave rotating joint

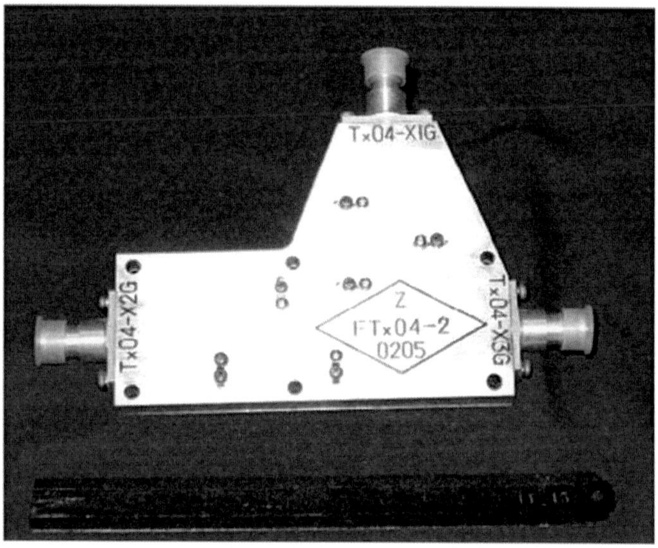

Fig. 11.15 C-band transceiver duplexer

reflection of the satellite, but still meets the requirements. All antennas of the satellite have been verified on orbit by eight satellites, meeting the specification requirements and are still operating on orbit.

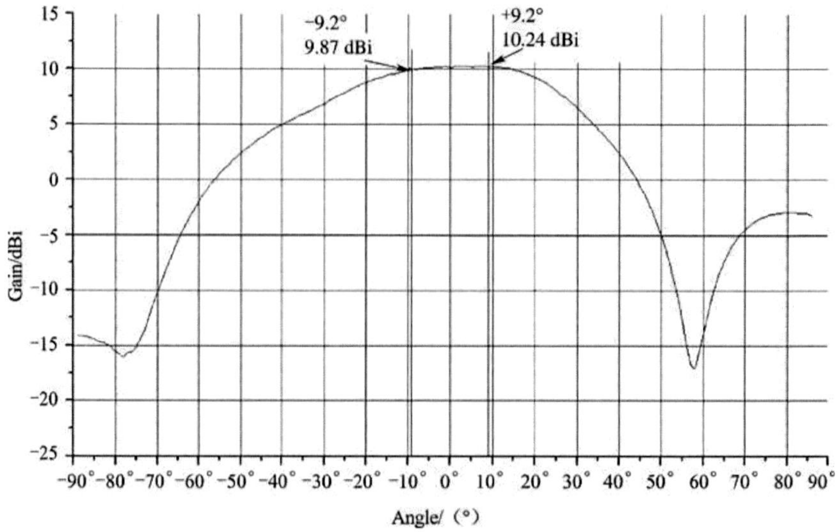

Fig. 11.16 Measured pattern and gain of UHF-band antenna

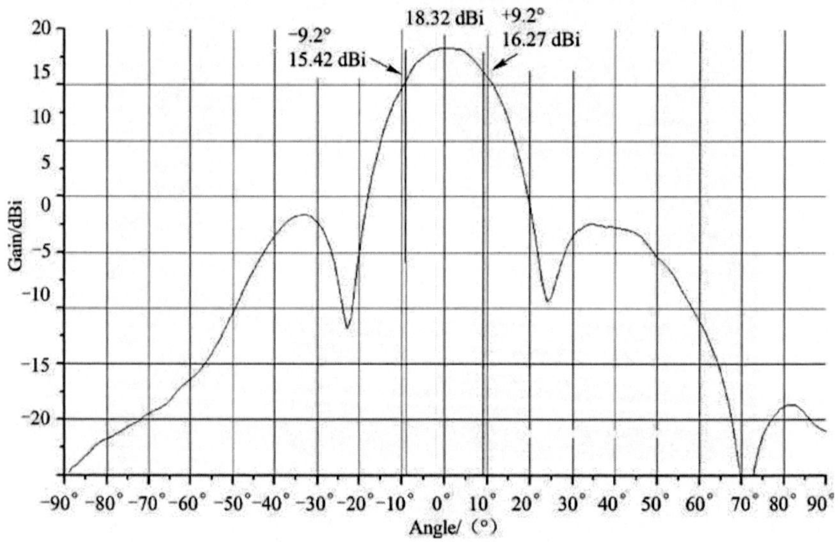

Fig. 11.17 S-band data and measured pattern and gain of transmitting antenna

11.2.4 Development Experience and On-Orbit Application

In the development of the antenna subsystem, due to the earlier time, some problems (such as passive inter-modulation, multipaction) are not identified and no conditions

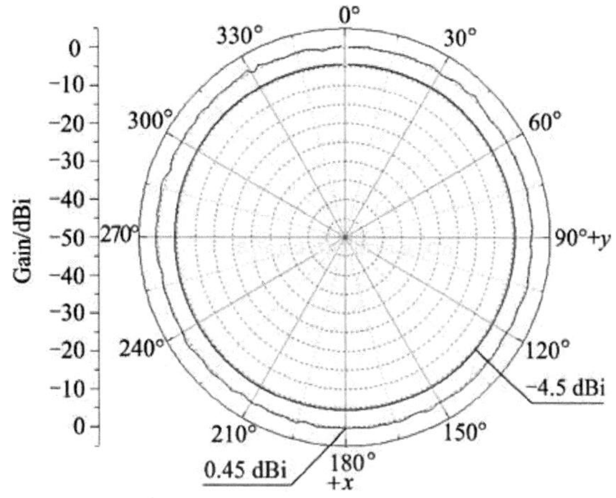

Fig. 11.18 Measured pattern and gain of C-band TT&C antenna

to be verified. Therefore, with the improvement of verification conditions in China, the antenna system has also been improved. The high-frequency connector has been changed from L16 to N-type and finally to TNC type. The impedance matching structure of the S-band antenna has been changed from split to continuous type, which greatly reduces the passive inter-modulation component. Quantitative regulation is made on the debugging process of rotating joint and duplexer, which reduces the risk of multipaction. At present, the TT&C, cloud map and data collection functions are all performed well in four satellites in use, which verifies the reliability and quality of the antenna system. The last one of the FY-2 series satellites was launched in 2018, which drew a happy ending for the development of the satellite.

11.3 Design Case of Space Science Satellite Antenna System

Hard X-ray Modulation Telescope (HXMT) satellite realizes space-borne hard X-ray high-resolution all sky survey. It will discover a large number of high-energy celestial bodies and high-energy radiation phenomena of celestial bodies and conduct high-sensitivity directional observation of important celestial bodies such as black holes and neutron stars. It is of important scientific significance for advancing human understanding of high-energy astrophysics, particle acceleration and radiation processes under extreme conditions.

The main functions of the HXMT satellite antenna are to realize remote control and telemetry, transmission and reception of ranging signal, GPS signal reception and scientific Data Transmission (DT) toward ground.

11.3.1 Mission Requirements and Technical Difficulties

According to the requirement of the payload to complete the flight mission, HXMT satellite works in four modes: all sky survey mode, fixed-point mode, small sky observation mode and orbit control mode.

Because of the different observation tasks, HXMT is quite different from the three-axis to-ground stable attitude satellite in remote sensing observation. Especially, the uncertainties of the toward-ground attitude caused by the satellite working mode increase the design difficulty of the antenna system. In the all sky survey mode, the HXMT satellite orients the -z-axis to the sun and rotates slowly around the $+z$-axis and completes the survey observation through the x-direction payload; the HXMT points to the fixed-point observation area in the fixed-point observation mode. The satellite has an arbitrary toward-ground attitude in the main working mode. The complex toward-ground attitude poses higher requirements for the full spatial coverage of the antenna.

Therefore, according to the working mode of the HXMT, besides the orbital control mode, the satellite is in arbitrary attitude toward Earth. In this case, in order to realize the functions of TT&C, GPS positioning and transmitting scientific observation data to the ground, it is necessary for the antenna system to realize the full-space coverage of the TT&C antenna, GPS antenna and DT antenna.

According to the above characteristics, the antenna system layout is as follows:

(1) USB TT&C antenna

The USB TT&C antenna subsystem achieves good near-full-space coverage with four antennas (each with two elements array, and complement each other), and each antenna element can achieve hemispheric coverage beams. Generally, satellite TT&C antennas were mostly distributed on the satellite's Earth-facing side and Earth-aft side, while the equipment of HXMT satellite's load mounting surface (i.e., the $+$ x-plane) are various, which bring great difficulties to the antenna layout. However, the $\pm z$-plane of HXMT has good installation conditions. In order to make full use of $\pm z$-plane resources and further improve the coverage of TT&C antennas, two sets of antennas are installed at $\pm z$-direction, one is orthogonal to plate, the other is tilting 25°, respectively, as shown in Fig. 11.19.

The FOV of the USB TT&C antenna installed in the loading cabin $+x$/$+z$-side and the service cabin $-x$/$-z$-side is without occlusion. For $\pm z$-plane USB TT&C antenna, the projecting device of the satellite surface will partly block its hemispheric FOV, and the solar wing will enter its FOV for a short time in the process of rotation, which still meets the requirements of the TT&C subsystem through analysis.

(2) GPS antenna

The design of the GPS antenna focuses on layout design. HXMT has an arbitrary attitude to the ground, so the GPS antenna cannot fully meet the requirements of satellite positioning if it is only placed on one surface of the satellite. In order to meet the reuirement of real-time and accurate positioning after acquiring four or more GPS

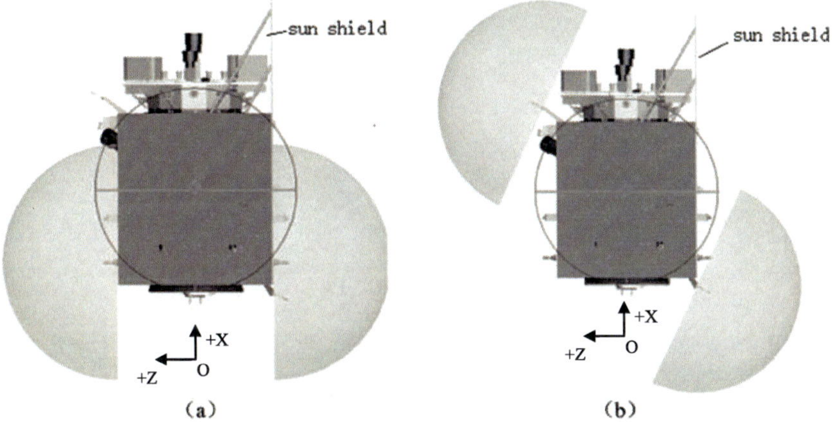

Fig. 11.19 FOV analysis of USB TTC antenna. **a** ±*z*-direction TT&C antenna; **b** ±*z*-direction 25° TT&C antenna

satellite signals, It should be considered that antenna elements are added on other satellite planes and two antenna arrays are used to increase the space coverage of GPS antenna by 0° and 180°.

The satellite consists of four GPS antennas, two in the service cabin ±*z*-plane and two in the hemisphere FOV.

Like the ±*z*-plane USB TT&C antenna, the devices of the satellite surface will have partial occlusion on its hemispherical FOV, and the solar wing will also enter its FOV in a short time during rotation (Fig. 11.20). After analysis, it meets the requirements for the use of the TT&C subsystem.

(3) Data transmission antenna

According to the analysis of data transmission tasks, in order to reduce the impact of satellite and on satellite payloads on data transmission antennas, two wide-angle hemispheric beam antennas with larger coverage are installed in the direction of ± z in 20° inclication, and data transmission to the ground is switched through waveguide switching units to ensure full-space coverage without interruption of observation in order to complete the task of scientific data download.

As shown in Fig. 11.21, the FOV of the data transmission antenna can satisfy the requirement of 90° half-cone angle of view without occlusion, and after reaching 100° half-cone angle, its +*y*- and −*y*-sides are partially occluded by the solar wing; for the data transmission antenna installed in the service cabin −*x*/−*z*-side, its FOV in the −*y*-direction will be partially occluded by the small sun shield on the satellite −*y*-side, while the data transmission antenna installed in the loading cabin +*x*/+*z*-side will be detected by X-ray photon probe. In the +*x*-direction, the antenna is mainly partially occluded by the low-energy detector under load. The RM satellite test results show that the layout of the data transmission antenna meets the application requirements.

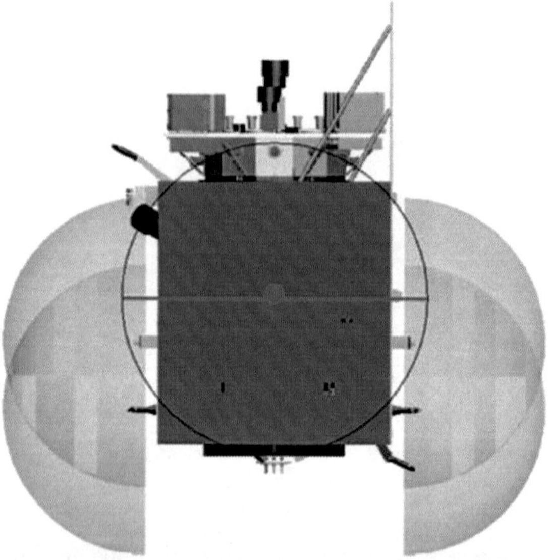

Fig. 11.20 FOV analysis of GPS antenna

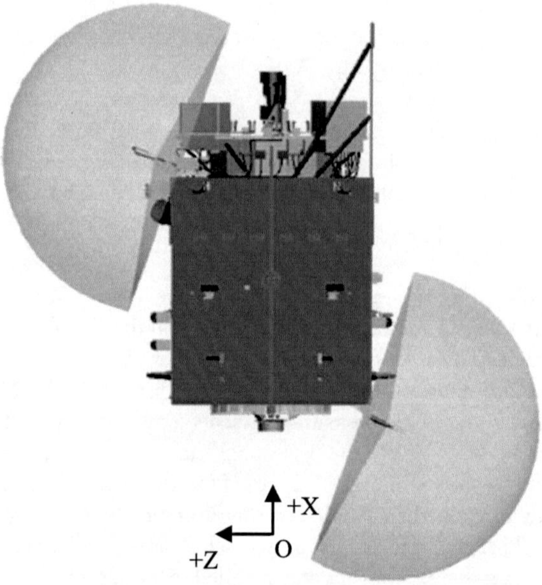

Fig. 11.21 FOV analysis of transmission antenna

Based on the above analysis, the design of HXMT satellite antenna system chooses mature antenna products as far as possible, and through reasonable working mode design and layout design, we strive to meet the satellite coverage requirements of the antenna subsystem with the most simplified system configuration.

11.3.2 Design Scheme of Antenna System

11.3.2.1 Design of USB TT&C Antenna

In the USB TT&C antenna, four broadband transceiver antennas are placed, respectively, in the $\pm z$-direction and in the direction of $\pm z$ with an inclination of 20°. Two antennas, respectively, on $\pm z$-plane and $\pm x$-plane, form a near-full-space coverage by the LHCP array and also play the role of mutual backup.

As shown in Fig. 11.22, the USB transponder *a* receives the uplink signal from the $\pm z$-direction of the TT&C antenna through the hybrid connector 1 and sends the downlink signal to the $\pm z$-direction through the hybrid connector 1 when the transmitter is powered on. The control antenna is connected; the USB transponder *b* receives the uplink signal from the $\pm x$-direction TT&C antenna through the hybrid connector 2 and sends the downlink signal through the hybrid connector 2 to the TT&C antenna in the $\pm$ x-direction when the transmitter is turned on. For on-orbit application, four antennas are simultaneously received for the uplink; for the downlink, only two antennas of one array are simultaneously transmitted.

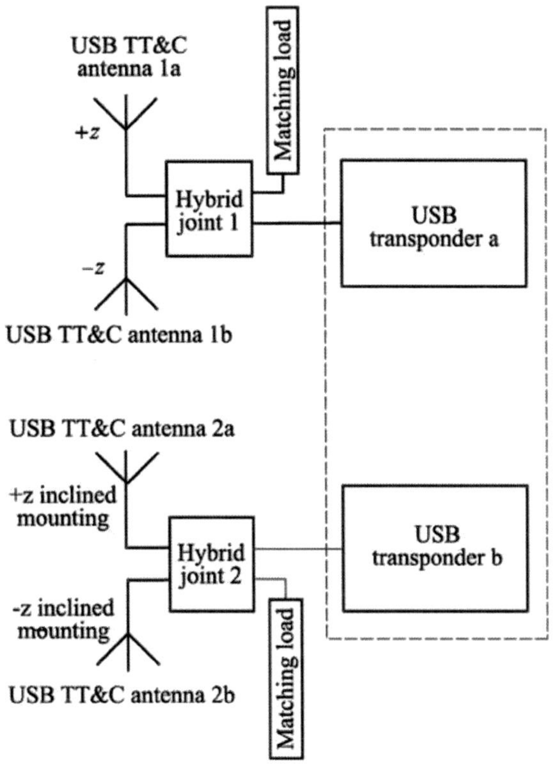

Fig. 11.22 Schematic scheme of TT&C antenna

The array interference area of the co-polarization array will affect the TT&C link to a certain extent, but the array method avoids the frequent switching of the upward change polarization of the ground station and is conducive to improving the reliability of the system application. The interference zone of the co-rotating array can be controlled within the allowable range of application by reasonable element and layout design. Most of the small LEO satellites adopt the method of co-polarization array, and there are many interference zones in the ground application process, which do not cause unacceptable impact on TT&C.

The USB TT&C antenna is a conical helical antenna, which has been successfully used in many LEO and MEO satellites. It can be directly used in the HXMT satellite USB TT&C antenna. Its appearance is shown in Fig. 11.23 and the simulation pattern is shown in Fig. 11.24. As can be seen from the simulation results, the gain of the element antenna is not less than 0 dBi in the range of the mechanical axis of the antenna ±70°.

Two LHCP array USB TT&C antennas *2a* and *2b* are installed in the ± *z*-direction of the HXMT satellite with an inclination of 25°. The test results of RM satellites with USB TT&C antennas array *2a* and *2b* show that the space with a gain not less

Fig. 11.23 S-band cone-column helical antenna

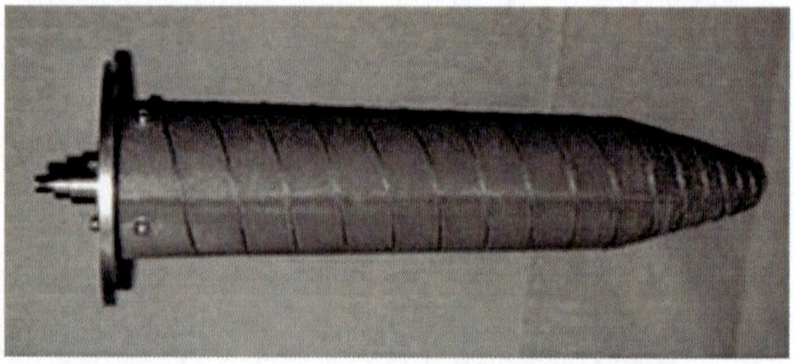

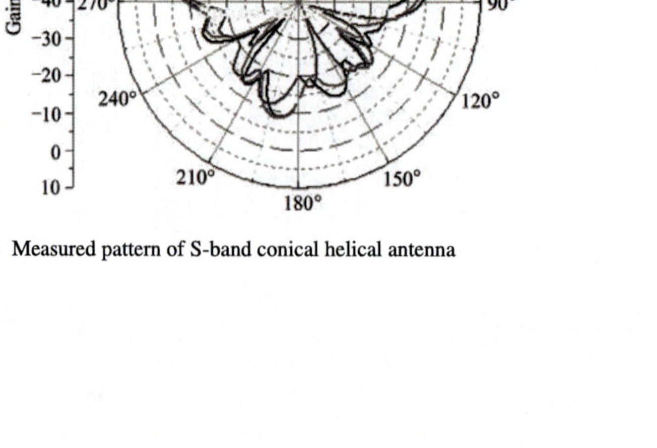

Fig. 11.24 Measured pattern of S-band conical helical antenna

than −18 dBi is more than 98.96% of the whole space (specification requirements ≥97%) and the space with a gain not less than −12 dBi is more than 94.88% of the whole space (specification requirements ≥90%). Therefore, it meets the requirement of on-orbit operation.

Two LHCP array USB TT&C antennas *1a* and *1b* are installed in the ± *z*-direction of the HXMT satellite. The test results of RM satellites with USB TT&C antennas array *1a* and *1b* show that the space with a gain not less than −18 dBi is more than 98.18% of the whole space (specification requirements ≥97%) and the space with a gain not less than −12 dBi is more than 91.97% of the whole space (specification requirements ≥90%), which is consistent with the simulation results, thus meeting the requirement of on-orbit operation.

11.3.2.2 GPS Antenna Design

After HXMT is on orbit, GPS antenna works continuously until the end of its life, covers nearly all space, receives GPS satellite data and implements orbit parameters and timing.

The GPS antenna consists of two independent channels, which connect the primary and backup GPS receivers through the combiner, respectively. The GPS antennas *a* and *c, b* and *d* constitute a set of working units, respectively. The two sets form a cold backup. The connecting block diagram of the GPS antenna subsystem is shown in Fig. 11.25. Two pairs of four GPS antennas are configured on the satellite ±*z*-plane, as shown in Fig. 11.25.

The HXMT satellite GPS receiving antenna uses a mature self-feeding phase four-arms helical antenna, as shown in Fig. 11.26. The antenna adopts the self-feeding principle, and the 3 dB/90° power split phase shifting network necessary for ordinary circularly polarized four-arms helical antennas is omitted, the structure is simple and compact, and the reliability is high. Satisfied application results have been achieved on many LEO satellites.

The comparison of GPS antenna element simulation and test result is shown in Fig. 11.27. It can be seen that the simulation and test result are consistent, satisfying the technical requirements of not less than −3.0 dBi ($0° \leq \theta \leq 75°$, $0° \leq \varphi \leq 360°$) and not less than −4.0 dBi ($0° \leq \theta \leq 85°$, $0° \leq \varphi \leq 360°$).

Figure 11.28 shows the net gain pattern of the GPS antennas a and c in the RM satellite state, which meets the satellite system requirements. The net gain direction of the GPS antennas b and d in the RM satellite state is the same.

11.3.2.3 Design of Data Transmission Antenna

Two Data transmission antennas are, respectively, installed on the +*x*/+*z*-side of the payload cabin and the −*x*/−*z*-side of the service compartment with 20° angle between the antenna axis and the *z*-axis to achieve the unobstructed 100° half-cone angle FOV. The to-ground data transmitting is switched by the waveguide switch to

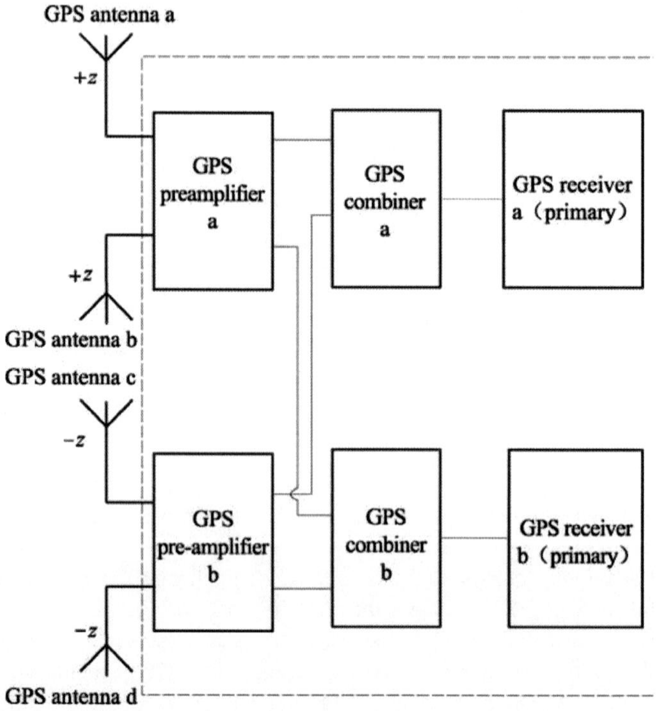

Fig.11.25 Block diagram of GPS antenna subsystem

ensure full coverage of the ground and complete the task of transferring the scientific data. The data transmission antenna connection block diagram is shown in Fig. 11.29.

The HXMT satellite data transmission antenna *a* and *b* are in the same state (the body and the bracket are the same), the data transmission antenna *a* is obliquely mounted on the $+z/+x$-side of the satellite body with 20° angle between the antenna axis and the $+z$-axis and the height of the inclined mounting frame of about 300 mm; the data transmission antenna *b* is inclined mounted on the $-z/-x$-side of the satellite body with 20° angle between the antenna axis and the -z-axis and the inclined mounting height of about 300 mm, as shown in Fig. 11.30.

The HXMT satellite data transmission antenna adopts a bifilar helical antenna scheme. By adjusting the electrical performance parameters of the antenna, the beam shape can be controlled to achieve a hemispherical coverage gain pattern. The antenna itself is made entirely of metal with high efficiency and reliable structure. The antenna fully inherits mature technology and can withstand 50 W power tolerance. The total height of the antenna is approximately 200 mm. Considering the mechanical properties, the dielectric block support is added between the outer conductor of the coaxial feed tube and the helix on the antenna (as shown in Fig. 34).

In view of the difference between the HXMT satellite and the three-axis to-ground-stabilized satellite platform, the helical parameters of the bifilar helical antenna are

Fig. 11.26 GPS antenna

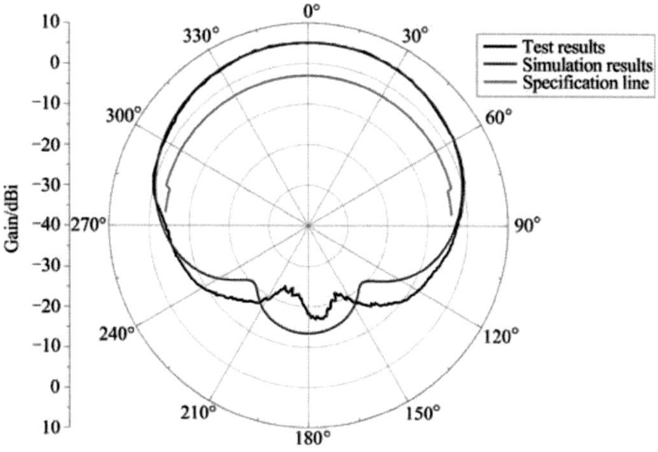

Fig. 11.27 Simulation pattern of GPS antenna element

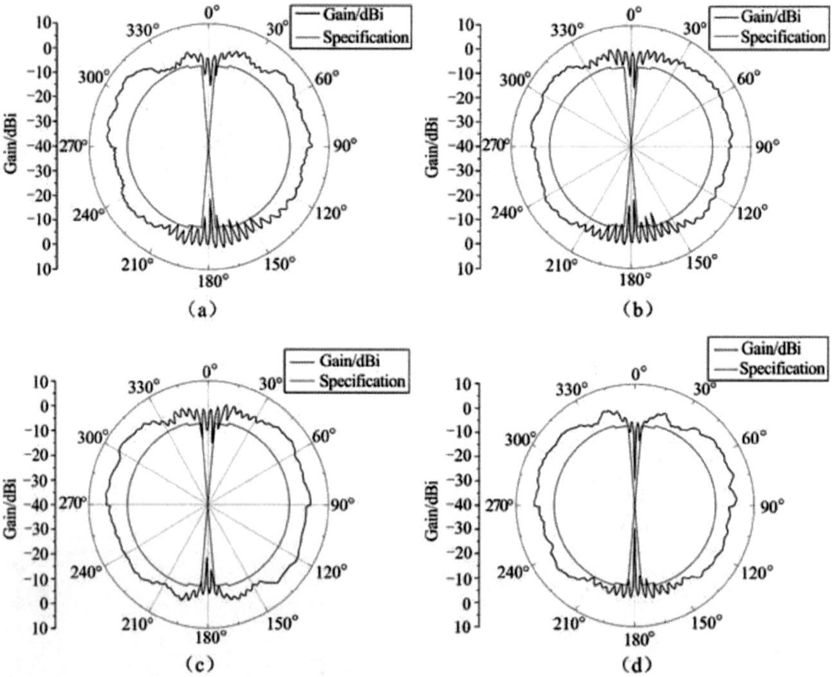

Fig. 11.28 GPS antenna a, c array gain pattern. **a** 0° plane gain pattern; **b** 45° plane gain pattern; **c** 90° plane gain pattern; **d** 135° plane gain pattern

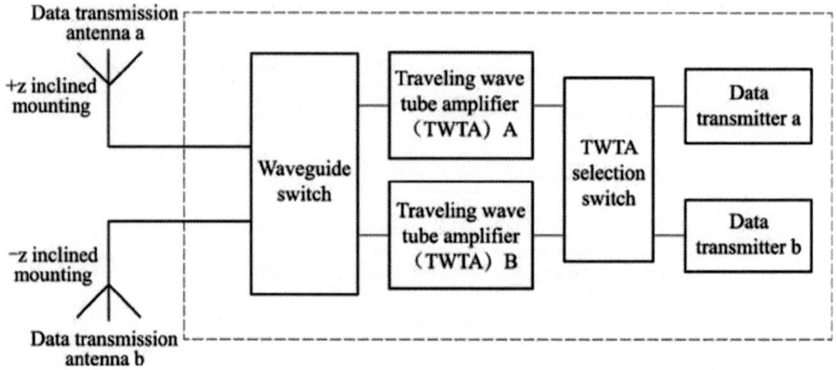

Fig. 11.29 Block diagram of data transmission antenna connection block diagram

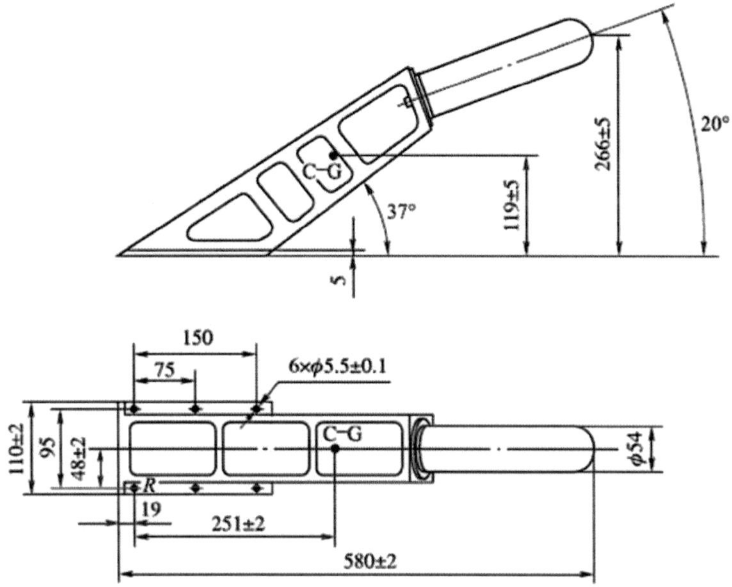

Fig. 11.30 Sketch of data transmission antenna

adaptively modified so that the pattern becomes a hemispherical beam. The unit antenna simulation pattern is shown in Fig. 11.31, whose gain is not less than −3 dBi

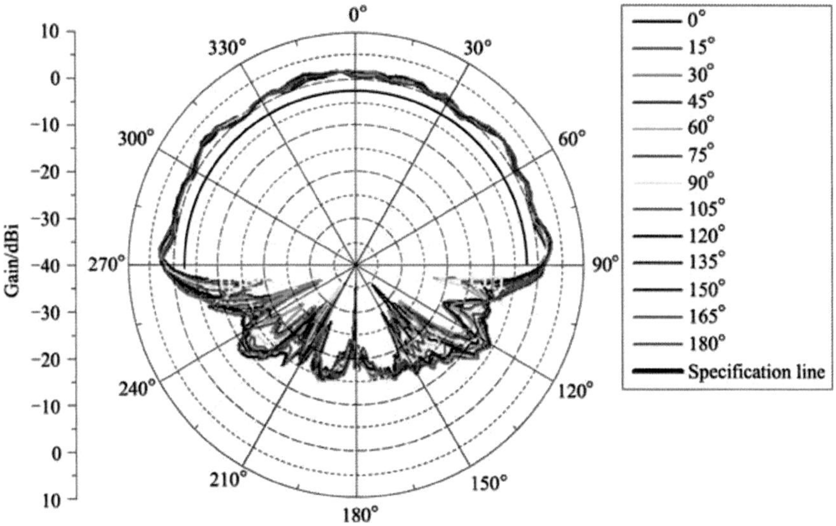

Fig. 11.31 Measured gain pattern of different sections of the data transmission antenna

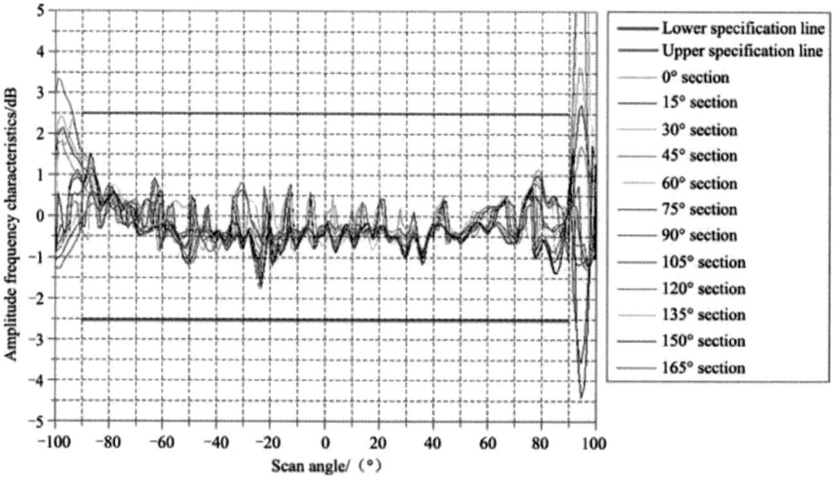

Fig. 11.32 Single-antenna amplitude-frequency characteristics of data transmission antenna

in the range of ±90° off-axis and with a gentle fluctuation (Fig. 11.32), which satisfies the specifications for the data transmission requirement of hemisphere coverage.

In the RM status, the net gains of the data transmission antennas a and b basically meet the requirement of not less than −3.0 dBi ($0° \le \theta \le 90°$, $0° \le \varphi \le 360°$, including polarization loss and satellite body influence), and there are individual points out of tolerance due to the influence of payload hood, but still meet the requirement of not less than −6 dBi.

Taking data transmission antenna with a slightly poor result and the solar wing near it to research the influence of the solar wing on data transmission antenna in the RM satellite, the test results of the RM status are shown in Fig. 11.33. It can be seen that the test results with and without the solar wing are basically the same, which shows that the solar wing has no significant influence on the DT antenna radiation pattern.

Since the launched on June 15, 2017, the HXMT satellite is normally operating with stable TT&C and GPS signal on orbit. The data transmission signal received by the ground station is normal at 60 Mb/s or 120 Mb/s, verifying the correctness of the design of the satellite antenna subsystem.

11.4 Design Case of Communication Satellite Antenna System

The communication satellite antenna system is an important part of the satellite payload, which should be selected and designed according to the user's requirements on coverage, frequency planning, gain requirements, antenna polarization

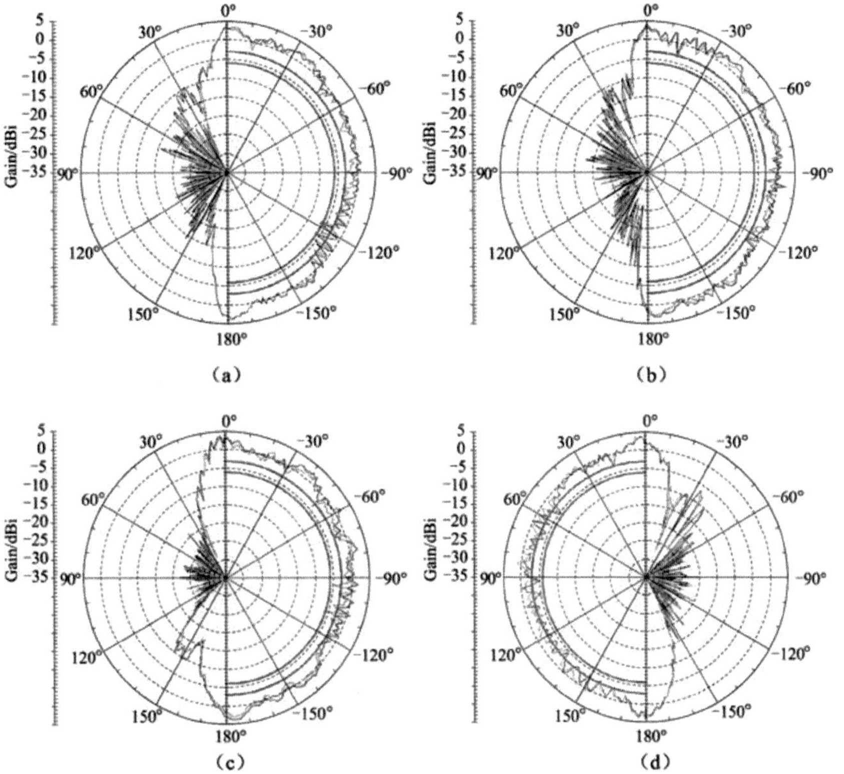

Fig. 11.33 Gain pattern comparison between data transmission antenna with and without solar wing. **a** 0° plane gain pattern; **b** 45° plane gain pattern; **c** 90° plane gain pattern; **d** 135° plane gain pattern

mode, satellite platform and carrying capacity after comprehensive analysis and consideration. The Nigerian communication satellite (NIGCOMSAT-1) is China's first export communication satellite with the DFH-4 satellite platform. Its payload includes L-, C-, Ku- and Ka-band communication systems covering most parts of Africa and parts of Europe. The following is an introduction to the antenna system design process with the NIGCOMSAT-1 as an example.

11.4.1 Technical Requirement

The main technical requirements of the satellite antenna system of the NIGCOMSAT-1are as follows:

(1) Satellite and delivery requirements

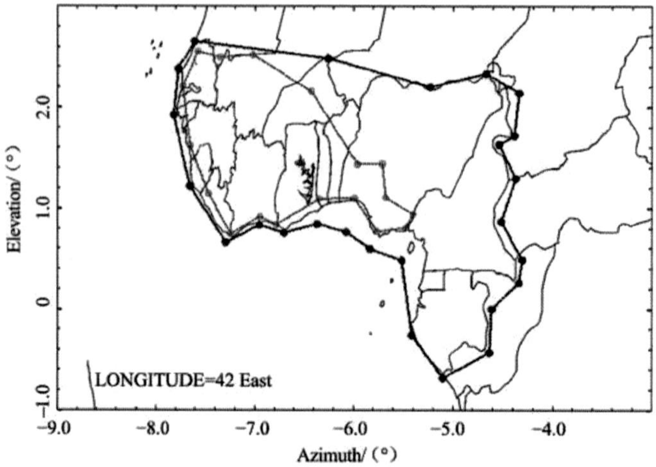

Fig. 11.34 Ku-band beam ECOWAS1 service area requirements

Orbit: 42°E (GEO).

Beam pointing accuracy: less than 0.1° (3σ) under normal conditions, less than 0.13°(3σ) under harsh conditions.

Satellite Platform: In terms of position keeping accuracy, the longitude is better than ± 0.05°, the latitude is better than ± 0.05°; the satellite platform is DFH 4.

Launch vehicle: CZ 3B.

(2) Coverage requirements

Ku-band beams require shaped beam coverage, as shown in Figs. 11.34 and 11.35.

The C-band beam requires shaped beam coverage, as shown in Fig. 11.36.

Ka-band beams require multi-beam coverage, as shown in Fig. 11.37.

The L-navigation beam requires global beam coverage, i.e., a ± 9° circular beam.

The C-navigation beam is required to cover North Africa and Europe, i.e., a ± 7.5° circular beam.

(3) Frequency planning

The frequency plans are shown in Tables 11.2, 11.3, 11.4, 11.5, 11.6 and 11.7.

(4) Gain requirement

The gain requirements are shown in Table 11.8.

(5) Cross-polarization (XPD) requirements

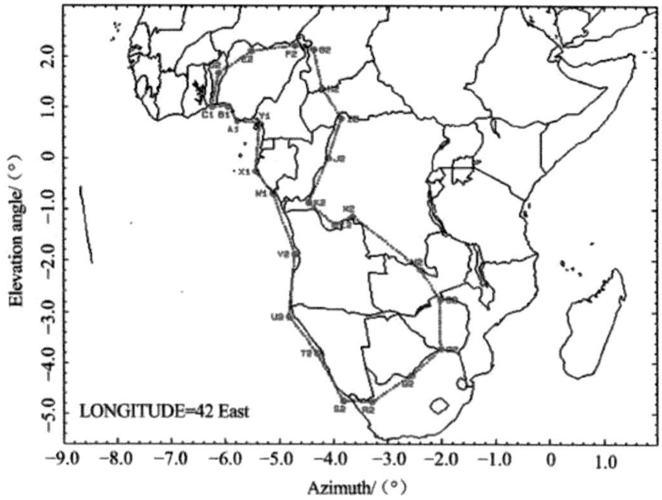

Fig. 11.35 Ku-band beam ECOWAS2 service area requirements

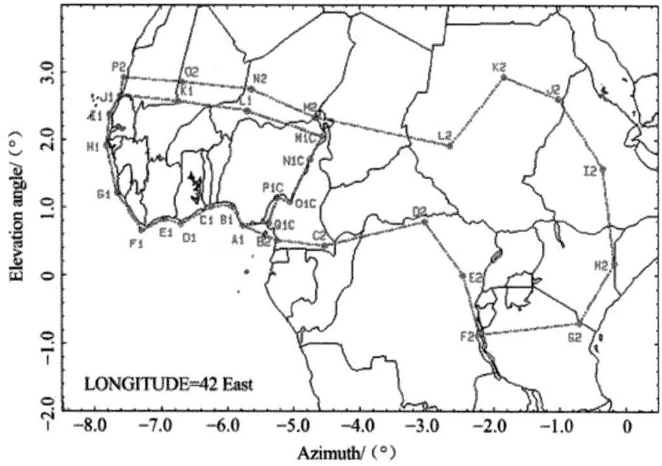

Fig. 11.36 C-band beam service area requirements

Ku-band: polarization isolation in coverage area ≥ 30 dB.

C communication frequency band: polarization isolation in 80% coverage area ≥ 30 dB; polarization isolation in 100% coverage area ≥27 dB.

Ka transmitting band: polarization isolation in coverage area ≥28 dB.

Ka receiving band: polarization isolation in coverage area ≥28 dB.

C-navigation antenna: polarization isolation degree in coverage area ≥27 dB.

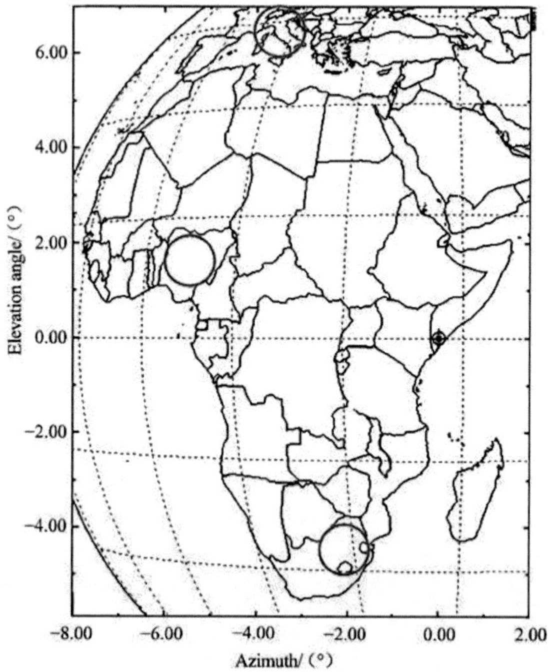

Fig. 11.37 Ka-band beam service area requirements

Table 11.2 Uplink frequency and polarization planning in Ku-band

Horizontal polarization uplink		Vertical Polarization uplink	
Channel	Frequency (MHz)	Channel	Frequency (MHz)
1	14,018.0	8	14,018.0
2	14,053.5	9	14,053.5
3	14,089.0	10	14,089.0
4	14,124.5	11	14,124.5
5	14,160.0	12	14,160.0
6	14,195.5	13	14,195.5
7	14,231.0	14	14,231.0

11.4.2 Scheme Design

11.4.2.1 Electrical Scheme Selection

According to the above-mentioned coverage area and technical requirements, for the antenna with high XPD requirements in Ku-band, the long focus single reflector or double-reflector antenna is preferred to obtain better XPD performance and good shaping effect; for the C-band antenna, the single reflector antenna is preferred

Table 11.3 Ku-band downlink frequency and polarization planning

Horizontal polarization downlink		Vertical Polarization downlink	
Channel	Frequency (MHz)	Channel	Frequency (MHz)
8	12,518.0	1	12,518.0
9	12,553.5	2	12,553.5
10	12,589.0	3	12,589.0
11	12,624.5	4	12,624.5
12	12,660.0	5	12,660.0
13	12,695.5	6	12,695.5
14	12,731.0	7	12,731.0

Table 11.4 C-band frequency and polarization planning

LHCP downlink		RHCP uplink	
Channel	Frequency (MHz)	Channel	Frequency (MHz)
1	3450	1	6475
2	3490	2	6515
3	3530	3	6555
4	3570	4	6595

Table 11.5 Ka-band downlink frequency and polarization planning

RHCP downlink		RHCP downlink	
Channel	Frequency (MHz)	Channel	Frequency (MHz)
1	19,075	5	19,675
2	19,225	6	19,825
3	19,375	7	19,975
4	19,525	8	20,125

Table 11.6 Ka-band uplink frequency and polarization planning

LHCP uplink		LHCP uplink	
Channel	Frequency (MHz)	Channel	Frequency (MHz)
1	28,875	5	29,475
2	29,025	6	29,625
3	29,175	7	29,775
4	29,325	8	29,925

Table 11.7 Navigation frequency and polarization planning

Channel	Frequency (MHz)	Polarization	Bandwidth (MHz)
L1-downlink	1575.42	RHCP	4
L5-downlink	1176.45	RHCP	20
C1-uplink	6697.42	LHCP	4
C5-uplink	6639.45	LHCP	20

Table 11.8 Antenna gain requirements

Beam	Downlink beam		Uplink beam	
	Gain (dBi)	Remarks	Gain (dBi)	Remarks
Ku-band beam ecowas1	32.3	EIRP = 52.0 dBW	32	$G/T = +\ 4.0$ dB/K
Ku-band beam ecowas2	28.3	EIRP = 48.0 dBW	29	$G/T = +\ 1.0$ dB/K
C-band service area I	26	EIRP = 41.0 dBW	29	$G/T = +\ 1.0$ dB/K
C-band service area II	20	EIRP = 35.0 dBW	23	$G/T =$ -5.0 dB/K
Ka-band European beam	38.0	EIRP = 52.0 dBW	41.2	$G/T = +\ 12.0$ dB/K
Ka-band Nigerian beam	39.1	EIRP = 52.0 dBW	42.0	$G/T = +\ 12.0$ dB/K
Ka-band Southern Africa beam	38.0	EIRP = 52.0 dBW	41.2	$G/T = +12.0$ dB/K
Navigation beam	13.5	1575.42 MHz	16.5	6697.42 MHz
	13.0	1176.45 MHz	16.5	6639.45 MHz

due to the low-frequency band (the large sub-reflector aperture of double-reflector antenna is not conducive to layout); for the antenna in Ka-band, in order to achieve good antenna performance, ensure beam coverage consistency and reduce component design difficulty, the antenna shall be designed as single-receiving and single-transmitting antenna for the large transmission and reception frequency gap and high gain, and the fixed spot beam coverage will be implemented, respectively, through three independent feeds; the L-band and C-band are separated by the navigation function, and the requirements can be met by simple unit antenna because the antenna gain requirements are low.

Based on the above analysis, it is preliminarily determined that the antenna subsystem consists of seven antennas, which, respectively, realize the following functions: two Ku-band reflector antennas, respectively, generate the ECOWAS1 beam and ECOWAS2 beam; one communication transceiver shared C-band antenna generates the C-beam coverage; one Ka-band offset parabolic antenna for transmitting to generate three-point beam areas of Nigerian beam, European beam and

South African beam, respectively; one Ka-band offset parabolic antenna for receiving to generates three-point beam areas of Nigerian beam, European beam and South African beam, respectively; one L-band helical antenna for navigation; one C-band multimode conical horn antenna for navigation covers North Africa and Europe.

11.4.2.2 Structural Layout Design

After the electrical scheme is preliminarily selected, the antenna system structure layout can be implemented under the consideration of the following constraints:

(1) Satellite platform conditions

The communication satellite adopts the three-axis stabilizing satellite platform DFH-4, and the available antenna installation surface on the satellite is the east and west cabin plates and the Earth-facing floor;

(2) Launch constraints

The satellite is launched by the rocket CZ-3B with the net envelope size inside the fairing of ϕ 3650 mm, which should not be exceeded for the complete antenna structure layout;

(3) Fixed position and coordinate system of the satellite

The fixed position and coordinate system of the satellite are closely related to antenna pointing direction and beam design, which should be considered in layout design;

(4) First-order fundamental frequency of satellite platform

In order to avoid the resonance between the first-order fundamental frequency of each structure and the satellite main body, the coupling between the first-order frequency of each antenna component structure and the fundamental frequency of the satellite should be avoided as far as possible in the structural design;

(5) Satellite telemetry and remote tele-control resources

In order to ensure the function of the antenna subsystem and on-orbit condition monitoring, it is necessary to acquire the remote control and telemetry resources provided by the satellite;

(6) Satellite power supply and distribution resources

The active equipment of the antenna system needs the power supply provided by the satellite platform to complete the corresponding functions. During the design, it is necessary to understand the power supply resources that the satellite can provide, such as the primary power supply, secondary power supply, line package power supply, interface circuit safety design and other requirements;

(7) Compatibility with platform products

Other devices on the satellite, such as Earth sensor, satellite lifting device, 10 N engine and its plume area, satellite fuel filling valve, radiation cooling tube heads, telemetry and tele-control antennas, shall be avoided during layout design, and no structural interference between components of antenna products is allowed;

(8) Technological process of satellite development

The tooling and the interface between tooling and antenna required to complete the functional performance test during the joint test of products will be planned through the technological process of satellite general assembly;

(9) Operability of general assembly integration on satellite platform

The antenna layout shall consider the product installation sequence on satellite, compatibility of antenna hanger and satellite products, implementability of final assembly and operation space, etc.;

(10) Analysis of scattering effects between antenna and satellite

In order to ensure the RF performance of each antenna, no obstructions within the FOV of the antenna is desired, and the structure on the satellite should not affect the performance of the antenna in the layout design;

(11) Antenna system weight

On the premise of meeting the function and performance of the antenna system, the optimization design and use of advanced materials shall be considered to minimize the design quality of the antenna system, improve the carrying capacity of the platform and extend the service life of the satellite.

11.4.2.3 Antenna System Design

On the basis of the above work, the antenna system scheme is finally formed by further refining, as shown in Fig. 11.38:

1. A 3.0 m × 2.2 m Ku-band deployable dual-reflector antenna with separate receiving and transmitting feeds is installed on the west wall panel of the satellite (the West antenna for short) to generate the ECOWAS1 beam.
2. A 2.6 m × 2.2 m Ku-band deployable dual-reflector antenna (the East antenna for short) is installed on the east wall panel of the satellite to generate the ECOWAS2 beam.
3. A 1.55 mC-band offset transceiver shared shaped parabolic antenna for communication (C communication antenna for short) is installed on the Earth-facing panel, and the beam coverage of the antenna meets the requirements of the coverage area.

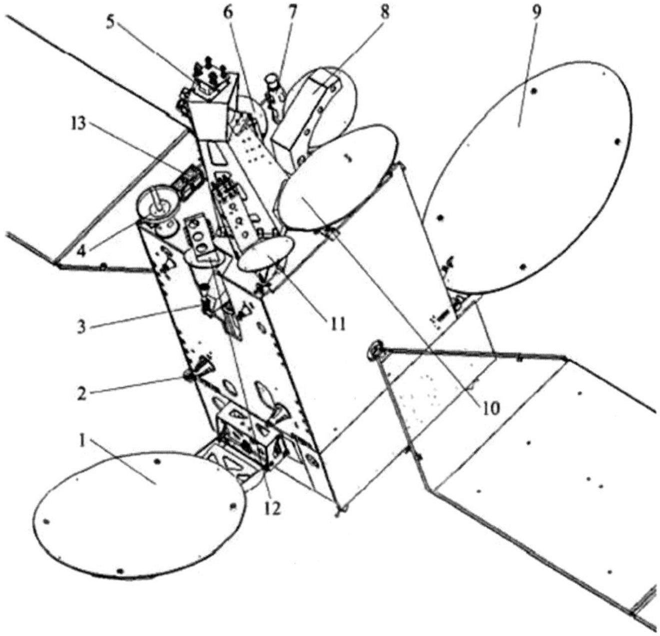

Fig. 11.38 Structure layout of antenna system

4. A Ka-band bias-feed standard parabolic antenna with 0.7 m aperture (Ka transmitting antenna for short) is installed on the Earth-facing panel and fed by three feeds placed on the focal plane of the reflector, generating three-point beam areas for Nigerian beam, European beam and South African beam, respectively.
5. A 0.65 m Ka-bandoffset3 standard parabolic antenna (Ka receiving antenna for short) is installed on the Earth-facing panel and fed by three feeds placed on the focal plane of the reflecting surface, generating three-point beam areas for Nigerian beam, European beam and South African beam, respectively.
6. A L-band backfire helical antenna for navigation, referred to as L-navigation antenna for short, is installed on the Earth-facing panel, and the antenna beam covers the whole world.
7. A C-band multimode conical horn antenna for navigation, referred to as C-navigation antenna for short, is installed on the Earth-facing panel, and the beam covers North Africa (including Nigeria) and Europe.
8. The ground TT&C antenna is installed on the top of C-antenna support tower to meet the TT&C business requirements of satellite platform.

(1) East antenna reflector assembly; (2) East antenna folding/unfolding device; (3) East antenna feed; (4) L-navigation antenna; (5) To-ground measurement and control antenna assembly; (6) Ka receiving antenna; (7) C-navigation antenna; (8) West antenna sub-reflector; (9) West antenna reflector assembly; (10) C communication

antenna; (11) Ka transmitting antenna; (12) East antenna sub-reflector; (13) Earth sensor.

The east and west antennas are, respectively, deployable antennas. During satellite launch, they are fixed on the satellite deck through the hold-down and release mechanism to achieve compatibility with the launch, as shown in Fig. 11.39. After the satellite is on orbit, the hold-down and release mechanism unlocks and releases. Under the action of the deployment mechanism and its controller, the antenna is unfolded and fixed at the working position, as shown in Fig. 11.40.

After analyzing and checking the compatibility of the antenna system and the carrier fairing, there are enough safety clearance between the antenna system and the fairing in *x*-, *y*-, *z*-direction, as shown in Fig. 11.41.

Taking the C communication antenna as an example to illustrate that there is no interference between the components of the antenna system, as shown in Fig. 11.42.

Fig. 11.39 Antenna system in the folded status

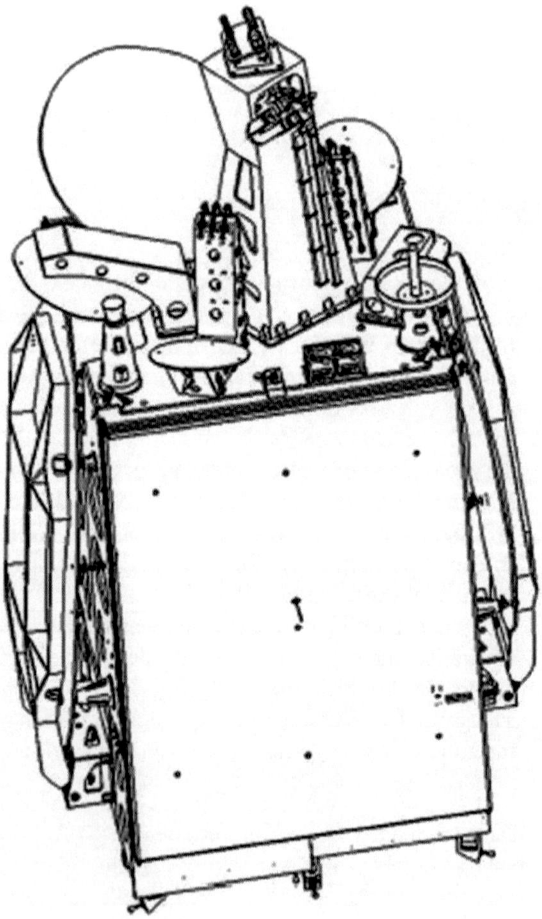

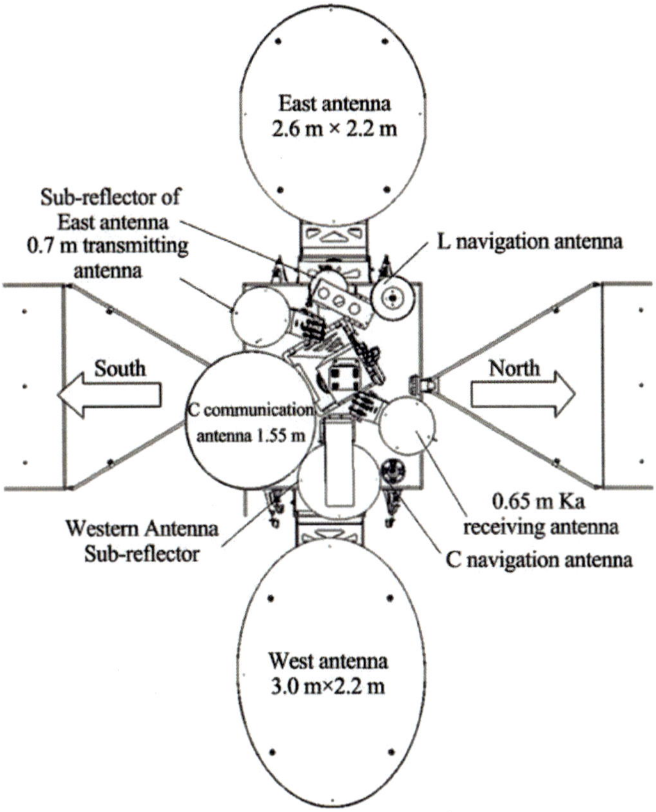

Fig. 11.40 Antenna system in the unfolded status (+*z*-direction)

Taking the East antenna as an example, the antenna beam is unobstructed, as shown in Fig. 11.43.

11.4.2.4 Electrical Performance Analysis Results

Taking the West antenna as an example, the design results of the electrical performance of the antenna are shown in Figs. 11.44 and 11.45.

After detailed design, the technical specifications of the antenna system can meet the technical requirements.

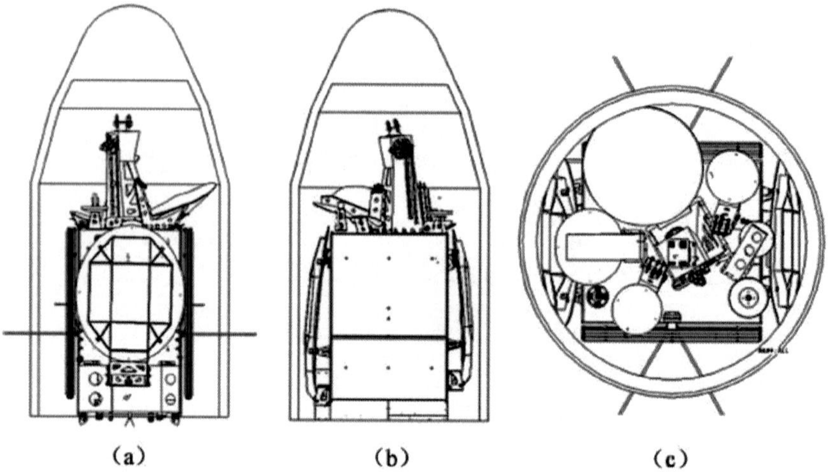

Fig. 11.41 Compatibility check of antenna system and fairing. **a** *x*-direction; **b** *y*-direction; **c** *z*-direction

11.4.3 Antenna System Application

At 0:01 on May 14, 2007, the NIGCOMSAT-1was successfully launched into space by the carrier rocket CZ 3B at Satellite Launch Center Xichang. The antenna is successfully unfolded and locked after the injection of the satellite. The results of on-orbit test show that the antenna beam pointing is correct, and the performance implements the specification and works normally. The antenna subsystem includes 7 communication antennas and 7 telemetry and tele-control antennas. The working frequency band covers L-band, C-band, Ku-band and Ka-band with the complex electromagnetic environment. By installing so many antennas in the limited space of the platform DFH-4, not only layout compatibility and structural achievability, field of view interference of communication antennas and feeds, but also high-power passive inter-modulation of antennas should be considered. The electrical performance test state of the antenna subsystem in the compact field is shown in Fig. 11.46, and the assembly state in the launching site is shown in Fig. 11.47. At that time, NIGCOMSAT-1 was the most complex antenna system with the largest number of antennas, the highest polarization isolation requirements, the largest power and the most frequency bands. Its success marks a new level in the research and development of space-borne antennas in China.

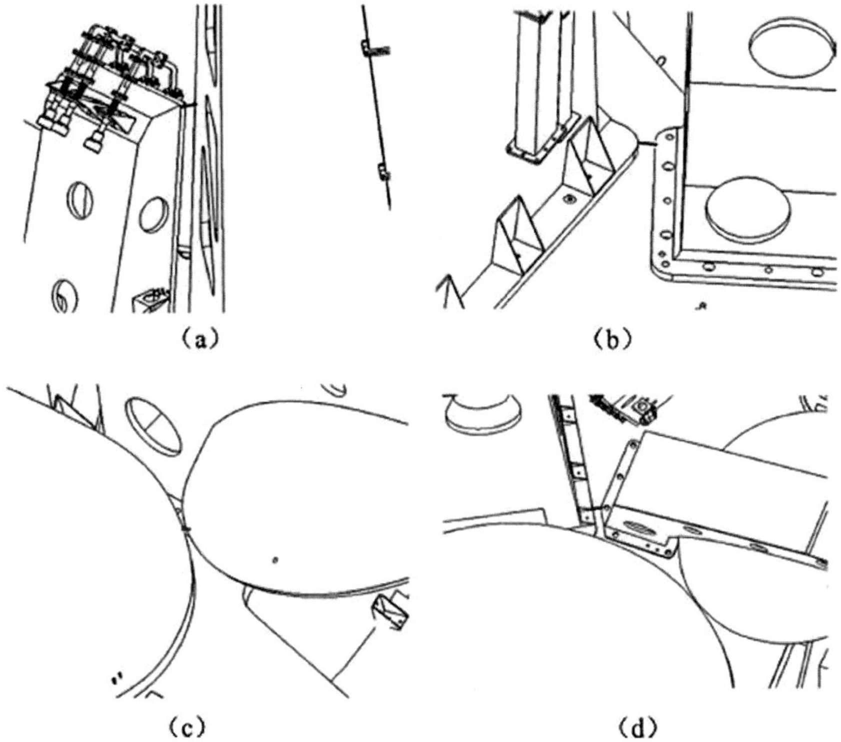

Fig. 11.42 Interference inspection for structure of C communication antenna and other components. **a** The gap with Ka receiving antenna is 46.2 mm; **b** The gap with East sub-reflector is 23.1 mm; **c** The gap with West sub-reflector is 6.3 mm; **d** The gap with West sub-reflector is 27.5 mm

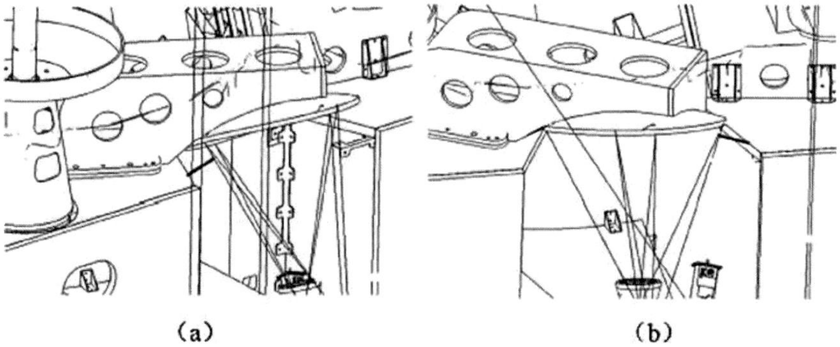

Fig. 11.43 Inspection for beam obstruction of East antenna. **a** The gap between the East antenna beam and the left side of the opening on the Earth-facing surface is 76.9 mm; **b** The gap between the East antenna beam and the right side of the opening on the Earth-facing surface is 95.8 mm

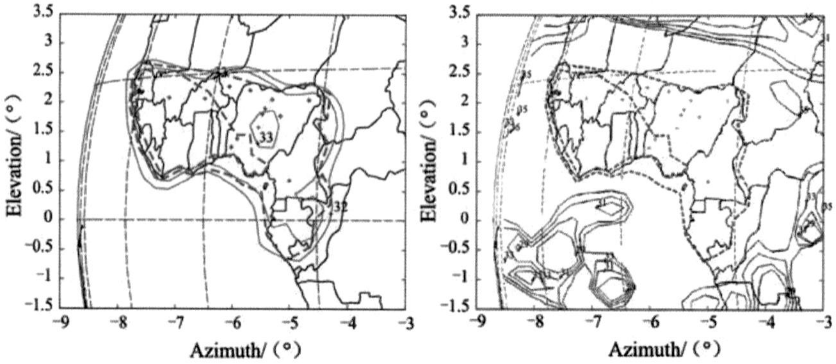

Fig. 11.44 West antenna downlink gain, cross-polarization contour map (12.625 GHz)

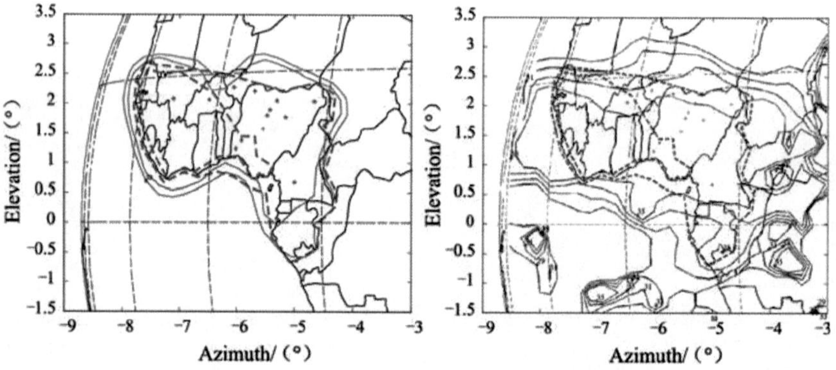

Fig. 11.45 Contour maps of uplink gain and cross-polarization of West antenna (14.125 GHz)

11.5 Design Case of Antenna System of the Recoverable Satellite

The scientific experimental satellite SJ-10 is a recoverable satellite specially used for space experimental research on "microgravity science and space life science", which is launched without fairing. Its main task is to make full use of the satellite orbital module and recoverable module to carry out several space science experiments in the fields of microgravity fluid physics, microgravity combustion, space material science, space radiation effect, gravity biological effect and space biotechnology.

Firstly, the recoverable spacecraft should be able to adapt to various environments during launch, flight on-orbit and reentry. These vehicles often have no fairing protection, so the difficulties of aerodynamic load environment in the active phase before on orbit must be got over. When the spacecraft enters the sound barrier area, the spacecraft is blocked, resulting in vibration, which will affect the spacecraft antenna.

Fig. 11.46 Electrical performance test of antenna subsystem in compact field

Fig. 11.47 Assembly status of antenna subsystem in the launching site

Secondly, when the speed of spacecraft reaches or exceeds the speed of sound, there is a great overload, which can be as high as dozens of g (acceleration of gravity). In addition, when the recoverable spacecraft is flying at a very high speed during reentry, the aerodynamic heating can make the surface temperature of the spacecraft as high as several thousand degrees Celsius.

The antenna system of the satellite SJ-10 plays a core role in the telemetry, telecontrol, receiving and sending of ranging signals and scientific data transmission between the satellite and ground station. The antenna should be able to withstand the aerodynamic load in the launch phase and the aerodynamic ablation environment in the reentry stage.

11.5.1 Mission Analysis

The flight process of the satellite SJ-10 is divided into four stages: launch phase, on-orbit operation phase, reentry and return phase and on-orbit test phase. The layout of the antenna subsystem is shown in Fig. 11.48. The mission requirements of the antenna subsystem in each stage are as follows:

In the **launch phase**, the launch vehicle will carry the satellite without fairing protection into the orbit as scheduled, and the satellite antenna shall be able to withstand the aerodynamic thermal load of the launch phase. In this phase, the C-band pulse transponder transceiver antenna, the S-band transponder antenna in the brake module and the S-band transponder antenna in the pressurized cabin are operating to complete the real-time orbit measurement, receive the uplink S-band measurement and control carrier, ranging and remote control signals, and transmit downlink TT&C carrier, ranging and telemetry signals.

In the **on-orbit operation phase**, the satellite enters three-axis stable attitude operation phase after separating from the rocket, and the transceiver antenna with C-band pulse transponder, S-band transponder antenna in brake module and S-band transponder antenna in pressurized cabin are operating, and the X-band data transmission antenna is operating when passing through the ground station.

In the **reentry and return phase**, the satellite returns according to the scheduled procedure after it completes the mission. The antenna of the recoverable capsule shall be able to withstand the mechanics and thermal environment during the reentry and return phase. In this phase, the C-band transponder transceiver antenna and the S-band transponder antennas in the brake module are operating. When the recovery beacon antenna is 17 km away from the ground, it is ejected and transmits the beacon to guide the ground searchers to search the recoverable capsule.

In the **on-orbit test phase**, the on-orbit test takes 3 days according to the requirements of payload test items if the onboard energy is available. In this phase, the S-band transponder antenna in the pressurized cabin is operating, and the X-band data transmission antenna is operating when passing through the ground station, and the test data is transmitted down (300 Mb/s in the this test phase).

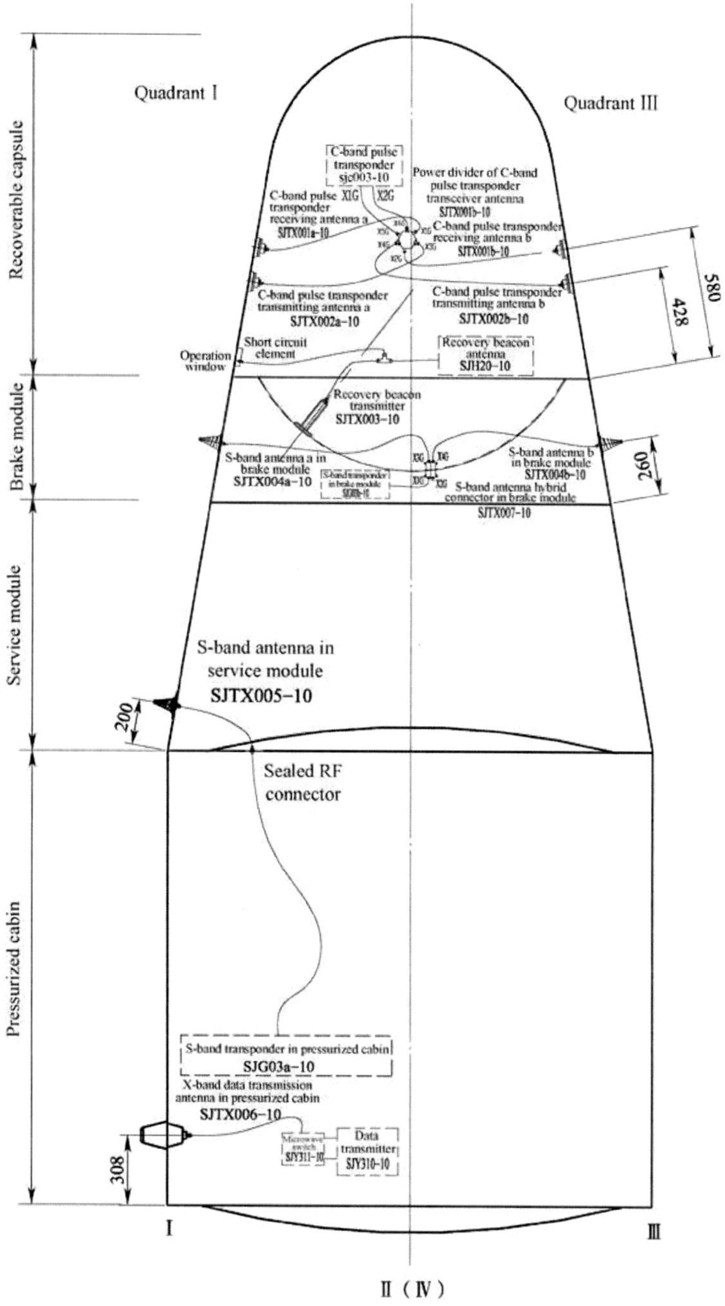

Fig. 11.48 Layout sketch of antenna system

11.5.2 Antenna System Design

Figure 11.48 shows the layout and connection of the antenna system on the scientific experiment satellite SJ-10. The C-band transponder transceiver antenna is arranged in quadrants I and III of the skirt of the recoverable capsule; the S-band antennas *a* and *b* in the brake module are arranged in quadrants I and III of the brake module, the S-band antenna in the service module is arranged in quadrants I of the service module; the data transmission antenna is arranged in quadrants I of the pressurized cabin.

The C-band pulse transponder receiving antenna in the recoverable capsule adopts two LHCP conformal anti ablation antennas and is installed on the bulkheads of the quadrants I and III of the capsule. The array mode forms a near-full-space coverage to meet the requirements of the recoverable capsule to receive the microwave pulse signals transmitted by the ground precision tracking radar. The C-band transmitting antenna with the pulse transponder in the recoverable capsule adopts two RHCP conformal anti ablation antennas array, which are installed on the bulkhead of quadrants I and III of the recoverable capsule, forming near-full-space coverage by array mode, meeting the requirements for the recoverable capsule to transmit the RF signal by the C-band pulse transponder.

In the reentry recovery phase, when the recoverable capsule descends to a certain height (17 km), the antenna cover will be flicked, the recovery beacon antenna will be automatically deployed in place by the spring device, and the recovery beacon will be transmitted. The recovery beacon antenna adopts the form of a quarter wavelength monopole to form an inverted “8” pattern.

The two S-band conical helical antennas are installed on the surface of quadrant I and quadrant III of the brake module, and the dual array is formed by the hybrid joint of equal power division to achieve near-full-space coverage. The service module adopts a single antenna with the same S-band antenna as that in the brake module, which is installed in quadrant I of the service module to realize half-space coverage to the ground.

The data transmission antenna in the pressurized cabin adopts the form of *X*-band bifilar backfire helical antenna to realize the Earth matching-shaped beam, which is installed in the first quadrant of the pressurized cabin to realize the data transmission to the ground; because the satellite does not have a fairing, the data transmission antenna is equipped with an antenna cover with a flat cone half-wave wall thickness that can withstand the aerodynamic environment of the active phase to realize the protection of the antenna itself.

11.5.2.1 C-band Transponder Transceiver Antenna

The C-band pulse transponder receiving antenna is a two-element array consisting of two unit antennas, which are, respectively, installed in the Earth-facing and the Earth-aft direction on the recoverable capsule. They are connected to the power divider of the C-band pulse transponder transceiver antenna and then connected to the receiving terminal of the C-band pulse transponder in sequence via the high-frequency cable.

The C-band pulse transponder transmitting antenna is a two-element array consisting of two unit antennas, which are, respectively, installed in the Earth-facing and the Earth-aft direction of the recoverable capsule. They are connected to the power divider of the C-band pulse transponder transceiver antenna and then connected to the receiving terminal of the C-band pulse transponder via the high-frequency cable.

A conformal anti-ablation antennas array is selected as the C-band pulse transponder transceiver antenna. Its working principle is that the circular polarization wave is formed by the jointly excitation by a pair of poles and groove joints of the slotted coaxial tube fed to it in the circular waveguide cavity of the antenna. LHCP or RHCP is realized by the different mounting angles of the poles designed. A quarter wavelength slotted coaxial tube is used for balanced feeding. Since the antenna is installed in the recoverable capsule, quartz glass and carbon fiber quartz reinforced composite materials are used for the electromagnetic window to prevent ablation.

The power divider has two layers, one is responsible for receiving and the other for transmitting. The power divider is fed by unequal power division, and the ratio of antenna power distribution in the direction of the Earth-facing to the Earth-aft is about 3:1.

In order to suppress surface wave, an aluminum foil with a width of 380 mm and a length of 710 mm shall be pasted on the outer surface of satellite around the radar transponder transceiver antenna and guide beacon antenna along the direction of bus.

The gain patterns of the C-band pulse transponder transceiver antenna array are shown in Figs. 11.49 and 11.50.

11.5.2.2 Recovery Beacon Antenna

The recovery beacon antenna working in UHF-band is compressed in the antenna base in the launch phase and orbital operation phase of the satellite. When the recoverable capsule passes through the atmosphere and reaches to about 17 km away from the ground, the antenna cover of the recovery beacon will be thrown away, so that the recovery beacon antenna will be ejected, and the antenna will be started up.

The antenna, the recovery beacon and the flange mounting socket with the short circuit element are, respectively, connected to the three ports of the coaxial tee element through the high-frequency cable.

The antenna consists of a coaxial feed section, a mounting flange, a radiation rod and a top load. A small disk is added to the top of the antenna pole to form the top load, so as to reduce the overall height of the antenna. The radiation rod is divided into three sections, which are covered with each other. When it is not in use, it is pressed into the conductor tube in the feed base with the aid of the antenna cover of the recoverable capsule; when it is in use, the antenna cover is thrown off, and the radiation rod is automatically ejected out of the cabin to a predetermined length

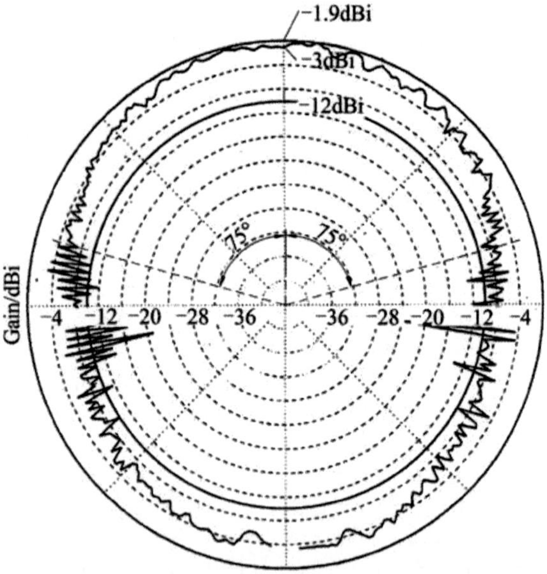

Fig. 11.49 Gain pattern of C-band pulse transponder receiving antenna array

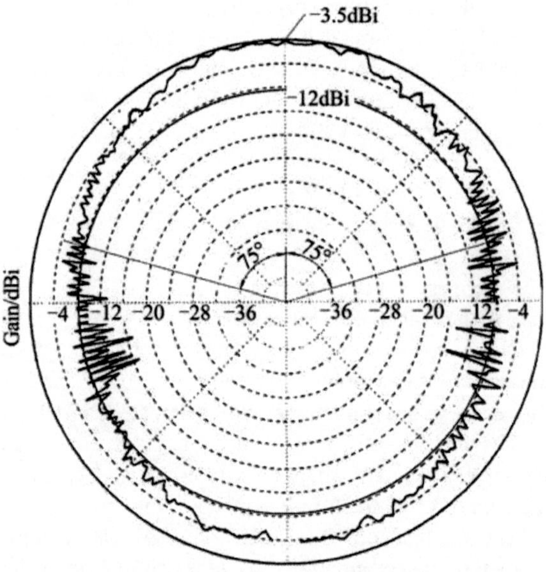

Fig. 11.50 Gain pattern of C-band pulse transponder transmitting antenna array

with the aid of the spring potential energy in the antenna, and the antenna starts up properly.

When testing on the ground, an external antenna is connected temporarily by a tee element, which is connected in series between the recovery beacon transmitter

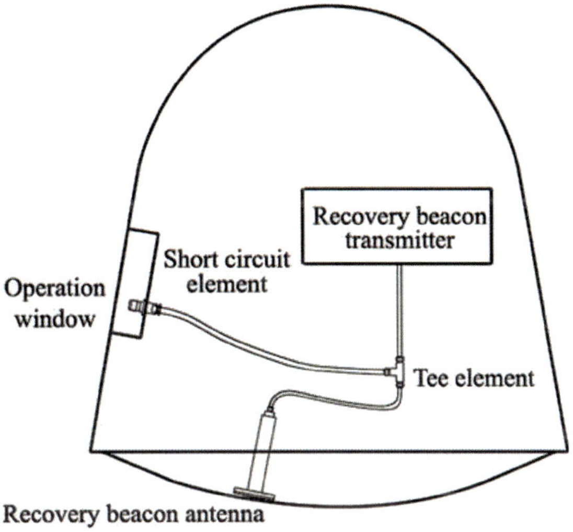

Fig. 11.51 Connection diagram of recovery beacon antenna

and the recovery beacon antenna. A coaxial cable is connected in parallel from the tee element to the short circuit element near the operation window of the recoverable capsule for the reliable short circuit. The distances between the center of the tee element and the short circuit point of the recovery beacon antenna or the short circuit element are all odd times of $\lambda/4$. During the test, the short circuit element can be replaced with an external antenna. When the antenna is not tested, the external cable can be replaced with a short circuit element, which does not affect the normal operation of the beacon antenna. The connection is shown in Fig. 11.51.

The pattern of recovery beacon antenna is tested with the scale model method, and the gain pattern is shown in Fig. 11.52, and the performance meets the overall requirements.

11.5.2.3 S-band Antenna in Brake Module

The S-band antenna in the brake module, a binary array consisting of two unit antennas, is installed in the Earth-facing and the Earth-aft direction of the brake module with a distance of 260 mm from the interface between the brake module and the service module, and the array pattern is shown in Fig. 11.53. They are connected to the S-band antenna power divider in the brake module and then the S-band transponder in the brake module in sequence through the high-frequency cable.

The conical constant angle helical antenna with good circular polarization performance belongs to broadband antenna, whose pattern width can be changed by changing the cone angle and the helix angle and can produce a single main lobe. The nearly rotationally symmetric pattern is with the maximum radiation direction

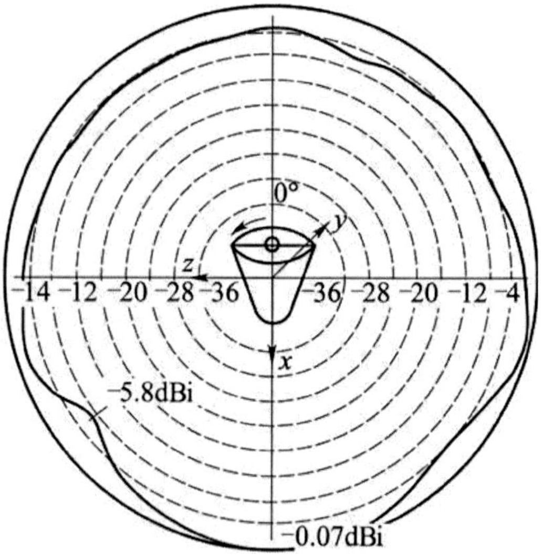

Fig. 11.52 Gain pattern of recovery beacon antenna

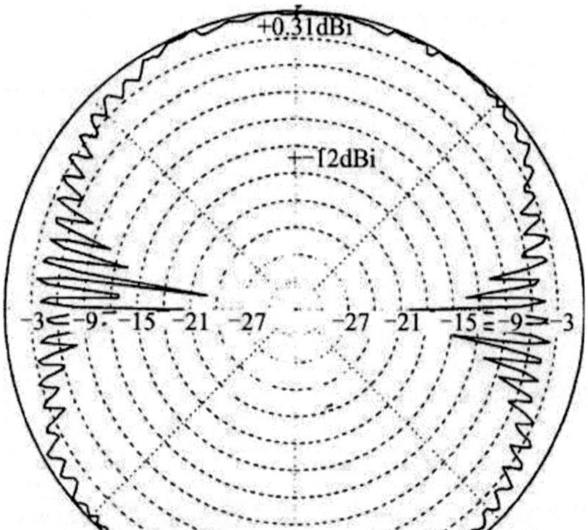

Fig. 11.53 Gain pattern of S-band measurement and control antenna array of brake module

from the cone bottom to the cone top. In the antenna, the feed is balanced by a slotted coaxial tube with quarter wavelength, and the antenna impedance is matched by a quarter wavelength impedance conversion section. The S-band antenna of the brake module has been successfully applied in a number of return satellites, which can meet the high-temperature requirements of upto 600°C (short term) in the launch phase, and the antenna performance meets the overall requirements.

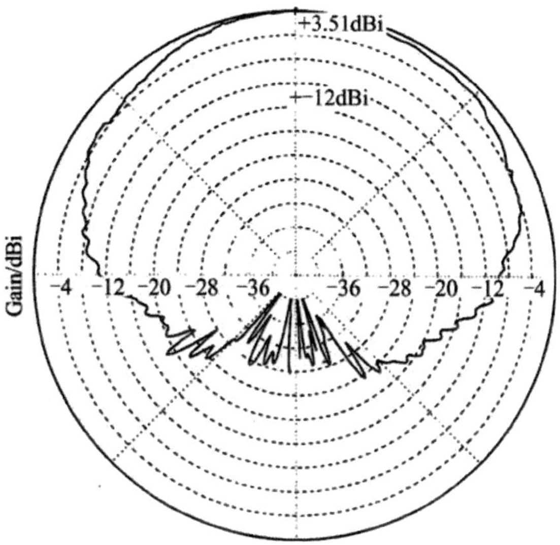

Fig. 11.54 S-band measurement and control antenna of service module

11.5.2.4 S-band Antenna in Service Module

The S-band antenna in service module consists of antenna, high-frequency cable assembly and sealed connector (XM1), which is connected to the sealed connector (XM1) and then the S-band transponder in the pressurized cabin in sequence via the high-frequency cable assembly.

The S-band antenna scheme of service module is the same as that in brake module, but with the slightly different installation interface sizes. The antenna axis is perpendicular to the satellite body surface. For the performance of S-band antenna in service module, please refer to S-band antenna in brake module. The antenna gain pattern on satellite is shown in Fig. 11.54. The antenna performance meets the overall requirements.

11.5.2.5 X-band Data Transmission Antenna of Pressurized Cabin

The data transmission antenna subsystem is composed of antenna and high-frequency cable assembly. The high-frequency cable assembly is used as TNC plug at the antenna end and SMA plug at the microwave switch end. One end of the high-frequency cable is connected with TNC plug for the antenna, and the other with SMA plug is for the microwave switch (for back up of data transmitter).

According to the mission requirements of the satellite SJ-10, the data transmission antenna has the following characteristics:

1. The antenna beam should be shaped to the ground,
2. The antenna can withstand the air scour of the launch phase, and

3. The antenna must be sealed.

From the above characteristics and overall specification, the following requirements on data transmission antenna can be derived:

1. After the installation of the antenna on satellite, the part protruding from the satellite body surface shall be minimized, and the antenna cover shall be used for protection, so as to overcome the scouring impact of the airflow in the launch phase;
2. In the saddle-shaped beam pattern of the antenna with the rotation axis to the ground, $G \geq 2$ dBi (including cable loss) in the $\pm 76°$ direction, and other points within the $\pm 76°$ range meet the requirements of the Earth matched shaped beam specification.

The X-band bifilar backfire helical antenna is used as the data transmission antenna. By adjusting the helix pitch and diameter, the shaped beam of the antenna can meet the requirements, and the antenna height can be adjusted. This product has been used on many low and medium orbit satellites. It works normally on orbit without failure. The longest on-orbit flight life has exceeded 7 years, and the product performance is stable and reliable.

The antenna is mainly composed of helical antenna, reflector, radome and sealed socket. The helical antenna itself is completely made of metal, with high efficiency and reliable structure. The antenna inherits the characteristics of mature products in material selection and structural form.

The reflector is of conical aluminum alloy structure with a wall thickness of 2 mm. In addition, the helical antenna is firmly connected with the reflector, and the bottom surface of the reflector is connected with the satellite body. The connecting surface of the reflector bottom surface is provided with a sealing groove, which is installed with an O-ring to realize the sealing between the antenna and the cabin.

The radome, a conventional half-wave wall wave-transparent structure with a wall thickness of 10.15 mm, is composed of a radome body and a flange. The radome is made of quartz fiber reinforced silica with good ablation resistance, wave-transparent and thermal load-bearing capacity. The antenna adopts a flat cone aerodynamic configuration, which is inverted over the antenna body, and the radome is connected with the satellite body bulkhead through the flange. The radome protects the antenna body from the scour of the airflow in the launch phase, thus ensuring the working temperature of each part of the antenna during the aerodynamic heating process in the rising stage and a stable and constant performance of the antenna.

Figure 11.55 shows the shape of the data transmission antenna and its connection with the satellite body. Figure 11.56 shows the gain pattern of the data transmission antenna on satellite, which meets the overall requirements.

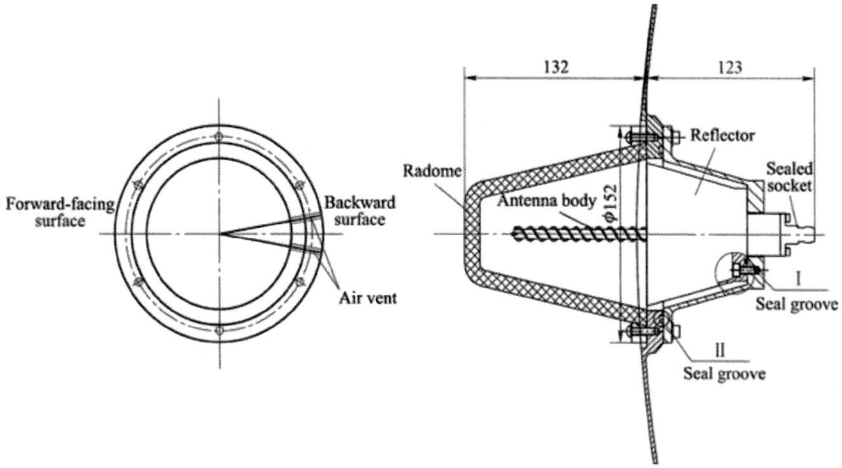

Fig. 11.55 Sketch of connection between X-band data transmission antenna and satellite body in pressurized cabin

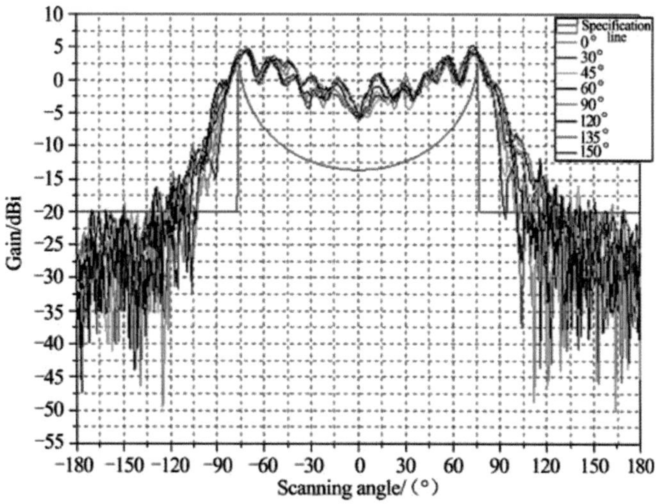

Fig. 11.56 Net gain pattern of center frequency of data transmission antenna ($f = 8212$ MHz)

11.6 Design Case of Antenna System for the Spacecraft Shenzhou

The spacecraft Shenzhou, which is equivalent to or even superior to the third-generation manned spacecraft technology in the world, is independently developed by China. The spacecraft Shenzhou is composed of a recoverable capsule, an orbital

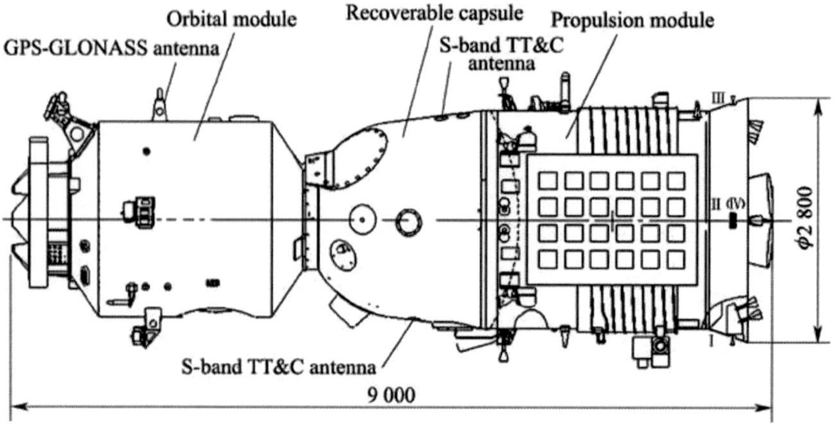

Fig. 11.57 Launch state configuration of the spacecraft Shenzhou

module and a propulsion module. The orbital module is in the front, the recoverable capsule is in the middle, and the propulsion module is in the back, the mechanical components of the docking mechanism are positioned in the front of the orbital module. The launching configuration of the whole ship is shown in Fig. 11.57. Compared with the third-generation spacecraft in the world, the spacecraft Shenzhou has the characteristics of staying on orbit.

The antenna system of the spacecraft Shenzhou is a crucial means for the spacecraft to communicate with the outside world. It undertakes the tasks of telemetry, remote control, voice communication, data transmission and the transmission and reception of various beacons in the whole flight process, such as launching phase, on-orbit operation phase, rendezvous and docking phase, return and reentry phase and landing phase.

Shenzhou spacecraft antenna system includes recoverable capsule S-band TT&C antenna, BD-GPS global positioning satellite receiving system antenna, VHF ultra-short-wave communication, beacon antenna and HF short-wave antenna, orbital module USB antenna, L-band marine relay phased array antenna, propulsion module S-band data transmission antenna, VHF antenna, C-band transponder antenna, Ka/S-band relay antenna, wide beam relay S-antenna, etc.

During the previous flight of the spacecraft Shenzhou, the design and implementation of antenna system have been tested by flight tests, and the scheduled test task has been successfully completed.

11.6.1 Engineering Design Requirements

In terms of design, the antenna of the spacecraft Shenzhou shall meet the following six requirements besides the coverage requirement of far-field radiation zone:

1. For the antennas that still need to work before and after the return of the spacecraft, it should be flush mounted and be thermal shock and ablation resistant.
2. The antennas in recoverable capsule and the orbital module must be sealed.
3. Antennas that still need to work after the return of the spacecraft should be designed in the form of a stretched antenna.
4. For low-frequency HF antennas, in principle, it should be designed in the form of electrically small antennas.
5. For the antenna that still needs to work in the recoverable capsule after landing, no matter what attitude it takes after landing, it should work properly and waterproof design measures should be taken.
6. The recoverable capsule and orbital module antennas must meet the requirements of manned environment such as fireproof, non-toxic, odorless, non-volatile and resistant to heat and humidity in material selection and design.

11.6.2 Recoverable Capsule Antenna

11.6.2.1 Design Characteristics of Recoverable Capsule Antenna

In view of the particularity of manned space engineering, the requirements of antenna's performance are extremely stringent. In addition to the high requirements on the electrical performance, the requirements on antenna layout, structure size, weight, sealing, thermal protection and ablation resistance, space mechanics test environment, electromagnetic compatibility between antennas are also very high, and it is required that the recoverable capsule antenna work reliably whether landing on the Gobi or desert or splashing down in the sea.

Recoverable capsule antenna has the following characteristics in design:

(1) A large number of antennas of various frequency bands and types are adopted in the recoverable capsule

The recovery capsule antenna includes S-band TT&C antenna, BD-GPS global positioning satellite receiving antenna, VHF ultra-short-wave communication, beacon system antenna and HF short-wave communication, beacon system antenna, etc. The feed network of the recoverable capsule antenna is complex and the antennas are in various forms. In the flight test, the whole antenna system should complete the specific work of different stages such as the launch phase, the on-orbit operation section, the rendezvous and docking section, the return reentry section and the landing section. The normal operation of the antenna system is an important prerequisite to ensure the successful completion of the flight test mission of the spacecraft. It should be emphasized that it is also a key issue to ensure that such a large number of antennas do not interfere with each other when working in a limited return capsule.

(2) Double seal of the antenna itself and the mounting surface

Recoverable capsule antennas are installed on the surface of the pressurized capsule of the spacecraft. In addition to sealing the connection of the bulkhead, the sealing performance of the antenna itself should be ensured. The sealing design of the antenna itself is very complicated, so it is necessary to consider all aspects of the antenna, such as mechanical interface, electrical performance, heat-proof performance and mechanical properties.

(3) Antenna flush mounting, miniaturization, heat-proof, anti-ablative, wide wave beam

There are many antennas on the return capsule, but due to the cramped installation space, the miniaturization design of the antenna is particularly meaningful. For this reason, many effective measures have been taken in the design of the antenna system. The flat-mounted antenna installed on the bulkhead of the return capsule should be heat-resistant and ablative resistant. By adopting the optimum design of mechanical, electrical and thermal integration, not only the electrical size of the antenna is reduced, but also the thermal resistance of the antenna is improved by 200 °C compared with similar antennas. Recoverable capsule antenna achieves wide beam and quasi-omni-directional coverage required by the overall TT&C requirements due to the optimization of the scheme design. Besides, in the system design, the distribution of transmitting and receiving power of antenna should be optimized under various working conditions, and the energy loss should be minimized.

(4) Conductive coating on the surface of return capsule heat-proof layer to eliminate the effect of heat-proof layer on the electrical performance of antenna

The influence of thermal medium layer on the design of S-band TT&C antenna and BD-GPS global positioning satellite receiving antenna must be taken into account. When the electromagnetic wave irradiates the heat-proof medium layer, the surface wave will be generated, thus causing the radiation pattern distortion, polarization torsion, impedance mismatch, etc. In the design, the thermal medium layer of the return capsule adopts honeycomb structure with the nonsmooth surface. So, it is difficult to be completely fitted between the metal film and the thermal medium layer. Therefore, the influence of surface wave must be eliminated by the scheme of spraying conductive coating around the antenna. The conductive coating is sprayed on the top of the heat-proof medium layer and under the heat-control coating, by which the heat-proof medium layer and the heat-control coating are firmly bonded.

(5) Double-layer microstrip antenna

The return capsule hatch is integrated in the antenna, and this has the advantages of small space occupation, lightweight and multi-frequency operation.

(6) After the return capsule is landed, the antenna can work properly under uncertain cabin condition

After the spacecraft returns, whether landing in the Gobi sand land or splashing down in the sea, both the onboard beacon antenna and the BD-GPS global positioning satellite receiving antenna should work reliably to ensure the safety of astronauts.

11.6.2.2 Types of Return Capsule Antennas

(1) S-band TT&C antenna

The S-band TT&C antenna of the return capsule is a unified S-band TT&C antenna used in spacecraft assembly, comprehensive test, the active launch, the on-orbit operation and the reentry phase. The antenna should maintain normal working state in the launch, on-orbit operation and reentry phase. Before the fairing is separated, the antenna is operated through the transparent wave port on the fairing; after separating and the reentry phase, the antenna should have quasi-omnidirectional beam coverage. So, only one antenna is not satisfactory for the above requirements. Through theoretical analysis and experimental verification, it is concluded that the receiving and transmitting are each implemented by two antennas according to the above requirements.

The working process of the S-band TT&C antenna in the return capsule is that two antennas are separately responsible for the transmitting and receiving of the spacecraft in the launch phase, the reentry section and the rendezvous and docking section. In addition, two pairs of transmitting antennas and two pairs of receiving antennas are connected to the transceiver through instructions, so that four antennas work in parallel. In the on-orbit section, when each one antenna works for transmitting and receiving, only one transmitting antenna and one receiving antenna are connected to the transceiver through instructions, so that one antenna is responsible for transmitting and one for receiving.

Due to the special use of the return capsule S-band TT&C antenna itself, it has the following features in design:

Firstly, according to the uniform S-band TT&C system of spacecraft, the S-band working frequency of the antenna can be divided into upstream and downstream bands. The upstream and downstream frequencies are coherent and the fixed forwarding ratio is adopted. The upstream antenna is LHCP and the downstream antenna is RHCP. The electrical performance of antenna mainly includes voltage standing wave ratio, polarization mode, pattern, gain and transceiver isolation.

Secondly, the S-band TT&C antenna of the return capsule is installed on the surface of the reentry module, and the installation space is cramped, so the miniaturization design must be adopted. In addition, the requirements of the whole working environment in depth should be considered in the design of the antenna. Since the spacecraft is carried by rockets, the antenna should be able to adapt to the various environments of the vehicle launch, flight and spaceship vehicle separation, cabin separation, return capsule reentry and landing impact. When the spacecraft returns, the return capsule flies at a very high speed, and aerodynamic heating may cause the antenna surface temperature to reach thousands of degrees. In order to effectively reduce the aerodynamic heating, the antenna should be so designed as to avoid causing discontinuities in the surface of the spacecraft. Due to the influence of high temperature, the strength and rigidity of the antenna metal material will be decreased, the dielectric constant ε_r of the dielectric material and the loss tangent $\tan\delta$ will become larger, that results in deterioration of the antenna performance.

Therefore, in the design of antenna, the metal material with high structural strength and the dielectric material with low loss ε_r and high heat stability should be selected. Because of the aerodynamic heating during reentry, it is required that the antenna should not be bulged on the surface of the spacecraft's heat shield as far as possible, and the antenna surface should be heat-proof and ablation resistant. In addition, the influence of plasma on the antenna should also be considered. The solution is to widen the VSWR bandwidth of the antenna as much as possible. Besides, electrical breakdown should also be considered in the design of antennas and feeding systems, and dielectric materials with small tanδ should be selected as far as possible.

Finally, because of the manned characteristics of the spacecraft, the antenna installed in the return capsule should be insulated and sealed on the basis of the structural strength and electrical performance to prevent thousands of degrees of high temperature outside the cabin from entering the return capsule through the antenna, thus ensuring the safety of the astronauts. In addition, the antenna body must be structurally sealed to ensure that the air in the return capsule is not leaked under vacuum. Due to the manned requirements, antenna material selection and design should meet the requirements of manned environment such as fireproof, non-toxic, odorless, non-volatile, humidity and resistant to damp heat alternation or constant damp heat, thus meeting the requirements of space electronic environment.

Based on an overall consideration of various factors, the S-band TT&C antenna element of the Shenzhou spacecraft return capsule adopts a flat-mounted embedded cross-cavity circular-polarized antenna design. The main technical requirements are shown in Table 11.9. The stand-alone antenna photo is shown in Fig. 11.58. The photo of the antenna after loading is shown in Fig. 11.59.

The whole spacecraft pattern test was carried out in the compact field, as shown in Fig. 11.60. The centroid of the reentry module is coordinate origin O, along the direction of the orbital module is the *Ox*-axis, and the connection between the S-band TT&C antenna and the centroid *O* is the *Oz*-axis. The typical pattern test results are shown in Fig. 11.61.

(2) BD-GPS antenna

BD-GPS antenna is used to receive signals from the satellites Beidou and global positioning satellites to determine the coordinate parameters of spacecraft. According to the requirements, BD-GPS antenna plays a core role in the whole process of flight, especially in the orbital operation stage, which requires precise orbital measurement. Therefore, the following requirements should be met in the design of the antenna:

1. In the transmitting/receiving phase, the antenna system can view more than four navigation satellites in its FOV within a certain period of time.
2. In the on-orbit operation phase, the BD-GPS antenna on the spacecraft needs and only needs to cover the upper half of the space.
3. After landing, the BD-GPS antenna can receive signals from the navigation satellite system to ensure positioning information regardless of the attitude.
4. The sealing of the antenna is ablative resistant and can work properly after atmospheric reentry.

Table 11.9 Main technical requirements of S-band TT&C antenna for return capsule

No	Item	Specification	
1	Working frequency (MHz)	Receiving frequency	2040 ± 20
		Transmitting frequency	2216 ± 20
2	Polarization mode	Receiving antenna	Left-handed circular polarization
		Transmitting antenna	RHCP
3	Axial axis ratio (dB)		AR less than 3
4	VSWR	Operating frequency point	VSWR ≤ 1.5
		The rest in band	VSWR ≤ 2.0
5	Pattern	Launch phase (after separating of fairing) return phase	In the range of ±76° of the antenna axis, the antenna gain is ≥−9.0 dBi In the range of ±60° of the antenna axis, the antenna gain is ≥−6.0 dBi
		On-orbit operating section	In the range of ±76° of the to-Earth angle, the antenna gain is ≥−6 dBi
6	Transceiver isolation (dB)		≥50
7	Tightness (P_a m^3 s^{-1})		≤1 × 10^{-7}
8	Operating temperature (°C)		−60 to +1400

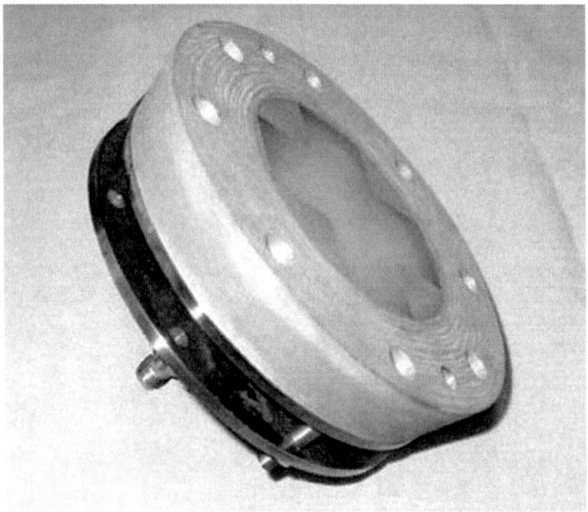

Fig. 11.58 Return capsule S-band TT&C antenna element

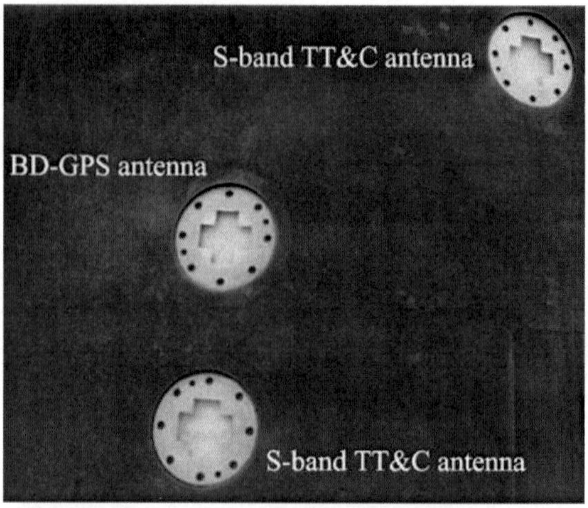

Fig. 11.59 S-band TT&C and BD-GPS antenna photos of the return capsule after loading

Fig. 11.60 S-band TT&C antenna test photo in the compact field

In order to balance various working conditions, the BD-GPS antenna system consists of two antennas, but only one antenna is in operation all along.

The working process of BD-GPS antenna is to connect and activate the antenna located near the quadrant III of the spacecraft in the transmitting phase, on-orbit operation phase, return and reentry phase until landing by command. After landing,

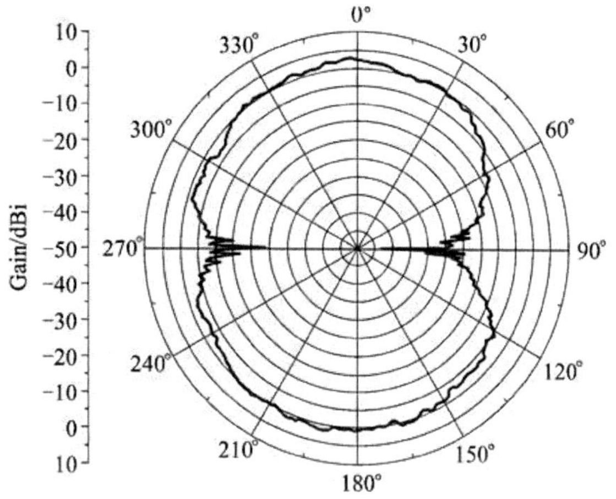

Fig. 11.61 Test results of typical pattern of S-band TT&C antenna

the gravity switch controls two antennas and activates the antenna which is higher than the horizontal plane.

Compared with the S-band TT&C antenna in the return capsule, the BD-GPS antenna has the same technical specifications except for the different electrical requirements. It also works at −60 to +1400 °C, so the both antennas have the same hard nut to crack. In addition, because the BD-GPS antenna has the same working frequency, but the transverse installation size is the same as that of the S-band TT&C antenna, so the installation space is relatively limited. Based on the electrostatic performance design, further miniaturization of the design should be considered, which greatly increases the difficulty of development. The miniaturization of antenna design can be satisfactorily realized by filling the cavity with dielectrics.

(3) VHF ultra-short-wave communication, beacon antenna

VHF antenna is used in ultra-short-wave communication, landing search beacon, international rescue and other systems, after the return capsule is reentered into the atmosphere.

The ultra-short-wave communication and landing search beacon antennas that start working after the return capsule passes through the Blackout are in the form of a two-layer microstrip antenna that is integrated in the return capsule door, i.e., the return capsule door antenna. The antenna features multi-frequency operation and high working environment temperature compared with similar antennas. Besides the antenna function, the door also has the function of heat insulation.

The reason why the double-layer microstrip antenna is adopted is that the bandwidth of common microstrip antenna is narrow, which cannot meet the needs of three frequency points working at the same time. Using double-layer microstrip, the lower layer works at a lower frequency. In addition to being used as a radiating element, it also serves as a grounding plate for the upper microstrip and adjusts the shape of the

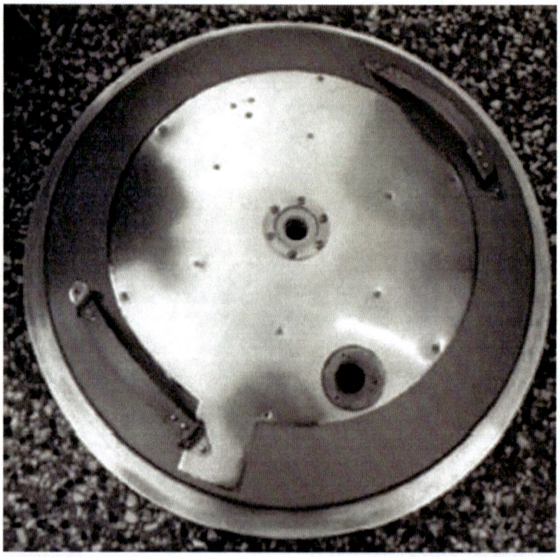

Fig. 11.62 Return capsule door antenna

upper microstrip to produce parasitic distributed capacitance and inductance, so that this microstrip resonates at two frequency points and three frequency points work simultaneously.

The double metal microstrip patch of the return capsule door antenna is processed by pressing the stainless-steel sheet with a specific thickness. The upper and lower dielectric substrates adopt a special foam sandwich structure skin made of quartz reinforced resin matrix composite. The test and flight results show that the performance of the return capsule door antenna is excellent, and the antenna appearance is shown in Fig. 11.62.

In order to make the VHF beacon antenna work normally no matter what the attitude is after landing, three deployable dipole antennas are installed on the side wall and the baseplate of the spacecraft. The three antennas and the return capsule hatch antenna are mutually arrayed. After landing, one higher than the ground plane is selected from the two arrays according to the gravity switch.

The VHF beacon antenna installed on the side wall and the base plate of the spacecraft is a compressible deployable dipole antenna. The wavelength of the antenna is long, so the longitudinal dimension of the antenna is large, and the installation space is very narrow. When the antenna is not working, it should be compressed in the cabin and ejected after landing. The electromagnetic environment around the antenna is complex. After analysis and a lot of experimental verification, a multi-section compressible deployable antenna with fan-shaped ground plate and increased top load is developed, which can implement both the electrical performance requirements and the general assembly requirements. The antenna satisfactorily solves the design problem of the ultra-slim spring and the requirement of miniaturization of the antenna. Another outstanding feature of VHF deployable dipole antenna design is that it uses flexible materials to design a compressible antenna transparent cover, which successfully cracks the problem of proper operation of the beacon antenna after the spacecraft splashes down in the sea.

VHF ultra-short-wave communication and beacon antenna participated in the low sea state comprehensive floating test along with the return capsule. According to the data received by the international rescue satellite, the carrier information of VHF ultra-short-wave communication and beacon antenna can be received as long as the satellite passes over the test area during the whole test process. Under the condition of good geometry of transit satellite measurement, the positioning deviation is within 1 km. The VHF ultra-short-wave communication and beacon antenna are always in good condition during the sea ranging test of the search and rescue ship. During the whole test, the search and rescue helicopter can successfully receive the carrier positioning information transmitted by VHF ultra-short-wave communication and beacon antenna. The antenna was manually watered with sea water for many times, or towed by a traffic boat to run in the sea. The antenna was submerged in the sea for a while and exposed to the sea for a while along with the return capsule. Under these two conditions, VHF ultra-short-wave communication and beacon antenna worked normally.

(4) HF short-wave communication and beacon antenna

HF short-wave communication and beacon antenna are composed of three antennas. One is parachute antenna; the other two are self unlocking spring extension antennas.

The working wavelength of HF self unlocking spring extended antenna is 15 m. If the $\lambda/4$ monopole is used, the antenna size is about 3.75 m. Because the antenna must be folded in the antenna cabin when the spaceship Shenzhou is launched and before returning, it can only be unlocked and deployed after landing, and the whole antenna installation space provided is only $\varphi 0.18\ \text{m} \times 0.12\ \text{m}$, the installation space is very limited, so great difficulties are faced in the structure design. On the basis of sufficient experimental verification and theoretical analysis, the scheme of normal mode spiral antenna is adopted, that is, the spiral antenna technology with the antenna diameter less than 0.18 is adopted, which greatly shortens the axial dimension of the antenna and makes matching easy. In order to make the antenna deploy independently, the helix should have certain elasticity, besides, it will not fall and deform due to its own weight, and there should be a support rod in the middle of the helix. A group of FRP hollow core sleeves with different thickness and a certain taper at both ends are chosen. The former can be sleeved together one by one. When the antenna is not deployed, the sleeve can be taken out from thin to thick and from inside to outside; when the antenna is deployed, the sleeve is pulled out section by section by helix, and the adjacent sleeves bite are engaged with each other to form a long rod with a thick head and a certain taper, which supports the helix. A locking/unlocking device is available to control the deployment of the antenna. When the antenna is not deployed, it is locked by the locking device. and the helix is folded in the antenna barrel. When the antenna needs to be deployed, the unlocking device unlocks and the antenna automatically deploys, which solves the problem of antenna folding and unlocking. After landing, the working state of HF extended antenna after deployment is shown in Fig. 11.63.

The parachute antenna is a λ/4 monopole antenna, which is integrated with the parachute hanging sling and is used when the parachute is deployed in the return phase.

The working procedure of HF antenna system is that before landing, the parachute antenna is pulled out together with the deceleration parachute, and the transceiver starts to work; after landing, the HF extension antennas in the return capsule and the base plate are, respectively, deployed, and one antenna with higher relative ground plane is selected by gravity switch to work.

(5) Mobile communications antenna

A micro-strip global mobile satellite communication antenna (Fig. 11.64) is separately equipped on the return capsule to ensure that the astronauts can make and answer satellite calls in the capsule after landing.

Fig. 11.63 After the spacecraft's return capsule landed, the unfolded HF extension antenna was unlocked

Fig. 11.64 After the spacecraft's return capsule landed, the unfolded VHF antenna and mobile communication antenna were unlocked

11.6.3 Orbital Module USB Antenna

The USB antenna of the orbital module is responsible for the TT&C, tracking, payload data transmission and other tasks of the orbital module. It starts to work before the unlocking and separation of the orbital module and the recoverable capsule. The unified S-band TT&C system is adopted, and data transmission antenna and TT&C antenna are shared.

The main technical requirements of the USB antenna in the orbital module are that the antenna gain is not less than −4 dBi within ±72° of the Earth flare angle. Therefore, it can meet the requirements to select the appropriate two antenna elements with separate receiving and transmitting to be installed in quadrant I of the orbital module. Due to manned demand, the antenna itself needs to adopt a sealed design. The antenna element adopts the cavity helical antenna technology, and the beam width of the conventional cavity helical antenna element is only about 100°. In order to implement the requirement of ±72° coverage specified in the technical specifications, the antenna designers have creatively designed an antenna beam broadening compensation element. On the aperture of the cavity helical antenna element, the antenna beam can reach ±72° by adding the antenna beam broadening compensation element, thus meeting the technical requirements. Figure 11.65 shows the appearance of USB antenna in the orbital module.

In addition, the orbital module is specially equipped with a communication antenna for spaceman outside the module for outbound activities, and an L-band phased array maritime satellite communication antenna is also configured for making/answering maritime satellite call on orbit.

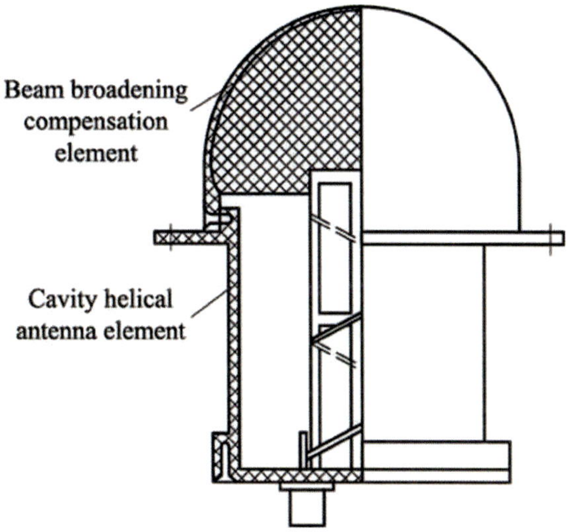

Fig. 11.65 Appearance of orbital module USB antenna

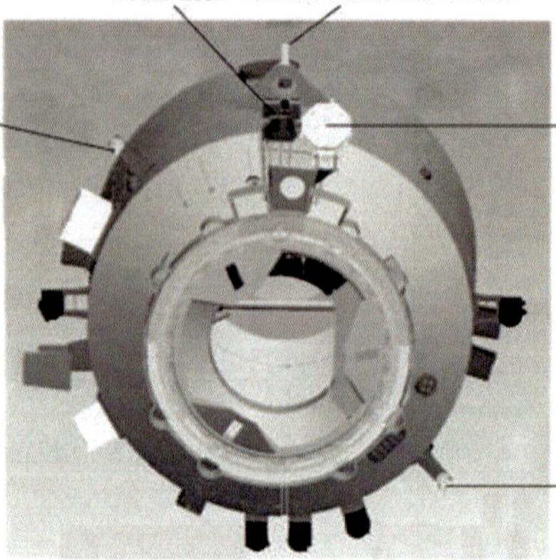

Fig. 11.66 Track module antenna equipment layout

For the requirements of the rendezvous and docking task, inter-module communication antennas, microwave radars and differential GNSS antennas are provided, as shown in Fig. 11.66.

11.6.4 Propulsion Module Antenna

The propulsion module antenna of the spacecraft Shenzhou mainly includes S-band data transmission antenna, C-band transponder antenna and VHF antenna, Ka/S relay antenna, wide beam relay-S antenna.

The S-band data transmission antenna of propulsion module is used for data transmission, the C-band transponder antenna for signal transmission of C-band transponder and the VHF antenna for voice signal transmission of spacecraft. All these antennas should work properly before the launch phase, on-orbit operation phase, propulsion phase and the separation of the propulsion with return module. Compared with the return capsule antenna, the mounting surface of the propulsion module antenna can protrude from the outer surface of the propulsion module, and there is no need to perform special heat and erosion resistant design and antenna self-sealing design.

11.6.4.1 S-band Data Transmission Antenna in Propulsion Module

In order to meet the requirements of quasi-omnidirectional coverage, the S-band data transmission antenna in the propulsion module adopts the form of two antenna element arrays, one in quadrant I of the propulsion module and the other in quadrant III. The antenna element form is the same as the USB antenna in the orbital module.

Before separating the fairing, it works through the electromagnetic window of the fairing.

In the launch phase of the spacecraft, two antennas should work at the same time. At this time, the two S-band antennas in the propulsion module are connected with the data transmitter through instructions, and the two antennas work in parallel. When only one antenna is needed to work in the on-orbit operation phase, only the quadrant I antenna is connected with the data transmitter through instructions.

11.6.4.2 Propulsion Module C-band Transponder Antenna

The C-band transponder antenna of the propulsion module adopts the form of separate receiving and transmitting, and the receiving and transmitting antennas are all installed in quadrant I of the propulsion module. The antenna element is the same as the USB antenna in the orbital module. The antenna is mainly used as the backup of the unified TT&C means for the S-band TT&C antenna of the return capsule.

11.6.4.3 Propulsion Module VHF Antenna

The VHF antenna in the propulsion module provides ultra-short-wave voice communication services for astronauts during the period from the takeoff to the separation of the propulsion module and the recoverable capsule. The antenna, a π-shaped unit dipole with a length of $\lambda/2$, is installed in the quadrant I of the propulsion cabin. For the impedance matching, the tuning bolt is so moved laterally along the antenna base that the antenna resonance is at the center frequency f_0. Before separating, the antenna works through the electromagnetic window of the fairing. VHF antenna in the propulsion module after installation is shown in Fig. 11.67.

Fig. 11.67 Propulsion module VHF antenna

11.6.4.4 Other Antennas in the Propulsion Module

Figure 11.68 shows a Ka/S relay antenna and a wide beam relay S-antenna for space-based TT&C.

11.6.5 Flight Test Results

During the previous flight of the spacecraft Shenzhou, the design and implementation of antenna system have been tested by flight tests, and the scheduled test task has been successfully completed.

The Shenzhou spacecraft antenna system works well according to the procedures in the flight test. The analysis of the flight results shows that the design of the spacecraft antenna is successful, and the spacecraft antenna has successfully completed the various tests.

11.7 Design Case of Chang'e-5 Lunar Probe Antenna System

The mission and goal of Chang'e-5 lunar probe are to realize the orbiting, landing and lunar samples returning, which will be another milestone in the field of deep space scientific exploration in China after orbiting of Chang'e-1 lunar probe, orbiting and landing of chang'e-3 lunar probe. Chang'e-5 lunar probe is composed of four parts, i.e., ascender, lander, returner and orbiter. As shown in Fig. 11.69, Chang'e-5 lunar probe will be carried by ChangZheng 5 and launched at the launch center Wenchang

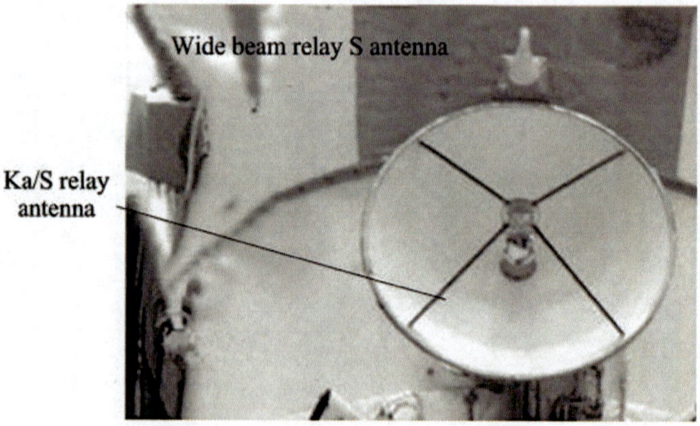

Fig. 11.68 Relay antenna

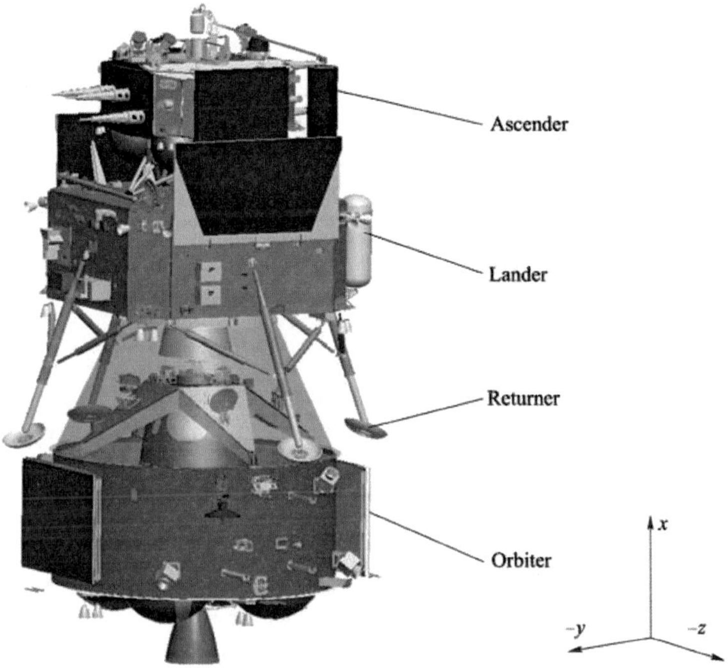

Fig. 11.69 Probe launch state configuration

in Hainan Province. After the Earth-moon transfer and circular lunar flight, it will land in the selected area of the lunar surface, collect lunar samples and send the lunar samples to the ground safely through the processes of takeoff of ascenders on the lunar surface, rendezvous and docking in the lunar orbit, Moon-Earth transfer and reentry and return.

11.7.1 Task Analysis of Antenna System

For spacecraft, the role of antenna system is equivalent to human eyes and ears, its importance is obvious, so backup measures must be taken when necessary. The flight process of Chang'e-5 lunar probe includes four modules joint flight state, flight state of combination of lander and ascender, flight state of orbiter and returner, and four modules independent flight state. In addition, the TT&C data transmission tasks of four modules are different. The orbiter undertakes TT&C and data transmission task, the ascender and the returner undertake TT&C task, and the lander undertakes data transmission tasks. In addition to the TT&C tasks, the returner also needs to send beacon signals for ground search and rescue. The above tasks of TT&C and data transmission need to be completed by establishing a communication channel through the antenna system. In other phases than the return phase, the unified carrier TT&C system of X-band is used for Chang'e-5 lunar probe TT&C, and X-band is used for data transmission. The unified S-band carrier TT&C system is adopted for the returner. Compared with other satellites, Chang'e-5 lunar probe is more complex in configuration, with multiple communication task modes and difficult antenna system design.

11.7.2 Antenna System Design

According to the flight status of Chang'e-5 lunar probe and the requirements of data transmission task for ground TT&C, the probe TT&C antenna is arranged in the orbiter, the ascender and the returner, the data transmission antenna is arranged in the orbiter and the lander, and the beacon antenna is in the returner.

TT&C antenna is absolutely indispensable for spacecraft, which is usually covered by two-antenna array with the same frequency or different frequencies. For the deep space probe, the requirements on the whole space coverage of TT&C are more stringent because of the complex flight process and the far away signal transmission distance. In order to avoid the interference area caused by the same frequency array, i.e., the TT&C blind area, the orbiter and ascender of Chang'e-5 lunar probe adopt different frequency array modes. The TT&C receiving antennas of the orbiter and the ascender are directly related to the success or failure of the probe task, so backup measures should be taken.

11.7.2.1 Orbiter Antenna Design

The orbiter with separate TT&C and data transmission capability must meet the requirements of TT&C data transmission in the state of single module, the state of combination of orbiter and returner and that of four modules. The TT&C antenna should achieve full coverage of TT&C, that is, two different frequency antenna arrays work at the same time. For array work, each antenna is responsible for half-space coverage. Under the consideration of the obstruction of the orbiter's solar wing to the antenna radiation as well as the flight attitude and available space constraints, the X-band TT&C omnidirectional antenna is installed on the $\pm z$-side of the orbiter, and the $\pm z$-antennas are responsible for the half-space TT&C coverage. The orbiter has four receiving antennas and two transmitting antennas, of which two receiving antennas are used as backups. An X-band directional antenna system with drive assembly mounted on the orbiter-z plane is used for orbiter data transmission to ensure that the orbiter can reliably track the Earth and return image. The biggest advantage of this antenna is that it can achieve wide-angle and high-gain coverage, greatly improving data transmission rate and reducing transmission power requirements.

11.7.2.2 Antenna Design of Lander and Ascender Combination

The combination includes the lander and the ascender. The ascender shall have independent TT&C capability to meet the TT&C requirements under the condition of single module, lander and ascender combination and four-module combination. The lander only has the function of data transmission, and its telemetry and remote control data are transmitted through the ascender. Generally, in order to avoid the radiation effect of solar wing on the TT&C antenna, the TT&C antenna is not installed on the same side as the solar wing, but in the vertical direction with the solar wing. To this end, considering the solar wing installation position of the lander and ascender, the X-band TT&C antenna is installed on the $\pm y$-side of the probe, and the $\pm y$-side antenna is responsible for half-space TT&C tasks, and the full-space TT&C coverage is realized by half-space different frequency arrays. The ascender has two transmitting antennas and three receiving antennas (one of which is used as backup).

The X-band medium-gain antenna in the lander, which is installed on the side plate of lander, is used for real-time transmission of images taken by landing camera in dynamic descent section. According to the coverage of the lander and ascender combination relative to the Earth in the dynamic descent section, and considering certain design margin, the medium-gain antenna in the X-band should achieve data transmission within the beam of about $\pm 20°$.

The X-band data transmission antenna of the lander adopts directional antenna, which is installed on the top plate of the lander. The antenna is similar to the data transmission antenna of the orbiter, which is driven to point to the ground station by drive assembly for high-speed data transmission to Earth.

11.7.2.3 Returner Antenna Design

The returner shall have independent TT&C function and provide search and rescue signal and are only used for the single module working mode after separation from the orbiter. According to the requirements of the TT&C mission, the S-band TT&C antenna starts up after the separation of the orbiter and returner combination. In order to achieve near-full-space coverage, the S-band TT&C antenna is installed at 115° and 295° positions on the side wall of the returner (Fig. 11.70), i.e., near quadrants

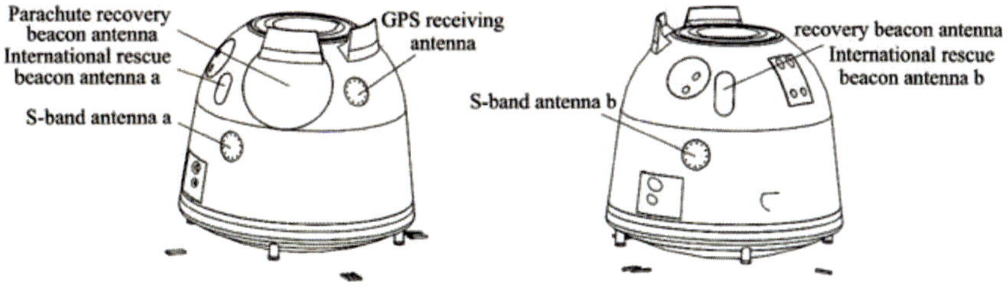

Fig. 11.70 Layout of returner antenna

II and IV, and near-full-space coverage is achieved through half-space co-frequency array. Considering the high-temperature ablation caused by airflow erosion during reentry, there should be no protuberance on the surface of the returner. To this end, the S-band TT&C antenna and GPS antenna should be installed flat with the surface of the module. The heat insulation design measures must be taken on the antenna surface to ensure that the antenna is not damaged by burning-through during reentry. During the reentry process, the returner beacon antenna is folded in the recoverable capsule, in which the parachute beacon antenna is deployed with the parachute during the falling process, and other beacon antennas operate only after landing. Due to the uncertainty of the landing attitude of the returner, in order to achieve better pattern gain coverage of the beacon antenna, two antennas are used for the recovery beacon antenna and the international rescue beacon antenna, respectively, and the antenna at the higher position is selected to transmit the beacon through the gravity switch.

11.7.3 Antenna System Operating Mode

In different flight phases, the Chang'e-5 antenna works in the following different modes:

(1) Launching phase

The X-band TT&C antennas of both the orbiter and ascender are working, but the antenna of orbiter is the main one. The X-band TT&C receiving antenna of the ascender works as the backup of the uplink TT&C. The directional antenna of the orbiter and lander is locked; the beacon antenna of the returner is folded in the returner cabin.

(2) From Earth-lunar transfer orbit phase to four modules separation in lunar orbit stage

When the detector in the $+x$-direction is favorable to the ground, the TT&C of the ground is completed by the ascender TT&C antenna. When the detector in the $-x$-direction is favorable to the ground, the TT&C of the ground is completed by the orbiter TT&C antenna. Under the control of the data management subsystem, the directional antenna of the orbiter is unlocked and deployed; the data transmission channel is established according to the ground command, and the surveillance images of the solar wing and the directional antenna are transmitted. The lander and the returner antennas do not work.

(3) From four modules separation of lunar orbit phase to Moon-landing phase

The combination of lander and ascender realizes TT&C through the ascender TT&C antenna. The TT&C of the orbiter and returner combination is realized by the X-band TT& antenna of the orbiter. The orbiter directional antenna establishes the downlink telemetry channel according to the ground command. The lander and the returner antennas do not work.

(4) Lunar surface work phase

After landing of the lander and ascender combination, the $-y$-side is to the Earth, and the TT&C is realized by the ascender $-y$-side TT&C antenna. Beside, the TT&C of the orbiter and returner combination is realized by the X-band TT&C antenna of the orbiter. The orbiter directional antenna establishes the downlink telemetry channel according to the ground command. The directional antenna of lander is unlocked under the control of the data management subsystem and points to the Earth for data transmission. The returner antenna does not work.

(5) Lunar ascent phase to rendezvous and docking phase

The ascender TT&C antenna realizes TT&C, the orbiter and returner combination realizes TT&C through the orbiter TT&C antenna. The orbiter directional antenna transmits the surveillance image of the process of the rendezvous and docking of the orbiter and returner combination with the ascender, and the process of lunar samples transfer after the lander antenna transmits the ascender takeoff image data, it stops working. Moreover, the returner antenna does not work.

(6) From the lunar orbit phase to the lunar Earth transfer orbit stage, the orbiter and returner combination separation

The orbiter/returner assembly realizes TT&C through the orbiter TT&C antenna. The orbiter directional antenna transmits the monitoring image during the separation of the orbiter and the returner, and the returner antenna does not work.

(7) Reentry stage

TT&C is carried out by the S-band TT&C antenna of the returner. After the parachute is deployed, the parachute antenna is ejected to transmit the recovery beacon signal, and the search helicopter tracks the recovery beacon signal. After landing, three recovery beacon antennas are ejected. The gravity switch automatically switches the beacon antenna which is favorable for the sky according to the landing stage, and tracks recovery beacon signal and international rescue beacon signal by the search helicopter.

11.7.4 Antenna Element Design

11.7.4.1 Design of Lander Directional Antenna

(1) Specifications

The technical specifications of directional antenna are shown in Table 11.10.

(2) Function and composition

Table 11.10 Technical specifications of directional antenna

Item	Requirement
Transmission frequency and bandwidth (MHz)	8450–8500
Antenna gain (dBi)	(net gain of directional antenna) ≥ +20.5 (±3.5° cone, including cable loss, feed and polarization loss)
Polarization mode	RHCP
Antenna coverage	1/4 space
System pointing accuracy (°)	(better than) ± 0.6
First mode (Hz)	(in folded state) > 70

The directional antenna of lander consists of the following parts: radio frequency part, including reflector antenna and high-frequency cable assembly; mechanism part of antenna, including two-axis drive mechanism and folded release mechanism; antenna structure part, including antenna support and system support, as well as low-frequency control wiring harness and thermal control components. The arrangement of the directional antenna is shown in Fig. 11.71.

(3) Antenna simulation

Feedforward reflector antenna is selected as reflector antenna. In order to improve the radiation efficiency of the antenna, the cross dipole is used as the feed of the antenna. The advantages of this feed are small aperture shielding, high efficiency, simple feed, lightweight and small mechanical complexity. The simulation results of gain pattern of reflector antenna are shown in Fig. 11.72.

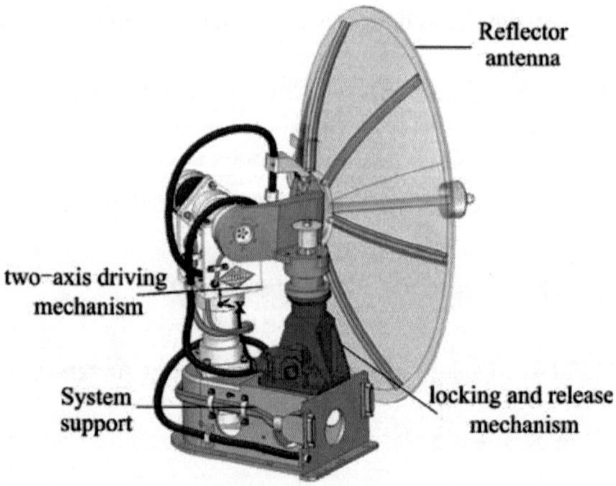

Fig. 11.71 Structure of directional antenna

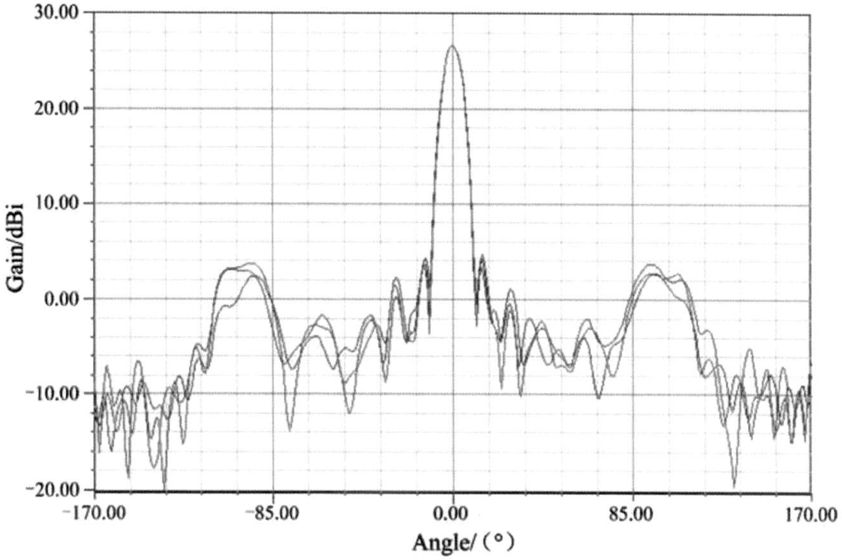

Fig. 11.72 Mesh reflector antenna gain pattern

11.7.4.2 Design of S-Band TT&C Antenna for Returner

(1) Specifications

The technical specifications of TT&C antenna are shown in Table 11.11.

(2) Function and composition

The S-band TT&C antenna of the returner shall be able to withstand the high-temperature ablation in the return phase and shall be installed on the side wall of the returner, in the form of a flat structure that conforms to the returner. The antenna body adopts cavity structure, which is composed of antenna heat insulated window,

Table 11.11 Technical specifications of TT&C antenna

Item	Requirement
Receiving frequency and bandwidth (MHz)	2025–2110
Transmission frequency and bandwidth (MHz)	2200–2290
Antenna gain (dBi)	After array formation (USB network influence is deducted after array formation): (gain is not less than) – ($0° \leq \theta \leq 78°$, $0° \leq \varphi < 360°$, 98% range, including feeder, polarization and other link losses, including the influence of probe shell)
Polarization mode	LHCP

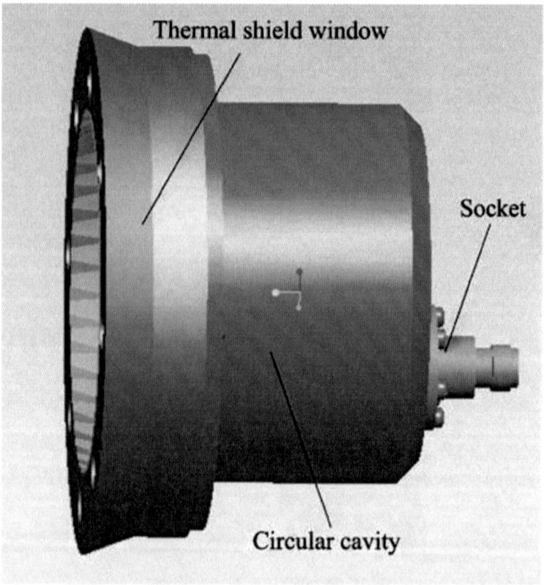

Fig. 11.73 Composition of S-band TT&C antenna

metal cavity, excitation structure and high-frequency socket. The excitation part of the antenna adopts single helix structure, the helix is fed by 50 Ω coaxial system and the end of the coaxial system is high-frequency socket (Fig. 11.73).

(3) Antenna simulation

The S-band TT&C antenna in the state of the returner model array is simulated by the simulation software. The simulation results are shown in Fig. 11.74.

11.7.4.3 Design of Returner Beacon Antenna

(1) Specifications

The technical specifications of the return beacon antenna are shown in Table 11.12.

The technical specifications of international rescue beacon antenna are shown in Table 11.13.

(2) Function and composition

Beacon antenna is a quarter wavelength monopole antenna, forming an inverted eight-shaped pattern. Beacon antenna can realize multiple folding and automatic deployment in structure design. The deployment status of beacon antenna is shown in Fig. 11.75. Beacon antenna mainly includes radiation pole, circular cover plate, installation flange of feeder base, tubular inner conductor of feeder base and spring

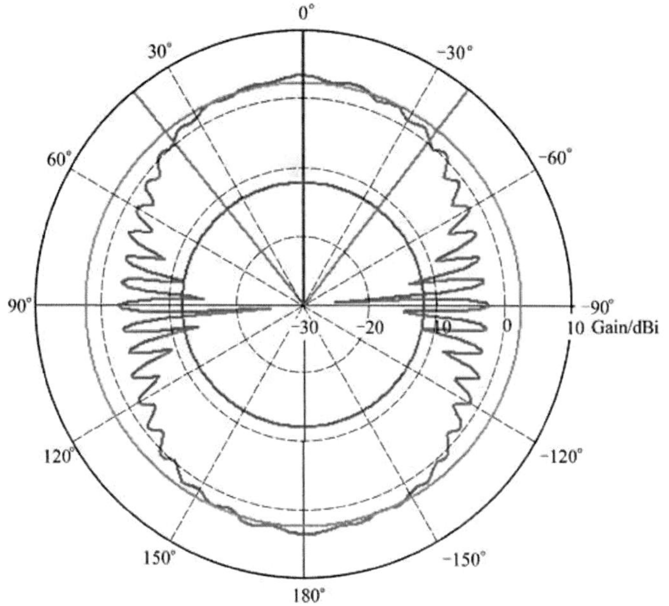

Fig. 11.74 Gain pattern of TT&C antenna array

Table 11.12 Technical specifications of return beacon antenna

Item	Requirement
Transmission frequency and bandwidth (MHz)	243 ± 1
Gain of parachute recovery beacon antenna (dBi)	(not less than) − 12 ($14° \le \theta \le 166°, 0° \le \varphi < 360°$)
Gain of ejection recovery beacon antenna (dBi)	(gain in 75% of space above the ground is not less than) − 12
Polarization mode	Linear polarization

Table 11.13 Technical specifications of international rescue beacon antenna

Item	Requirement
Transmission frequency and bandwidth (MHz)	406 ± 1
Antenna gain (dBi)	(75% of the space above the ground in landing state is not less than) − 8
Polarization mode	Linear polarization

device. The radiation pole is used to radiate electromagnetic signals. When it is folded, it is placed in the tubular inner conductor of the feeder seat. When it is used, it is ejected to a predetermined length by means of a spring device. The circular

Fig.11.75 Deployment status of beacon antenna

cover plate blocks the tubular inner conductor of the feeder seat. The spring device is placed in the tubular inner conductor of the feeder seat to ensure that the radiation pole can be repeatedly folded and ejected automatically.

(3) Antenna simulation

The simulation analysis pattern of beacon antenna in landing state is shown in Fig. 11.76.

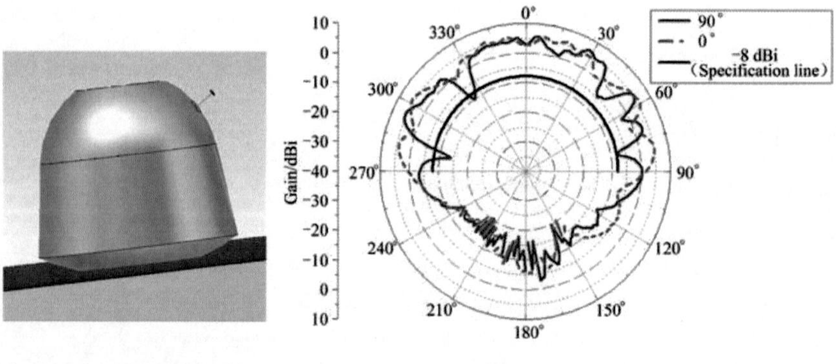

Fig. 11.76 Orthogonal section pattern of beacon antenna when the axis of the returner body is perpendicular to the ground

Chapter 12
Overview of New Technology of Spacecraft Antenna

Abstract With the development of new materials, new techniques and radio frequency microelectronics technology as well as the emergence of new requirements (such as space science exploration), spacecraft antenna technology has been kept improving. In this chapter, the application of smart reconfigurable antenna, multi-band reflector antenna, membrane phased array antenna, MEMS technology and terahertz technology in spacecraft antenna are briefly introduced.

12.1 Intelligent Reconfigurable Antenna

Reconfigurable antenna is a kind of antenna whose main performances (such as pattern, frequency, polarization) and parameters can be changed by command or auto-sensing. In a narrow sense, a phased array antenna is a reconfigurable antenna with variable pattern. The more intelligent adaptive nulling antenna is also based on the phased array, which can automatically sense the direction of the interference source, adjust the phase of the array elements, and form a concave area of the pattern in the direction of interference. Moreover, the regional power enhancement of the communication satellite Intelsat series is also achieved by the reconfigurable communication systems, including antennas.

On the demand side, with the cost of satellites becoming more and more expensive, users require that satellites not only meet the use requirements, but also be flexibly configured on demand. Especially for communication satellites, operators rely on flexible services to compete with ground systems, which require flexible on-orbit configurable (reconfigurable) functions of payloads. The antenna of the communication satellite, of course, stands in the breach, so the concept of reconfigurable satellite antenna emerges as the times require.

The reconfigurable antenna was first proposed by Schaubert et al. in the patent application “Frequency-Agile, Polarization Diverse Microstrip Antenna and Frequency Scanned Arrays” in 1983, and then applied in the base station Smart antenna for 3G mobile communication (auto-adaptive variable beam pattern).

R. You et al., *Technologies for Spacecraft Antenna Engineering Design*, Space Science and Technologies, https://doi.org/10.1007/978-981-15-5833-7_12

Currently, the antenna has become a "bottleneck" of high bit rate transmission, and reconfigurable antenna is one of the solutions for the large-capacity users' experience rate for 5G mobile communication (more than 10 times of 4G, upto 500 Mb/s). At present, the research of reconfigurable antenna can be divided into three directions.

12.1.1 Frequency Reconfiguration

In fact, frequency reconfiguration can be realized by switching among different frequency bands. The most basic method is to integrate radiation units of different frequency bands together (layered or horizontal layout), connect them with PIN diodes, and then use different radiation units through the on-off of PIN.

12.1.2 Pattern Reconfiguration

Phased array antenna is the most typical antenna with pattern reconfiguration mode, in which energy distribution in space is changed according to the demand, but its cost is very high. At present, the pattern is basically changed by three methods: phase control system, including using multi-feeds parabolic reflector antenna to achieve the spatial change of energy; adding parasitic branches connected with microwave switches in the unit to control its field distribution, thus achieving the pattern change; changing the shape of the antenna to achieve the pattern change.

12.1.3 Polarization Reconfiguration

The variable polarization is required to improve the anti-jamming ability. The transformation between circular polarization and linear polarization is basically realized by the switch network; and also, (vertical) linear-(horizontal) linear polarization transformation can be realized.

For the antenna reconfiguration methods mentioned above, the satellites after Intelsat V have been involved, but these terms have not been specifically mentioned. C-band regional beam coverage of the antenna can achieve an "equal gain" in the irregular continental region, then the EIRP adjustment of the area can be implemented with the adjustable gain of the transponder (generally within 6 dB) to achieve the regional enhancement required by user.

In the 1990s, Clarricoats et al. from Queen Mary College, University of London, supported by ESA, have studied the pattern reconfigurable antenna of the reflector for 10 years and carried it on the successor satellite of the European communication satellite Eutelsat-2. Over the past decade, ESA's European Space Research and Technology Centre (ESTEC), European Thales Corporation, the Space Department of Lockheed Martin Space Systems Company (LMT) in the United States and the NHK Science and Technology Laboratory in Japan have invested a lot of efforts in the in-depth study of reconfigurable antennas for satellites.

The basic principle of reconfigurable antenna studied by Clarricoats et al. is to change the shape of reflector with a large number of actuators for realizing dynamic-shaped reflector, so as to achieve reconfigurable pattern. Its prototype is an offset parabolic mesh antenna with a diameter of 0.85 m, Gold-plated molybdenum wire mesh and fifty-two excitation points. The step length of one pulse at each excitation point is changed to 0.025 mm. The antenna works at 10 GHz.

In 2011, Rodrigo from Cataluny University in Spain and Damgaci from Utah University in the United States proposed a reconfigurable antenna scheme [1], in which parasitic elements are added near radiation array units. Each parasitic element consists of a very small grid and an electronic switch. The parameters of the parasitic element are changed with the voltage of the electronic switch, thus changing the pattern. The prototype test of this scheme can change 10% of the working frequency band and the pattern change of $\pm 30°$.

A reconfigurable antenna with UWB tightly coupled array has been developed by the Beijing Institute of Spacecraft System Engineering (BISSE). The frequency and pattern of the antenna can be changed by switching the feeder network. As shown in Fig. 12.1, the pattern can be reconstructed within $\pm 45°$ in the range of 1–5 GHz, and VSWR can be less than 2 in this band.

The Grigori reflector reconfigurable antenna of Thales Corporation in Europe operates in Ku-band, and the shape of reflector is changed with a large number of actuators for realizing pattern enhanced (~+3 dB) [2].

In Japanese reconfigurable antenna scheme, the beam change and regional enhancement are achieved by using phased feed array. Its prototype is a Grigori parabolic reflector with a diameter of 1.8 m. It works at 21 GHz and shapes with 32 feeds for Japan's territory [3].

(a) (b)

Fig. 12.1 UWB tightly coupled array antenna. **a** Radiation aperture; **b** feed network

For the antenna, normal area communication and small area enhancement (3 dB) are realized by using power weighting and phase weighting. The normal beam has 36 dB edge gain for full Japanese and 40 dB gain for island. After reconfiguration, the edge gain is still 36 dB and the key region gain is 43 dB. The antenna achieves the design goal well.

The research on reconfigurable metasurface began in the 1990s. Parker et al. loaded PIN diodes on passive metasurface (Fig. 12.2) to switch the transmission and reflection of incident electromagnetic wave by metasurface in a specific frequency band. In recent years, some scholars combine reconfigurable metasurface with antenna to realize beam scanning and beam width control of antenna, aiming at the adjustable reflection amplitude/phase characteristics of reconfigurable metasurface. Reconfigurable metasurface is an emerging topic in the field of electromagnetic research, whose research started late in China, but it has developed rapidly in recent years. For example, Xu Jinping studied the reconfigurable metasurface loaded by PIN diode and its application in radar absorbing materials, which is insensitive to incident angle and polarization mode of electromagnetic wave. Jiang Jianjun studied the working mechanism and design method of reconfigurable metasurface used in radar absorbing materials and produced the absorbing structure that dynamically adjusts the absorbing performance by using the bias voltage of PIN diode. Zhang Liang studied an omnidirectional electric scanning antenna with reconfigurable metasurface based on a varactor. Qi Kainan et al. studied the optimal design of reconfigurable metasurface loaded by PIN diode using genetic algorithm.

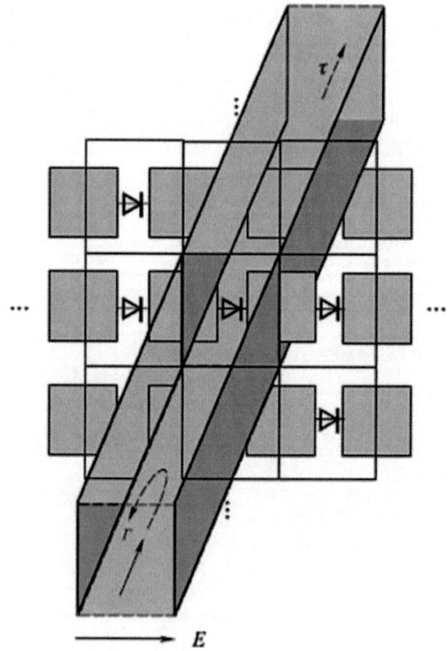

Fig. 12.2 Analysis model for reconfigurable metasurface loaded diode

The satellite-borne reconfigurable antenna is quickly developed under the increased market demand. Its technological theoretical basis is relatively mature, and the devices, such as MEMS, micro-motor, and highly integrated receiving module, are relatively new. However, great progress has been made in recent years. The development of satellite-borne reconfigurable antenna is characterized by demand traction and integration innovation, and oriented to the main markets such as communication and broadcasting satellite. Especially as a supplement to ground mobile communication, satellite communication has great advantages in desert, Gobi, ocean, polar and other sparsely populated areas, but these areas cannot be designed to cover all the time considering the cost, so reconfigurable antenna is preferred, which is another sharp weapon to expand the satellite market.

12.2 Multi-band Reflector Antenna

From the perspective of ground station and inter-satellite link applications, advanced multi-frequency reflector antenna is one of the development directions in the future. The advantages of this antenna include the reduced reflectors number, weight, and cost. The technical difficulties are mainly reflected in the reflector design, the feed design supporting multi-band, the feed line module design supporting separation of multi-band and better isolation, etc.

Traditional reflectors generally support single or dual bands. Although the performance of the parabolic reflector is independent of the frequency, the feed limits the bandwidth. In addition, the bandwidth of the reflector becomes narrow band application for the shape of the reflector designed for shaped beam. Figure 12.3 shows the dual-frequency antennas with two traditional methods. The first method uses frequency selective surface as a sub-reflector to separate high and low-frequency signals. Frequency selective surface reflects high-frequency signal, so high-frequency signal is realized by dual-reflector system. Low-frequency signal

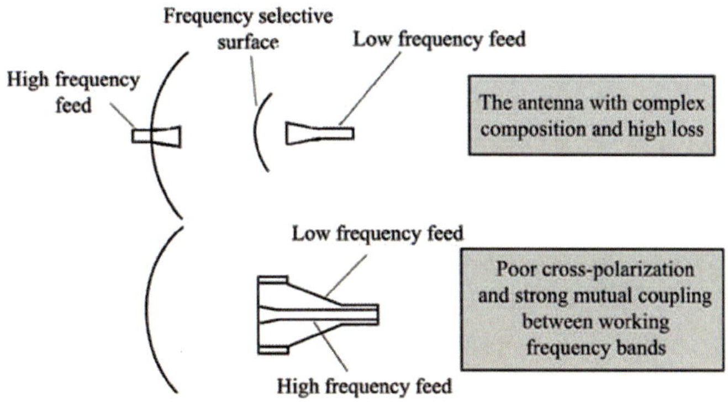

Fig. 12.3 Dual-band antenna using traditional reflector

transmits through frequency selective surface, so single reflector system is used in low-frequency band. The frequency selective surface used here utilizes the low-pass characteristic, which is realized by square ring, cross dipole or cross ring printed on the sub-reflector. This technique has been applied to the Cassegrain antenna Sailor S/X. The weakness of this design is loss of frequency selective surface, long coaxial connection of low-frequency band feed, two feeds and complex frequency selective surface.

The second method is realized by dual-band feed. The central waveguide is used in the high-frequency band and the signals in the low-frequency band are transmitted by the coaxial structure. This kind of structure requires two bands to be far enough, and high-frequency band passes through external coaxial structure, in which the coaxial mode is suppressed by filter or absorbing material. Its disadvantage is that the coaxial propagation mode has poor cross-polarization in low-frequency band, which is unacceptable for satellite applications and most ground applications.

The research progress of multi-band high-performance reflector antennas mainly includes dual-band antennas with single feed, dual-band antennas with multi-feed, multi-band antennas, stepped reflector antennas, antennas with full reflector and partial reflector.

With the increasing complexity of satellite payload and the development of high-power satellites (DC power upto 12–20 kW), antennas should work in more frequency bands at the same time. In addition, as satellites are equipped with the government and commercial payload, the cost reduction of the multi-band antennas is urgently needed. The following will mainly introduce multi-band feed technology based on the development of multi-band antennas.

12.2.1 Dual-Band Multipolar Feed

In 2010, Garcia et al. designed a C-band dual-frequency dual-circularly polarized feed based on the Tx/Rx circularly polarized Ortho-Mode Transducer (OMT) [4], as shown in Fig. 12.4. The specifications of the antenna are shown in Table 12.1.

In 2016, the ISSE of China Academy of Space Technology designed an K/Ka UWB dual-band dual-circularly polarized feed, as shown in Fig. 12.5. The feed operates at frequencies of 20 and 30 GHz and is used in a high-gain antenna for future communication satellite.

12.2.2 Triple-Band Multipolar Feed

In 2008, Rao et al. designed an antenna system supporting multiple frequency bands and multiple beams [5], which used a single antenna to simultaneously implement three frequency bands, covering 20, 30 and 45 GHz. The antenna has an offset feed reflector with a diameter of 1.1 m, a focal length of 1.4 m and a biasing distance of

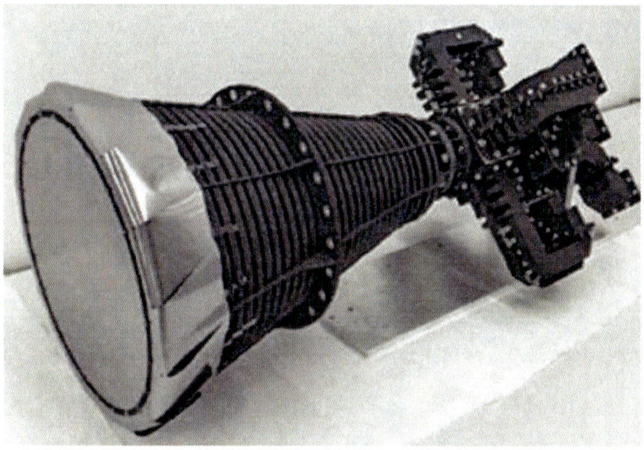

Fig. 12.4 C-band dual-band dual-circularly polarized feed antenna [4]

Table 12.1 C-band dual-band dual-circularly polarized feed antenna specification

Project	Transmitting	Receiving
Frequency (GHz)	3.625–3.945	5.850–6.170
Polarization	RHCP and LHCP	RHCP and LHCP
Irradiation angle (°)	±25	±25
Cross-polarization (dB)	<−35	<−35
Edge taper (dB)	>14.5	>20
Insertion loss (dB)	<0.3	<0.3
Standing wave ratio	<1.15	<1.15
Power capacity (W)	8 × 40	–
Receiving PIM (dBm)	−144 (two Tx tones of 40 W each)	

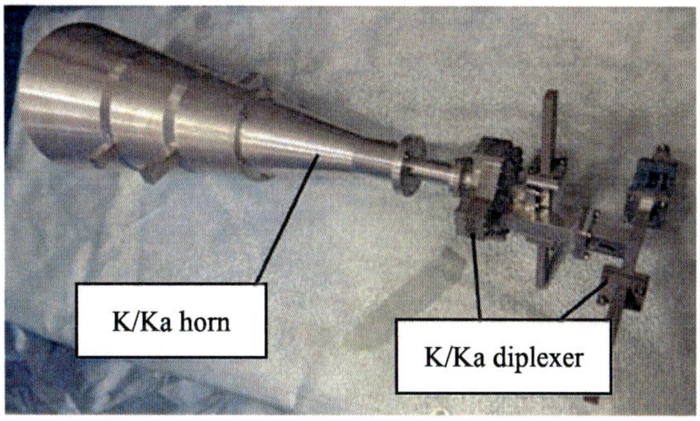

Fig. 12.5 X/Ka UWB dual-band circularly polarized feed antenna

Table 12.2 Key performance indices of K/Ka/EHF triple-band dual-circularly polarized antenna

Item	EHF transmitting	K transmitting	Ka receiving
Working frequency (GHz)	43.5–44.0	19.4–21.2	29.2–31
Polarization mode	LHCP and RHCP	LHCP and RHCP	LHCP and RHCP
Radiation pattern (°)	≤17.6	≤17.6	≤17.6
Cross-polarization (dB)	<−20	<−30	<−30
Insertion loss (dB)	<0.4	<0.3	<0.4
Return loss (dB)	≤−20	≤−20	≤−20
Port isolation (dB)	NA	>20	>20
Extra-band inhibition (dB)	>40	>40	>40
Power capacity (W)	NA	LHCP: 4 × 40 RHCP: 4 × 40 + 4 × 7.5	NA
Inter-modulation at 2 × 100 W	<−125 dBm in the 11th order	NA	<−125 dBm in the 4th order

0.45 m. It provides a 1° beam at 20 and 30 GHz and a 0.5° beam at 45 GHz. The feed is implemented by a triple-band multi-flare-angle horn.

In order to improve the overall performance of the reflector antenna in triple bands, the feed aperture is moved by 76 mm toward the reflector, because the phase center of the horn at 45 GHz frequency is inside the horn with the distance from the reflector of about 76 mm. The performance of the antenna is improved by changing the feed displacement. The cross-polarization is less than 20 dB, and the triple-band antenna obtains a good consistent pattern. Compared with the separated antenna, it has outstanding advantages.

In 2015, Ravanelli et al. designed two kinds of feed antenna topologies to meet the requirements of satellite communication for K/Ka/EHF triple-band dual-circularly polarized antenna [6]. Its specific indices are shown in Table 12.2, performance was very good.

12.2.3 Five-Band Feed

The classical ocean exploration microwave radiometer works in five frequency bands: 6.925, 10.7, 18.7, 23.8 and 37 GHz. CAST Xi'an developed a five-band reflector antenna with a wideband horn.

The five-band corrugated horn is shown in Fig. 12.6. Five-band performance of the antenna is shown in Table 12.3. Return loss is better than −21 dB, Cross-polarization is better than −25 dB and main beam efficiency is better than 90% in five bands [7].

Fig. 12.6 Configuration of five-band horn [7]

Table 12.3 Five-band reflector antenna performance

Frequency (GHz)	3 dB beamwide ° (Min)	3 dB beamwide ° (Max)	Main beam efficiency (%)
6.6	1.73	1.78	89.4
10.7	1.12	1.13	91.6
18.7	0.76	0.81	93.0
23.8	0.70	0.74	93.6
37.0	0.71	0.84	95.0

Hsu et al. designed a five-band multi-flare Horn, which contains seven different slopes [8]. The five-band antenna can support three kinds of satellite services, namely, Direct Broadcast Satellite (DBS), Reverse DBS (RDBS) and Personal Communication Satellite (PCS). With the approval of the RDBS band in the Federal Communications Commission (FCC), the future satellite services will be developed into RDBS, DBS and PCS integrated services.

With the development of communication satellites, there are many research achievements on high efficiency, low cost, lightweight and miniaturized multi-beam antennas in the United States and Europe. With the development of multi-function satellites, an antenna is required to achieve multi-function and multi-band work, which poses new challenges for multi-band reflector and is also one of the hot directions for the development of reflector antenna in the future. Many countries have conducted in-depth research on high-performance feeds operating in two or more different frequency bands and have obtained more research results. Some high-performance feeds have already verified by the space test.

12.3 Membrane Phased Array Antenna

With the rapid development of phased array radar and communication system, especially satellite-borne phased array radar based on space-based platform, in addition to large aperture and light mass, antenna also has the foldable and deployment feature, which should be folded to meet the platform requirements of small envelope volume when launching into orbit, and be deployed to a large aperture when working on orbit. At present, satellite-borne payload system with upto 700 m^2 phased array antennas should provide beam with high gain and wide-angle scanning. However, the mass area density of traditional phased array antenna technology is as high as 8–15 kg/m^2 (mainly including antenna array, electronic devices, structure, deployment servo mechanism, etc.), with large volume and high cost. It cannot meet the requirements of rocket carrying capacity and storage. Flexible membrane phased array antenna inherits the excellent performance of active phased array antenna and has the advantages of ultra-lightweight, small folded volume, low cost of fabrication and launching, so it becomes the development direction of the next generation of large satellite-borne phased array antenna.

Due to the advantages of membrane antenna, such as large aperture, lightweight and small folded volume, a lot of research on membrane antenna has been carried out by NASA and ESA since 1980s, but mainly concentrated on inflatable multi-layer microstrip membrane antenna and inflatable microstrip reflective array membrane antenna. These two kinds of membrane antennas cannot meet the application requirements of satellite-borne radar antenna because of no wide-angle beam scanning.

In 2003, JPL Laboratory in the United States, with the support of NASA, began to study membrane phased array antenna to obtain high gain and wide-angle beam scanning capability. JPL Lab has proposed the Global Earthquake Satellite System (GESS) scheme, in which 10 Geosynchronous Synthetic Aperture Radar (GEO SAR) satellites are distributed on five orbital planes, and two of them with the same ground track constitute a Differential Interferometric SAR (D-InSAR) system to be mainly used for seismic observation around the world with its imaging ability for most regions of the world within 10 min.

The key equipment of GEO SAR is large aperture, lightweight and low-cost active phased array antenna, which works in L-band with 30 m antenna aperture and two-dimensional beam phased scanning capability.

12.3.1 JPL Membrane Phased Array Antenna

In 2005, JPL developed an antenna structure with T/R modules. The prototype has an early form of large aperture membrane antenna array, which is composed of 2 × 4 L-band microstrip antenna elements. The frame is supported by rigid structure. T/R module is mounted on the bottom layer of membrane antenna. The necessary feeder network and control circuit are also constructed on this layer, so that the antenna has the characteristics of phased beam scanning.

T/R module is a hard nut in the design of the membrane phased array antenna. In order to maintain the flexibility of the thin film array, a small flexible T/R module is developed. By using PCB board made by flexible dielectric material, electric fitting by surface-mounting method, and shielding only some part of T/R module, not only the miniaturization and flexibility of T/R module can be realized, but also the crosstalk and self-excitation of radio frequency signal can be effectively prevented.

In 2011, JPL successfully developed a large membrane phased array antenna with 16 × 16 array and 2.3 m × 2.6 m in size [9]. Each four thin film antenna elements form a sub-array through a power division network, and then feed and amplitude and phase of the sub-array are controlled through a T/R module. A total of 64 miniaturized T/R modules are used to control the amplitude and phase of the 16 × 16 array, and the beam of the antenna scans at ±30°. A miniaturized rigid T/R module is designed by JPL. Compared with the flexible T/R module developed in 2005, the volume and mass of the T/R module have been dramatically reduced, but the power utilization and performance have been significantly improved.

12.3.2 Research Progress in China

Some representative achievements have been made in the research of large-scale membrane antenna. The research of membrane antenna in China is still in its infancy, and there is no mature report of system application. Further research can lay a foundation for the application of large-scale deployable membrane phased array antenna.

In 2012, China Electronic Science and Technology Group Institute No. 38 developed a membrane phased array antenna with a P-band 4 × 8 array, as shown in Fig. 12.7. The antenna is composed of a 4 × 8 array membrane array antenna and 8 T/R modules with a size of 10 cm × 5 cm. The test results of the membrane phased array antenna in a microwave anechoic chamber are as expected.

Aiming at the application requirement of satellite-borne SAR antenna, the BISSE of China Academy of Space Technology has carried out the related research work of membrane phased array antenna, completed and tested the development of membrane antenna prototype (Fig. 12.8) with the one-dimensional beam scanning ability, and its electrical performance test results met the design requirements.

Fig. 12.7 Membrane phased array antenna with P-band 4 × 8 array [10]

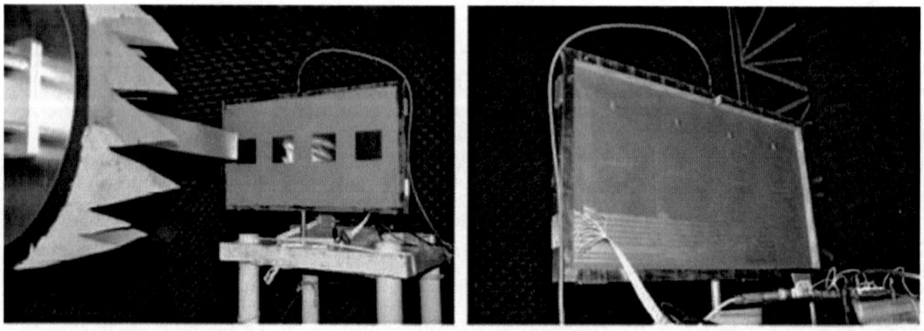

Fig. 12.8 Principle prototype of membrane antenna. **a** Near-field test status; **b** backside of feeder network layer

During the Twelfth Five-Year Plan period, the Beijing Institute of Spacecraft System Engineering (BISSE) carried out the research on the deployment structure of large-scale membrane phased array antenna for the application of satellite-borne SAR antenna, and developed the membrane structure principle prototypes with unfolded areas of 1 m × 3.5 m and 3 m × 11 m, as shown in Fig. 12.9.

From the current situation and trend of membrane antenna development, it can be seen that:

1. The membrane phased array antenna is the core technology to break through the development bottleneck of large-scale satellite radar antenna

Membrane phased array antenna is a kind of phased array antenna integrated by miniaturized T/R module and membrane antenna, with the large aperture, lightweight and small folded volume, which can make up for the shortcomings of the existing phased array antenna. It is regarded as the core technical means of breaking through the development bottleneck of the large satellite-borne SAR antenna by NASA and DARPA.

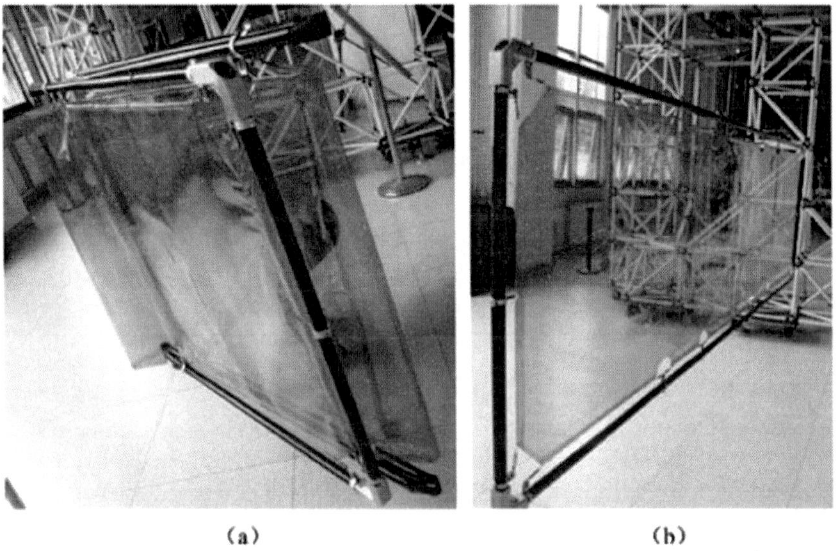

Fig. 12.9 Membrane structure prototype with unfolded area of 1 m × 3.5 m. **a** Before deployment; **b** after deployment

2. Membrane phased array antenna has wide application prospects

With large aperture, lightweight, high gain and wide-angle beam scanning, membrane phased array antenna can be used on the carrier platform, such as satellite and airship. Therefore, it has wide application prospects in radar imaging, ground target recognition and air moving target recognition.

3. The implementation of membrane phased array antenna faces many key technical challenges

Since the membrane phased array antenna is integrated into the flexible membrane antenna by miniaturized T/R modules, its implementation faces many technical challenges, such as membrane array design technology, miniaturized T/R module design technology, membrane array maintenance technology, array deployment technology and thermal control technology.

12.4 Application of MEMS in Antenna

Micro-Electromechanical System (MEMS) is an emerging multidisciplinary-crossed discipline involving precision machining, microelectronics materials science, as well as micro-fabrication process, system and control.

The technology combination of the MEMS and the antenna mainly includes the small antenna and the high-frequency antenna-oriented MEMS process, and the application of the RF MEMS devices in communication, such as the application of the MEMS phase shifter in the phased array antenna.

12.4.1 Brief Introduction to MEMS Machining Process

High machining precision and three-dimensional machining of the MEMS makes it possible that the design is not limited to the plane. Millimeter wave and terahertz antennas have increasingly higher requirements on processing accuracy with the increase of frequency. However, traditional mechanical processing faces many technical problems, such as gap size control, deformation control, burr cleaning, assembly accuracy assurance, welding accuracy and deformation assurance. The difficulty of traditional technology limits the ability of flexible antenna design to obtain the best operating characteristics. The emergence of MEMS technology makes antenna design break through the limitation of traditional processing technology to a certain extent.

MEMS technology is developed on the basis of micro-electronic integrated circuit technology. Because microelectronics technology is a kind of planar technology, it is difficult to fabricate three-dimensional structure of MEMS. In order to achieve high depth to width ratio of three-dimensional micro-fabrication, the methods such as LIGA and laser processing have been developed through the cross-penetration of multi-disciplines. In addition, a series of special technologies must be used to construct various special structures of the MEMS, including bulk micro-fabrication technology, surface micro-fabrication technology, LIGA technology, photolithography technology and three-dimensional non-silicon micro-fabrication technology.

12.4.1.1 Comparison of MEMS Process with Other Processes

1. Comparison of MEMS process with IC process (silicon processing)

In silicon processing technology, the main difference between the MEMS process and IC process is that the main goal of IC process is to achieve electrical performance, and the mechanical and electrical performance can be realized simultaneously in the MEMS process. The differences between the two processes in the selection of specific processes are shown in Table 12.4.

2. Comparisons of MEMS technology with machining technology (non-silicon machining)

In non-silicon fabrication, compared with the traditional mechanical fabrication, the MEMS technology has more advantages in processing accuracy, micro-fabrication and three-dimensional fabrication, especially for the fabrication of small antennas and millimeter wave/terahertz band antennas. Traditional mechanical processing is facing

Table 12.4 Comparison of MEMS with IC processes

Process type		MEMS process	IC process
Lithography technology		Double-sided lithography technology	One-sided lithography technology
Corrosion technology	Dry process	Deep and high depth to width ratio corrosion	General film corrosion
	Wet process	Anisotropic corrosion, self-stopping technology, deep silicon corrosion	Isotropic corrosion, anodic corrosion, electropassive corrosion, limited to surface processing
Sacrificial layer technology		Surface micro-fabrication process, compatible with IC process, for manufacturing surface active structures	Not commonly used
Bonding		Silicon/silicon direct bonding, silicon/glass anodic bonding	Silicon (SOI) material on insulator by high-temperature bonding
LIGA		High depth to width ratio structure, high cost	Not in use

some difficulties in ensuring the processing accuracy of high-frequency antenna, while the MEMS technology breaks through the limitation of traditional processing technology to a certain extent, making it easier to ensure that the processing accuracy of high-frequency antenna reaches the sub-wavelength level.

12.4.1.2 Selection of the Antenna Machining Oriented MEMS Process

In order to meet the processing requirements of miniature high-frequency (millimeter wave/terahertz) antennas, the MEMS technology provides a selective advantage for ensuring the processing accuracy. Especially, the general micro-fabrication technology of non-silicon membrane materials basically overcomes the key technical problems of integrated manufacturing of various non-silicon materials and makes it possible to fabricate small antenna oriented to complex structure design. When facing different thickness or multi-layer superposition processing, the selection criteria of the MEMS process are as follows:

1. For thin-layer microstructures with thickness less than 0.2 μm, LIFT-OFF or mask electroplating process can be used to ensure 1–2 μm machining accuracy.
2. For several micron-thick micro-structure layers, it is better to choose the mask electroplating process. When the mask electroplating cannot be realized, the universal processing flow is preferred to ensure the processing accuracy.
3. For thicker layers with tens of microns thicker thickness, direct lithography and general processing technology are inevitable choices. However, the requirement of the machining accuracy of 1–2 μm is unrealistic.

4. The multi-layer superposition technology can provide a feasible way to improve the processing accuracy of thicker microstructures, but it also puts forward higher requirements for lithography equipment and techniques.

12.4.2 MEMS Phase Shifter

12.4.2.1 Overview

The technology of RF MEMS aims to produce the RF devices with the size of micron to millimeter. Compared with traditional RF devices, RF MEMS devices are not only smaller in size and easier to integrate with monolithic circuits, but also better in performance (such as linearity, broadband, low power consumption and low insertion loss). They can replace traditional PIN diodes and similar ferrite products.

DARPA provided support to the first RF MEMS projects in the 1990s, aiming at phase shifters and microwave switches for high-performance phased array radar for anti-missile applications. Compared with the phase shifter realized by PIN diode and GaAs FET electronic control switch, it has the advantages of low loss and driving power (microwatt level). The microwave parallel switch developed by the company Raytheon represents the highest level nowadays. Typical performances are that driving voltage is 30 V, switching delay is 10 μs, insertion loss is less than 0.20 dB at 30 GHz, isolation is greater than 40.00 dB and switching length is only 0.1 mm. Typical performances of microwave series switch are that driving voltage is 30 V, switch delay is 10 μs, insertion loss is less than 0.20 dB at 30 GHz, isolation is greater than 50.00 dB, and contact area only is 0.2 mm $\times$ 0.2 mm. Compared with the existing PIN and FET electronic switch, the advantages of microwave switch are low insertion loss, low driving power consumption and monolithic integration. In order to deploy the phased array radar to space for anti-missile, the satellite-borne test of microwave switch was carried out in 1998 in the United States. A pair of tethered "nanosatellites" launched in 2000 also used microwave switches to switch radio frequencies.

The principle, architecture, technology and performance of the traditional phase shifter have been changed dramatically when the MEMS technology is applied to the design and processing of the traditional phase shifter, resulting in a new type of MEMS phase shifter. The performance comparison of the MEMS phase shifter with the traditional phase shifter is shown in Table 12.5.

From Table 12.5, a conclusion can be reached that the MEMS phase shifter has obvious advantages in terms of cost and DC power consumption. It also has good reliability and low RF loss. However, the current processing technology of the MEMS is not mature enough, which results in a larger size of the MEMS phase shifter. Therefore, how to miniaturize the phase shifter is one of the key issues in the design.

Table 12.5 Performance comparison of various phase shifters

Type	Ferrite phase shifter	Solid-state phase shifter	MEMS phase shifter
Cost	Very high	High	Low
Reliability	Best	Good	Satisfied
RF loss	X-band (<1 dB/360° s)	Ka-band (2 dB/bit)	Ka-band (2.3 dB/360° s)
DC power consumption	High (10 W)	Low (<10 mW)	Ignored

Table 12.6 Performance comparison of MEMS phased array with reflector antenna and traditional phased array antenna

Item	Reflector antenna	Traditional phased array antenna	MEMS phased array antenna
Pattern performance	The pattern characteristics depend on the antenna array, which needs to be guaranteed by the design and processing of the antenna array	It is easy to realize the pattern characteristics of different demands by weighting the array	It is easy to realize the pattern characteristics of different demands by weighting the array
Beam switching speed	Slow (depending on servo or satellite attitude adjustment)	Fast (currently more than 100 ns of beam switching time can be achieved)	Faster (can achieve 20 μs of beam switching time)
Power waste	Low (RF loss can be neglected)	High (T/R module efficiency is generally less than 30%, RF heat consumption is more)	Lower
Reliability	High	The system is complex, there are many electronic devices, and the reliability is not easy to guarantee	Higher
Cost	Low	High	Lower

12.4.2.2 Types of MEMS Phase Shifters

According to the structure characteristics, the MEMS phase shifters can be roughly divided into three categories: switching line type, distributed type and reflective type. Switch-line MEMS phase shifter is mainly composed of transmission line and MEMS switch in series and parallel. Distributed MEMS phase shifter is composed of periodically loaded MEMS metal bridges in Coplanar Waveguide (CPW). Reflective

MEMS phase shifter is also composed of transmission lines and switches in series and parallel and is connected with switch-line MEMS phase shifter, and the difference with the switch-line MEMS phase shifter is a coupler added.

The biggest advantage of switch-line MEMS phase shifter is that it can achieve broadband and low loss in parallel with simple structure, high phase shifting accuracy and high isolation [11, 12]. However, the processing technology of the MEMS switch which constitutes the phase shifter is facing the higher requirements. Distributed MEMS phase shifter features miniaturization, low loss and high integration. It can be used in millimeter wave band, but its phase shifting accuracy is poor. Reflective MEMS phase shifter has the advantages of smaller size, wider frequency band and no requirement on impedance matching. However, in the high-frequency band, the loss of coupler and switch will increase, which leads to the increase of the loss of reflective MEMS phase shifter.

12.4.2.3 MEMS Phased Array Antenna

At present, large-scale antenna is basically in the form of reflector or phased array. The phased array antenna based on the MEMS phase shifter can make a certain application space between them. The characteristics comparison of the MEMS phased array antenna with the reflector and the traditional phased array antenna is shown in Table 12.6.

The MEMS phased array belongs to array antenna in system. It has the advantage of array antenna, which can realize beam scanning by weighting the array. However, since the mechanical response of the micro-electromechanical structure is slower than that of the electromechanical structure, the response time of the beam switching of the micro-electromechanical phased array is longer than that of the traditional electro-scanning phased array. Therefore, the MEMS antenna is not suitable for military phase-controlled communication with nanosecond beam agility requirements, but its microsecond beam switching time can meet the needs of civil communications, SAR beam scanning and other applications.

In addition to the analysis and comparison of the above-mentioned principles, in terms of technology maturity, reflector antenna and traditional phased array antenna start early and their technologies are relatively mature; although the development of micro-electromechanical technology is currently a hot spot, it mainly focuses on the field of micro-electromechanical sensors and inertial units of MEMSs, and radio frequency micro-electromechanical devices are still being developed. For example, MEMS phase shifter still has a certain gap in phase bits and phase shifting accuracy. The RF MEMS technology products are relatively not rich and mature, thus restricting the technical specifications of MEMS phased array to some extent.

Table 12.7 THz location in electromagnetic spectrum

Frequency (Hz)	10^5–10^8	10^9–3×10^{11}	~3×10^{12}	~10^{14}	3.75×10^{14}–7.5×10^{14}	10^{15}–10^{16}	>3×10^{17}
Range	Radio/TV	Microwave	Terahertz	Infrared	Visible	Ultraviolet	X/γ ray

12.5 Application of Terahertz Technology in Antenna

12.5.1 *Terahertz Wave Definition and Characteristics*

The combination of radio wave and aerospace technology is like the icing on the cake, the communication satellite, navigation and positioning satellite, remote sensing satellite, and meteorological satellite greatly improved the quality of human life. Lunar exploration, lunar landing, deep space exploration and space exploration have broadened the scientific horizon of mankind. The radio wave has become an indispensable part of human life and scientific exploration. However, in the context of this prosperity, Terahertz-THz Wave at 300–3000 GHz (0.3–3 THz corresponding wavelength is 0.1–1 mm) is still a less developed area, namely, sub-millimeter wave for most radio scientists and engineers. Table 12.7 shows the location of terahertz in the electromagnetic spectrum, it can be seen that the THz wave is between microwave and far infrared.

The single photon energy in terahertz is 1.2–12.4 meV and its equivalent black body temperature is 14–140 K. The feature that this energy band is generally consistent with the vibrational magnitude of some light molecules and atoms can be used for spectral analysis, detection and recognition of the thermal radiation and absorption characteristics of these light elements. This is also the principle of most terahertz detectors.

Due to the short wavelength, the great attenuation in the air, the interaction with water, oxygen and carbon molecules, terahertz waves show obvious absorption lines at several frequencies, i.e., the maximum attenuation point [13, 14]. Table 12.8 is shown the main absorption spectrum lines of THz in atmosphere. Using this characteristic, the content of water, oxygen, carbon, ozone, methane and other components in gas can be detected, and air pollution, atmospheric change, meteorological information, etc. can be indirectly detected.

12.5.2 *Application of Terahertz Technology in Space Field*

The first high-end microwave (millimeter wave) was used in space on the U.S low-orbit meteorological satellite (NOAA), whose microwave radiometer (SSM/T) is still in use. Three channels at 183 GHz are available to measure the vertical profile of

Table 12.8 Main spectrum lines of THz in atmosphere

No	Molecule	Absorption peaks (GHz)
1	H_2O	22.35/183.31/325.15/380.1974/439.15/439.15/445.00/448.00/474.68/488.49/556.936/620.7/658.0/752.0/859.865/916.17/970.315/987.93
2	O_2	49.7/119.75/368.4983/424.76/487.25/715.39/773.84/834.14
3	CO	346/0461/576 …
4	SO_2	362/456/556/662/765/858 …
5	H_2S	740/1000/1290/1610 …
6	NH_3	575/662/825 …

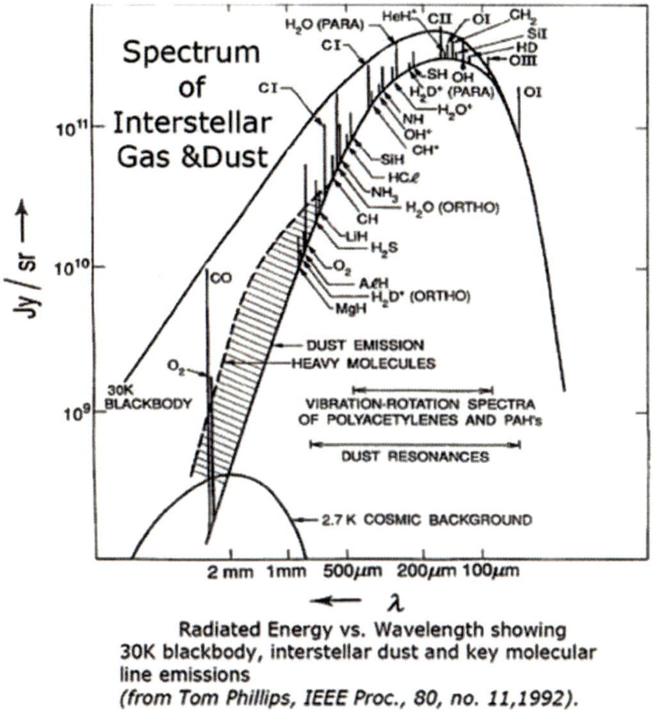

Fig. 12.10 THz spectral lines in cosmic exploration [15]

atmospheric water vapor and one channel at 92 GHz to measure the vertical profile of atmospheric temperature. Terahertz technology is mostly used for space exploration, and the detection materials are shown in Fig. 12.10. As can be seen in Fig. 12.10, a large number of cosmic molecules and dust can be detected by terahertz wave.

Table 12.9 shows typical terahertz space detectors in Europe and the United States. A large number of new phenomena in the universe have been detected by using terahertz waves, which provides strong evidence for the origin theory of the Big Bang of the universe.

12.5.3 Technological Characteristics of THz Antenna

Terahertz detection is mainly used for receiving and detecting, which includes direct detection and heterodyne detection. Direct detection with the characteristics of broadband, high-sensitivity and low-spectral resolution is used in spectrophotometer, cooled bolometer (0.01–4 K), photoconductor (100 GHz–30 THz), etc. Heterodyne detection with the characteristics of narrow band, medium to high sensitivity

Table 12.9 Typical terahertz space detectors in Europe and America

No	Detector	Launch year	Scientific objectives	Terahertz equipment	Major Performance Achievements
1	COBE, NASA	1989	Measuring the cosmic microwave background and infrared background to provide evidence for the big bang theory	Differential microwave radiometer (DMR), 31.5, 53, 90 GHz, resolution 7°	Its data analysis results play a decisive role in supporting big bang theory. Two U.S professors won the Nobel Prize in Physics in 2006
2	Odin, Sweden	2001	Study of galaxy formation/early solar system, ozone layer changes in the upper atmosphere and coupling between the upper and lower atmosphere	Φ 1.1 m antenna, a 119 GHz fixed-frequency heterodyne radiometer, a 495, 549, 555, 571 GHz four-frequency adjustable heterodyne radiometer	First discovery of nebulae containing oxygen molecules
3	Rosetta Orbiter, ESA	2004	Measurement of temperature and electrical characteristics of micrometeoroids and comets	Φ 0.3 m antenna, two heterodyne radiometers operating at 190 GHz (bandwidth 0.5 GHz) and 562 GHz (bandwidth 1 GHz) can measure the contents of H_2O, CH_3OH and CO	Rosetta Orbiter was successfully awakened in January 2014 for detection of the comet 67P/CHC, and a clear image of the small and medium-sized stones was taken. On August 12, 2015, the comet entered 185,000 km nearest to the sun
4	Planck, ESA	2009	High precision measurement of cosmic microwave background (CMB) inhomogeneity	Φ 1.5 m antenna; LFI: 30, 44, 70 GHz; HFI: 100, 143, 217, 354, 545, 850 GHz; the lowest temperature of the bolometer is 0.1 K and $\Delta T/T$ ~ $2 * 10^{-6}$ The cooling system is superfluid He	The dark matter, proportion of dark energy to normal matter was determined accurately. Dark energy was 68.3%, dark matter was 26.8%, normal matter was 4.9%, Hubble constant was 67.3 km/s/MPC

(continued)

Table 12.9 (continued)

No	Detector	Launch year	Scientific objectives	Terahertz equipment	Major Performance Achievements
5	Herschel, ESA	2009	Origin of galaxies/milky way galaxy	Φ 3.5 m antenna; infrared and terahertz, wavelength coverage of 60–650 μm (0.5–5 THz); two infrared spectroscopes (PACS, SPIRE) and one heterodyne terahertz spectrometer HIFI; 6 terahertz channels: 0.4–0.64 THz, 0.6–0.8 THz, 0.8–0.96 THz, 0.96–1.12 THz, 1.12–1.25 THz, 1.41–1.91 THz; sensitivity: 90–650 K The cooling system is superfluid He	Many achievements have been made, such as the discovery of supermassive black holes and the presence of H_2O and CO molecules around the sun-like Helix Nebula (NGC 7293) star at the end of its lifetime

and high-spectral resolution is mainly used in Schottky diode mixers, hot electron bolometer (4 K), superconducting tunnel junction (4 K).

Due to the short wavelength, terahertz antenna is basically an aperture antenna, including dielectric lens antenna and reflector antenna. The large aperture reflector antenna is suitable for space systems because of the long detection distance. Dielectric lens antenna is mainly used for secondary beam focusing. According to the theory of reflector antenna, the shape accuracy of terahertz antenna is 0.01–0.1 times that of centimeter wave antenna under the condition of given phase error, which is basically micron level. Therefore, the commonly used carbon fiber composites and processes are not applicable, only optical reflector materials (such as SiC) and processes can be used, but the absolute aperture of the antenna is much larger than that of light lens, which brings great difficulties to processing. The geometric installation accuracy of the main reflector, the secondary reflector and the feed and other components is also greatly improved, so the requirement on the system debugging is relative d to high. Herschel telescope antenna includes a 3.5 m primary parabolic reflector lens with SiC, its surface accuracy must reach at 6 μm. In order to reduce noise, the whole antenna is in a low temperature of 80 K on orbit. This accuracy is a great challenge to the process. The detector PLANK, which is also launched with Herschel, has a maximum frequency of nearly 1 THz. The temperature of the detector PLANK is 0.1 K, approaching absolute 0 K. The temperature of the next generation of the detector PLANK will reach the magnitude of mK, which is a serious challenge to the antenna, radiometer and space thermal control technology. Rui proposed a method to measure the vertical distribution of atmospheric temperature and humidity in a small area (within 10 km of transverse direction) by using a large aperture reflector with a diameter of more than 3 m, 256 feed arrays and on-chip radiometers (ROS) to predict the approaching (in 1–2 h) of rainfall in a small area (within 10 km of transverse direction). However, it is difficult to realize the scheme of near prediction of rainfall in a small area.

The terahertz antenna is closely related to its detector technology. The terahertz signal source is still a difficult problem, and the most important one is the efficiency problem. At present, the efficiency is less than 5%. If the efficiency can be increased to 20%, the application of terahertz will be greatly expanded, such as non-destructive testing and personnel safety inspection. For antenna, the processing technology is still a difficult problem. For the proximity prediction of rainfall in a small area (10 km) (in 1–2 h), an antenna with a diameter of more than 3 m should be placed on the GEO orbit to detect the water vapor and temperature absorption lines in 183–600 GHz. The proximity prediction of rainfall in a small area can be achieved by numerical calculation of the measured data, but the antenna processing complexity and cost limit its development.

Stated thus, terahertz technology development may have the following trends:

1. Improve the efficiency of terahertz signal generator.
2. Miniaturization of the front end of the normal temperature receiver.
3. Promote of material and technology of large aperture antenna (more than 2 m).

References

1. Rodrigo D, Damgaci Y, Unlu M et al (2011) Antenna reconfigurability based on a novel parasitic pixel layer. In: 2011 5th European conference on antennas and propagation (EuCAP)
2. Mizzoni R, Noschese P, Campana P et al (2014) Antenna technologies for science missions in Thales Alenia Space Italia. In: The 8th European conference on antennas and propagation (EuCAP 2014), pp 3321–3325
3. Charistodoulou CG, Tawk Y, Lane S et al (2012) Reconfigurable antennas for wireless and space applications. Proc IEEE 100(7):2250–2261
4. Garcia R, Mayol F et al (2011) Circular polarization feed with dual frequency OMT-based turnstile junction. IEEE Antennas Propag Mag 53:226–236
5. Rao SK, Hsu C-C, Chan KK (2008) Antenna system supporting multiple frequency bands and multiple beams. IEEE Trans Antennas Propag 56(10):3327–3329
6. Ravanelli R, Cecchini P, Mizzoni R et al (2015) A K/Ka/EHF feed chain for dual-use telecom. In: 9th European conference on antennas and propagation (EuCAP)
7. Cao D, Ye C, Lv Q (2016) Design of five bands horn for wide band microwave radiometer. Space Electron Technol (1):61–65
8. Hsu C-C, Rao SK (2008) Advanced multiple beam antenna system supporting multiple satellite communications services. In: Asia-Pacific microwave conference, APMC 2008

9. Moussessian A (2011) Large aperture, scanning, L-band SAR (membrane-based phased array). In: 2011 earth science technology forum, NASA, JPL
10. He X (2012) Development of flexible active array antenna T/R components. Silicon Valley 11:66–67
11. Psychogiou D, Li Y, Hesselbarth J, Kühne S, Peroulis D, Hierold C, Hafner C (2013) Millimeter-wave phase shifter based on waveguide-mounted RF-MEMS. Microw Opt Technol Lett 3:465–468
12. Anitha G, Usha Kiran K (2018) Miniaturized switched line MEMS phase shifter. Microw Opt Technol Lett (6):1526–1531
13. Rui Y, Jianlou Z (2011) High resolution precipitation sensing in GEO orbit using multi-beam radiometer of millimete wave. In: 62nd international astronautical congress 2011, IAC 2011
14. Rui Y, Tao Z, Ni R, Zhang L (2018) Space applications of terahertz present and potential futurre. In: 11th UK-Europe-China workshop on millimeter waves and terahertz technologies (UCMMT), Hangzhou
15. Phillips TG, Keene J (1992) Submillimeter astronomy. Proc IEEE 80:1662–1678

Appendix

A.1 Basic Equation of Wire Antenna

Wire antenna is the oldest and most common antenna form. Wire antenna refers to the antenna composed of wires (such as dipole, symmetric vibrator, asymmetric vibrator, ring antenna, helical antenna, folded vibrator, Yagi-Uda antenna). In the analysis of the wire antenna, the current distribution on the conductor can be accurately assumed, and the impedance characteristics and radiation field in space can be calculated with the current distribution on the conductor. This kind of antenna can be used in ultra-short-wave and microwave frequency bands, and widely used in various radio equipments in other frequency bands.

In order to get the strict solution of the wire antenna, it is necessary to solve the current distribution which satisfies the boundary condition that the electric field is equal to zero along the tangent of the wire, so as to derive an integral equation which satisfies the current distribution—Hallen integral equation or Pocklington integral equation. These classical solutions are lengthy and limited to the shape with a few simple wires. The numerical method is effective and convenient for solving the integral equation and can be applied to the shape of many wire antennas.

A.1.1 Current Distribution and Radiation Field of Symmetrical Dipole

Symmetrical dipole is the simplest form of wire antenna (Fig. A.1), which consists of two wires or metal bars of the same size. An alternating voltage source is connected to the two corresponding ends of the two arms to excite the dipole at the center, thus generating current on the two arms. In the figure, L is the length of the dipole, and

R. You et al., *Technologies for Spacecraft Antenna Engineering Design*, Space Science and Technologies, https://doi.org/10.1007/978-981-15-5833-7

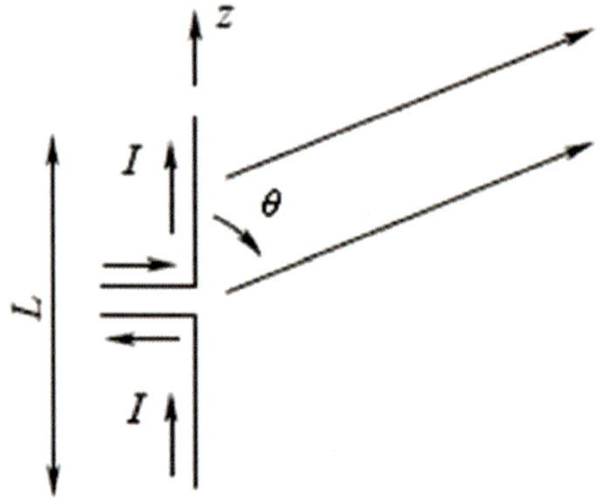

Fig. A.1 Symmetrical dipole antenna

the gap between the middle end points is far less than the wavelength, which can be ignored. The symmetrical dipole can be used as an independent antenna or as a feed of paraboloid antenna. In addition, it is widely used as radiation element of array antenna.

In Fig. A.1, the symmetrical vibrator is oriented along the z-axis, which can be regarded as an open dual-wire transmission line bent from $L/2$ of the end length, assuming that the current distribution of the bending part is not changed, as shown in Fig. A.2. The current distribution on the dipole is a sinusoidal function and can be written as.

$$I(z') = I_m \sin\left[k\left(\frac{L}{2} - |z'|\right)\right] \quad |z'| < \frac{L}{2} \tag{A.1}$$

The current distribution of dipoles with different lengths is shown in Fig. A.3. The current at the end of the vibrator is zero, and sine wave is sent to the feeding point. The distance λ/4 from the end is the maximum value of the current I_m, and the distance λ/2 is zero. The current is reversed every λ/2, and the phase difference is 180°. The input current depends on the dipole length $I_{in} = I_m \sin(kL/2)$. The arrows indicate the direction of the current. The current direction of the corresponding part of the two arms is the same, and the radiation field is enhanced with each other. The direction of the transmission line is opposite, so the radiations almost cancel out.

The vector potential can be expressed as follows:

$$\mathrm{A} = \hat{\mathbf{z}} \frac{\mathrm{e}^{-\mathrm{j}kr}}{4\pi r} \int_{-L/2}^{L/2} I(z') \mathrm{e}^{\mathrm{j}kz'\cos\theta} \mathrm{d}z'. \tag{A.2}$$

Then, the radiation electric field is

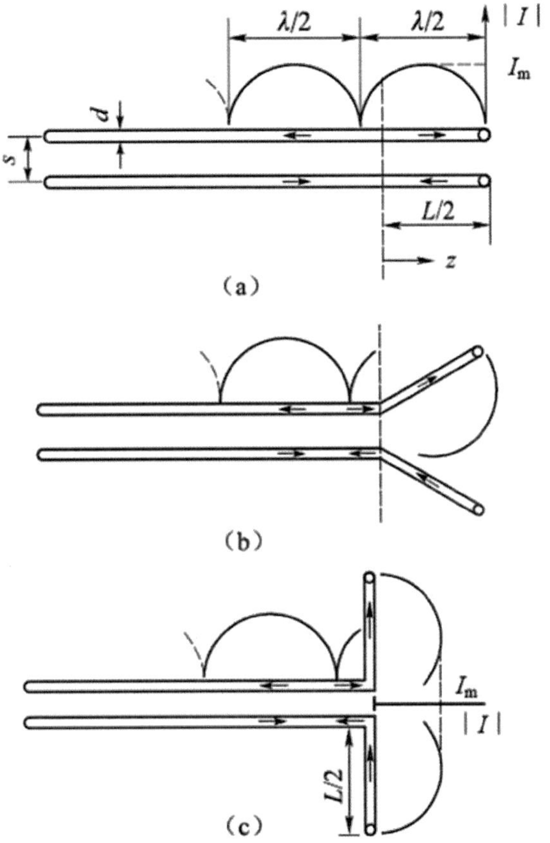

Fig. A.2 Current distribution. **a** Lossless double wire transmission line. **b** The end flares out. **c** Symmetrical dipole

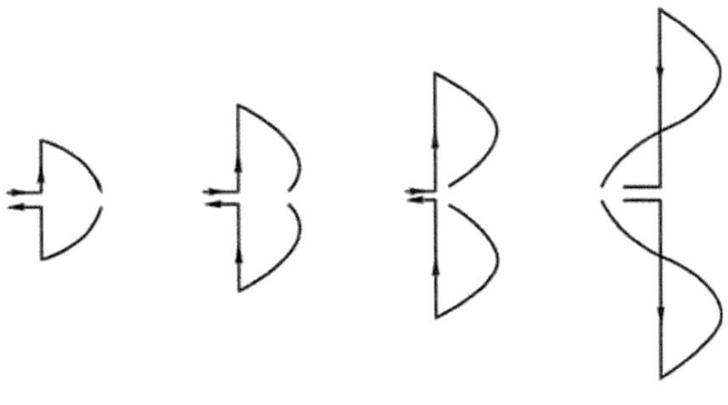

Fig. A.3 Current distribution of dipoles with different lengths. **a** $L = \lambda/2$; **b** $L = 3\lambda/4$; **c** $L = \lambda$; **d** $L = 3\lambda/2$

$$\mathbf{E} = \hat{\theta}\,\mathrm{j}\omega\mu\;\sin\theta\;A_z$$

$$= \hat{\theta}\,\mathrm{j}\omega\mu\frac{\mathrm{e}^{-\mathrm{j}kr}}{4\pi r}\sin\theta\int_{-L/2}^{L/2} I(z')\mathrm{e}^{\mathrm{j}kz'\cos\theta}\mathrm{d}z', \tag{A.3}$$

where $\sin\theta$ is the element factor, which is the pattern function of the basic vibrator located at the origin of the coordinate. Besides, the integration in the above formula is normalized to obtain the spatial factor, which is only determined by the current distribution. The pattern of symmetrical dipole is equal to the product of element factor and space factor.

The symmetrical dipole can also be divided into several basic vibrators with a length of $\mathrm{d}z'$ and a current of $I\ (z')$, and the sum of the radiation fields of all the vibrators is equivalent to the radiation field of the symmetrical dipole. In the limiting case, sum operation should be changed into integral operation. In a sense, a symmetrical dipole can be regarded as a continuous array. The pattern of a continuous array is equal to the product of element factor and space factor, which is called the pattern product theorem of a continuous array.

A.1.2 *The Pattern of Symmetrical Dipole*

Unnormalized spatial factor is calculated as

$$f_{\mathrm{un}} = \int_{-L/2}^{0} I_{\mathrm{m}}\sin\left[k\left(\frac{L}{2}+z'\right)\right]\mathrm{e}^{\mathrm{j}kz'\cos\theta}\mathrm{d}z' + \int_{0}^{L/2} I_{\mathrm{m}}\sin\left[k\left(\frac{L}{2}-z'\right)\right]\mathrm{e}^{\mathrm{j}kz'\cos\theta}\mathrm{d}z'. \tag{A.4}$$

The following integral formula can be used for integral of each term in the formula

$$\int \sin(a+bx)\mathrm{e}^{cx}\mathrm{d}x = \frac{\mathrm{e}^{cx}}{b^2+c^2}[c\sin(a+bx) - b\cos(a+bx)], \tag{A.5}$$

where $a = kL/2$, $b = \pm k$, $c = \mathrm{j}k\cos\theta$,
then, we have

$$f_{\mathrm{un}} = \frac{2I_{\mathrm{m}}}{k\sin^2\theta}\left[\cos\left(\frac{kL}{2}\cos\theta\right) - \cos\left(\frac{kL}{2}\right)\right]. \tag{A.6}$$

Substituting the above formula into Formula (A1.5), and then we have the far-field component

$$E_\theta = \mathrm{j}\omega\mu \frac{\mathrm{e}^{-\mathrm{j}kr}}{4\pi r} \frac{2I_\mathrm{m}}{k\sin\theta}\left[\cos\left(\frac{kL}{2}\cos\theta\right) - \cos\left(\frac{kL}{2}\right)\right]. \tag{A.7}$$

Since $\omega\mu/k = \eta$, the above formula can be simplified as

$$E_\theta = \mathrm{j}\ 60\ I_\mathrm{m} \frac{\mathrm{e}^{-\mathrm{j}kr}}{r\sin\theta}\left[\cos\left(\frac{kL}{2}\cos\theta\right) - \cos\left(\frac{kL}{2}\right)\right]. \tag{A.8}$$

Unnormalized pattern function for symmetric dipoles is

$$f(\theta) = \frac{1}{\sin\theta}\left[\cos\left(\frac{kL}{2}\cos\theta\right) - \cos\left(\frac{kL}{2}\right)\right]. \tag{A.9}$$

With $L = \lambda/2$, the normalized pattern function of a half-wavelength dipole is

$$F(\theta) = \frac{1}{\sin\theta}\cos\left(\frac{\pi}{2}\cos\theta\right) \quad \left(L = \lambda/2\right). \tag{A.10}$$

A.2 Basic Equation of Reflector Antenna

A.2.1 Common Methods for Analyzing the Radiation Characteristics of Reflector Antenna

Two approximate methods were initially proposed and are still commonly used to analyze the radiation characteristics of the reflector antenna, or to design the geometric profile and size of the reflector:

1. Aperture field distribution method is also known as Geometrical Optics (GO), with which the field distribution of the antenna reflector aperture is determined by the feed pattern according to the principle of geometrical optics and the law of conservation of energy, and then the radiation characteristics of the antenna is calculated with the Fresnel Kirchhoff diffraction formula.
2. Mirror current method (induced current method) is also known as Physical Optics (PO). According to the electromagnetic field radiated by the feed on the reflector, the mirror current excited by the reflector is calculated approximately by the PO method, and then the radiation characteristics of the antenna are obtained by the standard radiation integral formula.

Comparing the above two methods, the mirror current method is more accurate than the aperture field distribution method. Both methods are applicable to the assumption that the plane wave is projected onto the infinitely large ideal conductor plane, but in fact the radiation field of the feed is a spherical wave rather than a plane wave, so the calculated radiation characteristics are not very accurate, especially for the far side lobes. In order to correct the calculation results of the far side lobes, the above two methods are extended by the additional ray diffraction, forming Geometrical Optics/Geometrical Theory of Diffraction (GO/GTD), Physical Optics/Physical Theory of Diffraction (PO/PTD).

However, the radiation field calculated by PO method is more accurate in the main lobe and the side lobe near the main lobe. In general, for the main constrained area of reflector antenna beam is main lobe and its adjacent side lobe, the requirement of detail shape of far side lobe is not strict. Therefore, the following will go deep into the introduction of PO as the analysis theory of reflector antenna radiation field.

A.2.2 PO Method

Physical Optics (PO) method is a common high-frequency approximation method. It can only approximate the induced current (equivalent field source) on the surface of the scatterer as an ideal electrical conductor, and then integrate the induced current to obtain the radiation field. However, when the size of the scatterer and the radius of curvature of the surface is greater than 5 times the wavelength, the accuracy of radiation field calculated by PO is satisfied well. According to the PO method, the induced current on the surface of the scatterer (reflecting surface) is

$$\mathbf{J}_s(\mathbf{r}') = \begin{cases} 2(\hat{\mathbf{n}} \times \mathbf{H}_i) & \text{illumination surface} \\ 0 & \text{shadow surface} \end{cases}, \tag{A.11}$$

where $\boldsymbol{r}'$ is the position vector at any point on the surface of the scatterer, $\hat{\mathbf{n}}$ is the unit external normal vector at $\boldsymbol{r}'$ on the surface of the scatterer, and $\boldsymbol{H}_i$ is the incident magnetic field at $\boldsymbol{r}'$on the surface of the scatterer.

After calculating the surface induced current, we can get the radiation field as follows:

$$\mathbf{E}(\mathbf{r}) = -j\omega\mu\,\mathbf{A}(\mathbf{r}) + \frac{1}{j\omega\varepsilon}\nabla\,[\nabla\cdot\mathbf{A}(\mathbf{r})] \tag{A.12}$$

$$\mathbf{H}(\mathbf{r}) = \nabla \times \mathbf{A}(\mathbf{r}), \tag{A.13}$$

where $\boldsymbol{r}$ is the vector of observation point, and the vector magnetic potential $\boldsymbol{A}(\boldsymbol{r})$ is

$$\mathbf{A}(\mathbf{r}) = \iint_{S'} \mathbf{J}_s(\mathbf{r}')\,G(\mathbf{r}|\mathbf{r}')\,\mathrm{d}s' = \iint_{S'} \mathbf{J}_s(\mathbf{r}')\,\frac{e^{-jk|\mathbf{r}-\mathbf{r}'|}}{4\pi|\mathbf{r}-\mathbf{r}'|}\,\mathrm{d}s', \tag{A.14}$$

where free space wave number is $k = \omega\sqrt{\varepsilon\mu}$,$G(r|r')$ is a scalar Green function and ds' is a surface element on the reflecting surface.

Formula (A.14) is substituted into Formula (A.12) and Formula (A.13). After mathematical derivation, the radiation field calculation formula of the scatterer can be obtained:

$$\mathbf{E}(\mathbf{r}) = -\mathrm{j}k\eta \iint\limits_{S'} \left\{ g_1 \mathbf{J}_s(\mathbf{r}') - g_2 \left[\mathbf{J}_s(\mathbf{r}') \cdot \hat{\mathbf{R}} \right] \hat{\mathbf{R}} \right\} \frac{\mathrm{e}^{-\mathrm{j}kR}}{4\pi R} \mathrm{d}s' \tag{A.15}$$

$$\mathbf{H}(\mathbf{r}) = -\mathrm{j}k \iint\limits_{S'} g_3 \left[\hat{\mathbf{R}} \times \mathbf{J}_s(\mathbf{r}') \right] \frac{\mathrm{e}^{-\mathrm{j}kR}}{4\pi R} \mathrm{d}s' \tag{A.16}$$

$$g_1 = 1 - \mathrm{j}\frac{1}{kR} - \frac{1}{(kR)^2} \tag{A.17}$$

$$g_2 = 1 - \mathrm{j}\frac{3}{kR} - \frac{3}{(kR)^2} \tag{A.18}$$

$$g_3 = 1 - \mathrm{j}\frac{1}{kR} \tag{A.19}$$

$$\mathbf{R} = \mathbf{r} - \mathbf{r}', \quad R = |\mathbf{R}|, \quad \hat{\mathbf{R}} = \mathbf{R}/R, \tag{A.20}$$

where η is the free space wave impedance, $\eta = \sqrt{\mu/\varepsilon}$.

When $kr \gg 1$ or $r \gg \lambda$, and $r \gg r$, the far-field approximation is introduced, and the $(1/\mathrm{j}kR)$ and its higher order terms in g_1, g_2 and g_3 are ignored

$$R = |\mathbf{r} - \mathbf{r}'| \approx r - \mathbf{r}' \cdot \hat{\mathbf{r}}, \quad \frac{1}{R} \approx \frac{1}{r}, \quad \hat{\mathbf{R}} \approx \hat{\mathbf{r}}. \tag{A.21}$$

For the far field, the above approximate method can be explained by the following simple geometric relationship: it is assumed that every ray from any point of the scatterer (field source) to the observation point is parallel to each other, as shown in Fig. A.4. Then, we can get the Formula (A.21) from the figure. However, the assumption of parallel rays is correct only when the observation point is located at an infinite distance.

The radiation far field is calculated as

$$\mathbf{E}^{\text{far}}(\theta, \phi) = -\mathrm{j}k\eta \frac{\mathrm{e}^{-\mathrm{j}kr}}{4\pi r} \left(\overset{\rightrightarrows}{\mathbf{I}} - \hat{\mathbf{r}}\hat{\mathbf{r}} \right) \iint\limits_{S'} \mathbf{J}_s(\mathbf{r}')\, \mathrm{e}^{-\mathrm{j}k\mathbf{r}' \cdot \hat{\mathbf{r}}} \mathrm{d}s' \tag{A.22}$$

$$\mathbf{H}^{\text{far}}(\theta, \phi) = -\mathrm{j}k \frac{\mathrm{e}^{-\mathrm{j}kr}}{4\pi r} \hat{\mathbf{r}} \times \iint\limits_{S'} \mathbf{J}_s(\mathbf{r}')\, \mathrm{e}^{-\mathrm{j}k\mathbf{r}' \cdot \hat{\mathbf{r}}} \mathrm{d}s', \tag{A.23}$$

where $\overset{\rightrightarrows}{\mathbf{I}}$ is unit dyad, $\hat{\mathbf{r}}$ is the unit vector at the observation point:

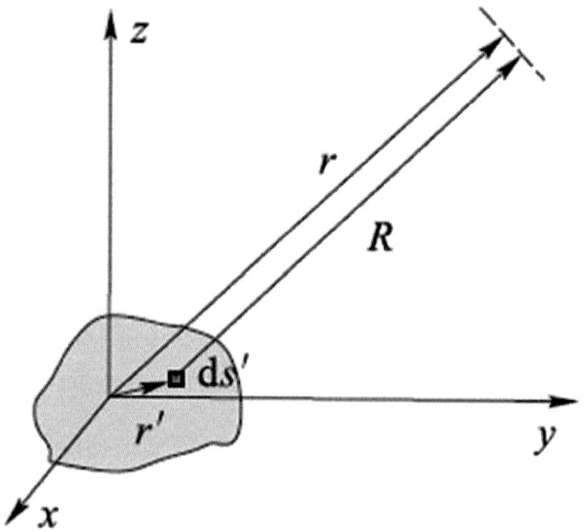

Fig. A.4 Radiation far field of scatterer

$$\hat{\mathbf{r}} = \hat{x}\sin\theta\cos\phi + \hat{y}\sin\theta\sin\phi + \hat{z}\cos\theta. \tag{A.24}$$

When calculating the radiation field of the reflector antenna, the integral in Formula (A.22) or Formula (A.23) is the surface integral on the reflector, which can be converted into the two-dimensional integral on the projection plane of the reflector by proper transformation.

We usually use following two methods to calculate surface integral, the first is to express the integrand function and *z*-coordinate of any point on the surface of the reflecting surface as a function of *x* and *y*, and then project the surface integral into the xOy-plane; the second is to express the integrand function and *x*-, *y*-, *z*-coordinate of any point on the surface of the reflecting surface as a function of the other two parameters *u* and *v*, and then project the surface onto surface corresponding to the two newly defined parameter.

1. If the surface *S* is a smooth two-sided surface

$$z = z(x, y) \quad (x, y) \in \sigma, \tag{A.25}$$

where $\sigma$is the projection of surface *S* on the *xOy*-coordinate plane, *z* (*x*, *y*) is a single-valued continuous differentiable function, function *f* (*x*, *y*, *z*) is defined and continuous on each point of surface *S*, then this integral is independent of the direction of surface *S* (normal direction)

$$\iint\limits_S f(x, y, z)\mathrm{d}S = \iint\limits_\sigma f[x, y, z(x, y)]\sqrt{1 + \left(\frac{\partial z}{\partial x}\right)^2 + \left(\frac{\partial z}{\partial y}\right)^2}\,\mathrm{d}x\mathrm{d}y. \tag{A.26}$$

2. The surface S is given by a continuous differentiable function,

$$\begin{cases} x = x(u, v) \\ y = y(u, v) \ (u, v) \in \Omega \\ z = z(u, u) \end{cases} \tag{A.27}$$

then

$$\iint\limits_{S} f(x, y, z)\mathrm{d}S = \iint\limits_{\Omega} f[x(u, v), y(u, v), z(u, v)]\sqrt{EG - F^2}\mathrm{d}u\mathrm{d}v, \tag{A.28}$$

where

$$E = \left(\frac{\partial x}{\partial u}\right)^2 + \left(\frac{\partial y}{\partial u}\right)^2 + \left(\frac{\partial z}{\partial u}\right)^2 \tag{A.29}$$

$$F = \frac{\partial x}{\partial u}\frac{\partial x}{\partial v} + \frac{\partial y}{\partial u}\frac{\partial y}{\partial v} + \frac{\partial z}{\partial u}\frac{\partial z}{\partial v} \tag{A.30}$$

$$G = \left(\frac{\partial x}{\partial v}\right)^2 + \left(\frac{\partial y}{\partial v}\right)^2 + \left(\frac{\partial z}{\partial v}\right)^2. \tag{A.31}$$

The radiation integral formula of reflector antenna can be calculated by either of the above two methods.

A.3 Basic Equation of Planar Array Antenna

A.3.1 Radiation Field of Antenna Unit

It is hard to treat the radiation problem of the array mathematically and thoroughly (including the analysis of the interaction between the elements and the accurate analysis of the element pattern in the array). Even the mathematical analysis of a single element, the calculation of the trivial vector field will be involved, and the interaction between the elements should also be considered in the analysis of the element in the array. The array theory provides a tool, with which the approximate array structure can be determined according to a given application, and the most arrays can also be synthesized and designed.

Figure A.5 depicts the general distribution of array elements. Each element is a small radiation surface. For a very long distance from the element, the radiation field is spherical wave multiplied by angle vector function (i.e., element pattern). Assuming that any point in the array is taken as the coordinate center (generally,

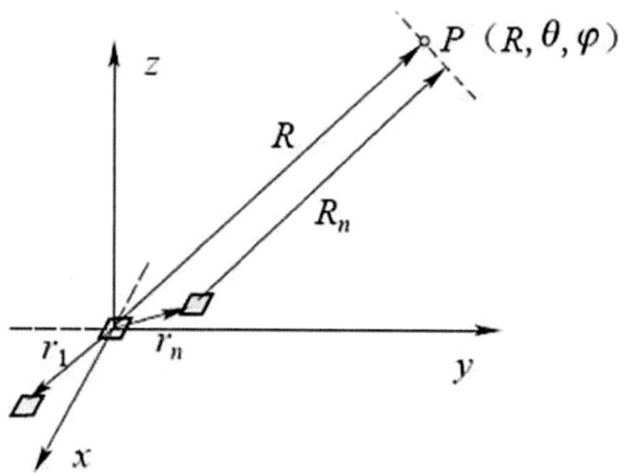

Fig. A.5 Generalized array structure

the center of the array is taken as the reference center), and the vector connecting the reference center and the far-field observation point is $\boldsymbol{R}$, the vector between the n_{th} element of the array and the reference center is $\boldsymbol{r}_n$, and the vector relative to the observation point is $\boldsymbol{R}_n$, then the electric field F_n generated by the n_{th} element of the array at the observation point P (R, θ, φ) is

$$F_n(u, v) = A_n \frac{\mathrm{e}^{-\mathrm{j}k R_n}}{4\pi R_n} E_n(u, v), \tag{A.31}$$

where $u = k\sin\theta\cos\varphi$, $v = k\sin\theta\sin\varphi$, k is the wave number of free space, and (θ, φ) is the far-field observation point in spherical coordinate system. The excitation current (amplitude and phase) is A_n, and E_n (u, v) is the radiation pattern of the n_{th} element, i.e., element factor.

When $R \gg \lambda$ and $R_n \gg r_n$, the far-field approximation is introduced, i.e.,

$$R_n = |\mathbf{R} - \mathbf{r}_n| \approx R - \mathbf{r}_n \cdot \hat{\mathbf{R}}, \quad \frac{1}{R_n} \approx \frac{1}{R}, \quad \hat{\mathbf{R}}_n \approx \hat{\mathbf{R}}. \tag{A.32}$$

Then, the far-field produced by the n_{th} element of the array at the observation point P (R, θ, φ) is

$$F_n(u, v) = A_n \frac{\mathrm{e}^{-\mathrm{j}k R}}{4\pi R} \mathrm{e}^{\mathrm{j}k\mathbf{r}_n\hat{\mathbf{R}}} E_n(u, v), \tag{A.33}$$

where the radiation pattern of the n_{th} element can be regarded as the Fourier transform of its aperture field distribution, namely,

$$E_n(u,\ v) = K \iint_{x,y} e_n(x, y)\ \mathrm{e}^{\mathrm{j}(ux+vy)} \mathrm{d}x\mathrm{d}y, \tag{A.34}$$

where e_n (x, y) is the excitation distribution of the n_{th} element in the local coordinate system. The normalized constant K can be approximated as

$$K = \frac{1}{\lambda}\sqrt{\frac{4\pi}{\iint\limits_{x,y} |e_n(x, y)|^2 \mathrm{d}x\mathrm{d}y}}. \tag{A.35}$$

The above normalization can make $E_n(u, v)$ a good approximation of the pattern of the viewing axis region. Here is an example to verify the correctness of the above statement. When e_n $(x, y) = 1$, for uniform distribution, $|E_n(0, 0)|^2$ is equal to $4\pi S_n/\lambda^2$, where S_n is the aperture area of the n_{th} element, which is the directivity coefficient of the larger array with uniform excitation. However, in order to accurately calculate the array pattern and consider the mutual coupling effect between elements, E_n (u, v) should take the active pattern of the n_{th} element and can simulate the pattern of elements in the array through simulation software (such as HFSS, CST), or use the measured pattern of elements in the array.

A.3.2 Radiation Field of Linear Array

Suppose that there are N elements arranged in a linear array along the Z-axis (Fig. A.6), the center of the nth element is at $Z = Z_n$, and the excitation current (amplitude and phase) is A_n. Select the first cell as the coordinate origin (reference point). The transverse electromagnetic spherical wave is generated in the far field of each element, and the field strength is proportional to the exciting current.

The electric field generated by the n_{th} element at the observation point P (R, θ, φ) is

$$F_n(\theta, \phi) = A_n \frac{\mathrm{e}^{-jkR_n}}{4\pi R_n} E_n(\theta, \phi), \tag{A.36}$$

Fig. A.6 Arbitrary N-element linear array

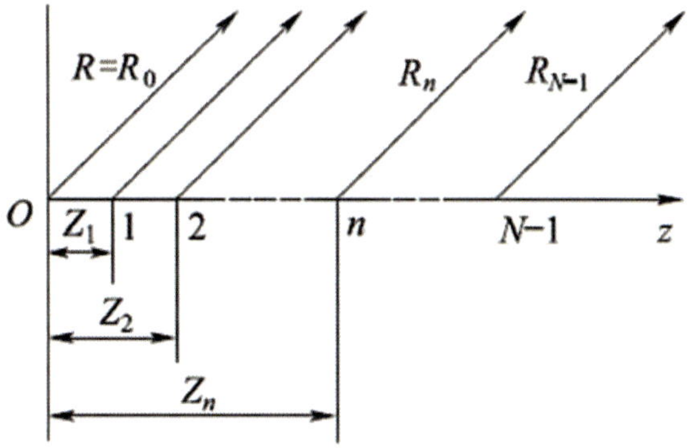

where $E_n\ (\theta, \varphi)$is the radiation pattern of the n_{th} element, i.e., the element factor, which can be approximated as the Fourier transform of its aperture field distribution.

Substituting into the far-field approximation condition,

$$R_n \approx R - Z_n \cos\theta, \quad \frac{1}{R_n} \approx \frac{1}{R}, \tag{A.37}$$

we have

$$F_n(\theta, \phi) = \frac{\mathrm{e}^{-\mathrm{j}kR}}{4\pi R} E_n(\theta, \phi) A_n \mathrm{e}^{\mathrm{j}k Z_n \cos\theta}. \tag{A.38}$$

The last factor represents the relative phase produced by the spatial position and observation angle of the element. According to the superposition principle, the field produced by the linear array at the observation point is equal to the vector sum of the field produced by each element at the observation point

$$F_{\text{linear}} = \sum_{n=0}^{N-1} F_n = \frac{\mathrm{e}^{-\mathrm{j}kR}}{4\pi R} \sum_{n=0}^{N-1} A_n E_n(\theta, \phi) \mathrm{e}^{\mathrm{j}k Z_n \cos\theta}. \tag{A.39}$$

The linear array pattern is

$$F(\theta, \phi) = \sum_{n=0}^{N-1} A_n E_n(\theta, \phi) \mathrm{e}^{\mathrm{j}k Z_n \cos\theta}. \tag{A.40}$$

Let

$$F_a(\theta, \phi) = \sum_{n=0}^{N-1} A_n \mathrm{e}^{\mathrm{j}k Z_n \cos\theta} \tag{A.41}$$

be an array factor. The pattern of antenna array is equal to the product of element factor and array factor.

Assuming that the spacings d of the elements are equal, the coordinates of the n_{th} element are (0, 0, nd). The corresponding linear array pattern is

$$F(\theta, \phi) = \sum_{n=0}^{N-1} A_n E_n(\theta, \phi) \mathrm{e}^{\mathrm{j}knd \cos\theta}. \tag{A.42}$$

A.3.3 Radiation Field of Plane Array

Considering the array antenna with N elements in the xOy-plane, the radiation pattern is expressed as

$$F(u,v)=\sum_{n=1}^{N}A_nE_n(u,v)\ \mathrm{e}^{\mathrm{j}(x_nu+y_nv)},\tag{A.43}$$

where $u=k\sin\theta\cos\varphi$, $v=k\sin\theta\sin\varphi$, k is the wave number of free space, and (θ, φ) is the far-field observation point in spherical coordinate system. A_n, (x_n, y_n), E_n (u, v) are the complex excitation factor, position and the radiation pattern of the nth element, respectively. In Formula (A3.13), the term $\mathrm{e}^{-\mathrm{j}kr}/r$ that vary with distance and the time change factor $\mathrm{e}^{\mathrm{j}\omega t}$ are ignored.

A.3.4 Array Antenna Pattern Synthesis Method—Projection Matrix Method [1, 2]

Taking the plane array model as an example, we can deduce it to the linear array model. It is assumed that F_{d} (u, v) is the desired far-field gain pattern. The aim is to find an incentive set A_n ($n=1, 2,\ldots, N$), which satisfies the following equation:

$$F_{\mathrm{d}}(u,v)=\sum_{n=1}^{N}A_nE_n(u,v)\ \mathrm{e}^{\mathrm{j}(x_nu+y_nv)}.\tag{A.44}$$

In order to solve A_n, M far-field points are selected and M constraint equations are obtained. The matrix form is as follows:

$$\begin{bmatrix}F_{\mathrm{d}1}\\F_{\mathrm{d}2}\\\vdots\\F_{\mathrm{d}M}\end{bmatrix}=\begin{bmatrix}T_{11}&T_{12}&\cdots&T_{1N}\\T_{21}&T_{22}&\cdots&T_{2N}\\\vdots&\vdots&\ddots&\vdots\\T_{M1}&T_{M2}&\cdots&T_{MN}\end{bmatrix}\begin{bmatrix}A_1\\A_2\\\vdots\\A_N\end{bmatrix},\ \mathbf{F}_{\mathrm{d}}=\mathbf{TA},\tag{A.45}$$

where $F_{\mathrm{d}m}$ is the abbreviation of $F_{\mathrm{d}}(u_m, v_m)$, $\boldsymbol{F}_{\mathrm{d}}$ is the desired pattern, $\boldsymbol{T}$ is the pattern under the excitation of each element in the array and $\boldsymbol{A}$ is the complex excitation coefficient to be calculated.

$$T_{mn}=\ E_n(u_m,v_m)\mathrm{e}^{\mathrm{j}(x_nu_m+y_nv_m)},\qquad 1\le m\le M,\ 1\le n\le N,\ M\ge N.\tag{A.46}$$

In order to get the phase distribution of the desired far-field pattern, an iterative method is used. First, assuming that a trial excitation vector $\boldsymbol{A}^{(1)}$ is normalized with the input power; using $\boldsymbol{A}^{(1)}$ instead of $\boldsymbol{A}$ on the right side of Formula (A.45), then the corresponding far-field vector $\boldsymbol{F}^{(1)}$ is obtained. Obviously, at most far-field points, the element of $\boldsymbol{F}^{(1)}$ is different from that of $\boldsymbol{F}_{\mathrm{d}}$. The error pattern is calculated as follows:

$$\Delta\mathbf{F} = \mathbf{F}_{\max}[|\mathbf{F}_n| - P_M\{|\mathbf{F}_n|\}]\left(\mathbf{F}^{(1)}/\left|\mathbf{F}^{(1)}\right|\right)\mathbf{F}_{\text{weight}} \tag{A.47}$$

$$P_M\{|\mathbf{F}_n|\} = \begin{cases} \mathbf{M}_U & |\mathbf{F}_n| > \mathbf{M}_U \\ |\mathbf{F}_n| & \mathbf{M}_L \le |\mathbf{F}_n| \le \mathbf{M}_U, \\ \mathbf{M}_L & |\mathbf{F}_n| < \mathbf{M}_L \end{cases} \tag{A.48}$$

where $\boldsymbol{F}_{\max}$ is the maximum value of the current pattern, $\boldsymbol{F}_n$ is the normalized pattern and $\mathrm{P_M}$ operator projects the normalized pattern to the specified area of the mask. $\mathrm{P_M}\{|\boldsymbol{F}_n|\}$ is a projection operator defined by a mask, $\boldsymbol{M}_{\mathrm{U}}$ is the upper bound mask of the desired pattern, $\boldsymbol{M}_{\mathrm{L}}$ is the low-bound mask of the desired pattern. $\boldsymbol{F}_{\text{weight}}$ is a weight factor, which is used to strengthen the constraint on the specific observation area pattern.

Then, let

$$\Delta\mathbf{A} = \left(\mathbf{T}^H\mathbf{T}\right)^{-1}\mathbf{T}^H\Delta\mathbf{F}, \tag{A.49}$$

we have incremental excitation coefficient $\Delta\mathbf{A}$ vector, which is updated as

$$\mathbf{A}^{(2)} = \mathbf{A}^{(1)} + C\cdot\Delta\mathbf{A}, \tag{A.50}$$

where C is the step factor. In order to achieve better control effect, refer to the step selection rule in the steepest descent method, Let the value of step length be related to the incremental excitation coefficient, then

$$C = \frac{\delta l}{\left\|\Delta\mathbf{A}/\mathbf{A}^{(1)}\right\|_2}, \tag{A.51}$$

where δl is a preset quantity, and $\|\ \|_2$ represents the 2-norm of the matrix, that is, the Euclidean length of the vector. The step factor C calculated by Eq. (A.51) may be too large or too small, so it needs to be further projected into an appropriate range (such as [0.001, 0.5]).

After normalizing the new excitation vector, the whole process is repeated until a satisfactory far-field pattern is obtained or the number of iterations reaches the preset maximum.

When the method of projection matrix is used to synthesize the beam of phased array antenna, the convergence of the algorithm can be guaranteed by selecting the range and interval of the angle of observation point reasonably. Moreover, the observation angle range must be larger than the beam width of each element, and the angle interval must be smaller than the beam width of the whole aperture uniform distribution of phased array.

References

1. Xiaomeng L, Wenjun G et al (2012) Synthesis technique of array-fed shaped-reflector antenna for DBF-SAR application. IEEE Antennas Wirel Propag Betters 11:30–33
2. Bhattacharyya AK (2007) Projection matrix method for shaped beam synthesis in phased arrays and reflectors. IEEE Trans Antennas Propag 55(3):675–683